T0132576

# Materials and Expertise in Early Modern Europe

# *Materials and Expertise in Early Modern Europe*

**Between Market and Laboratory**

EDITED BY URSULA KLEIN AND E. C. SPARY

The University of Chicago Press
Chicago and London

Ursula Klein is a senior research scholar at the Max Planck Institute for the History of Science in Berlin and a professor at the University of Konstanz. She is the author of *Experiments, Models, Paper Tools: Cultures of Organic Chemistry in the Nineteenth Century.*

E. C. Spary is a lecturer in the history of eighteenth-century medicine at the Wellcome Trust Centre for the History of Medicine at University College London. She is the author of *Utopia's Garden,* published by the University of Chicago Press.

The University of Chicago Press, Chicago 60637
The University of Chicago Press, Ltd., London

Printed in the United States of America

18 17 16 15 14 13 12 11 10 09 1 2 3 4 5

ISBN-13: 978-0-226-43968-6 (cloth)
ISBN-10: 0-226-43968-2 (cloth)

Library of Congress Cataloging-in-Publication Data

Materials and expertise in early modern Europe : between market and laboratory / edited by Ursula Klein and E. C. Spary.
p. cm.
Papers of two workshops which were held at the Max-Planck-Institut für Wissenschaftsgeschichte, Berlin, in July 2004 and December 2006.
Includes bibliographical references and index.
ISBN-13: 978-0-226-43968-6 (cloth : alk. paper)
ISBN-10: 0-226-43968-2 (cloth : alk. paper)
1. Materials—Europe—History—Congresses. 2. Technological innovations—Europe—History—Congresses. 3. Laboratories—Europe—History—Congresses. 4. Expertise—History—Congresses. 5. Knowledge, Theory of—Europe—History—Congresses. I. Klein, Ursula, 1952– II. Spary, E. C. (Emma C.)
TA403.6.M369 2010
670.94′0903—dc22

2009026122

∞ The paper used in this publication meets the minimum requirements of the American National Standard for Information Sciences—Permanence of Paper for Printed Library Materials, ANSI Z39.48-1992.

# Contents

# *Acknowledgments*

This book is the result of two workshops which were held at the Max-Planck-Institut für Wissenschaftsgeschichte, Berlin, in July 2004 and December 2006. As always, the editors have accrued many debts along the way. In particular, several people participated in the original workshops and contributed extensively to the discussions which shaped the form this book has taken, as well as the individual contributions. First and foremost, both editors and contributors wish to acknowledge the input of Larry Epstein (Stefan Epstein). His untimely death before this project was complete deprived us of a valued contributor.

In addition, we wish to thank Andy Pickering, Robert Fox, and John Pickstone for their involvement in the workshops and for their commentaries. Lissa Roberts and an anonymous reader for the University of Chicago Press helped us to reformulate the volume as a whole and provided valuable suggestions for the introduction, for which we are extremely grateful. Our thanks also go to Karen Darling for her enthusiastic support in the final stages of preparation.

*Ursula Klein*
*E. C. Spary*
*Berlin and Cambridge, August 2008*

1

# Introduction: Why Materials?

URSULA KLEIN AND E. C. SPARY

The dramatis personae of this book are materials such as metals, gunpowder, pigments, and foods. *Materials and Expertise in Early Modern Europe* takes useful material substances, "materials," as a route into mixed artisanal and learned practices, which contributed to artisanal innovation, the development of the consumer market, and the formation of the observational and experimental sciences of the early modern period. It examines the ways in which innovative practices involving materials, technical expertise, and learned natural knowledge converged and considers the experts who occupied themselves with materials. In certain types of early modern workshops, laboratories, and marketplaces, materials continually resisted routine manufacture, caused consternation, or stimulated curious inquiries. Central to this volume are just such challenging materials and ingenious practices, developed between the early sixteenth and the late eighteenth centuries at sites where commerce intersected with the production of learned natural knowledge and where hybrid experts united hands-on experience with higher learning.

Each chapter in this volume analyzes practices and forms of knowledge associated with a particular material or class of materials: metals painstakingly extracted from the earth, the red pigment vermilion redolent of blood and life, beautiful simulacra of plants and animals molded in clay, ink produced through prolonged and arduous labor, chemical remedies stimulating endless experimental trials, distilled liqueurs attracting polite connoisseurs, milk eluding chemical analysis, mineral waters converted into commodities through chemical knowledge, vegetable materials explored by economic societies, dyestuffs submitted to continual quality control, and gunpowder perpetually resisting standardization. By taking a particular material as the focal point of their analysis, the essays in this volume place emphasis upon things while at the same time seeking to explore practitioners' knowledge claims and ontologies. The chapters thus attempt to move beyond existing studies of material culture and to raise new questions about the interconnections between practical expertise and learned natural inquiries in the early modern period.

The materials we are investigating were plural in nature, at once the stuff of ingenious labor, mundane consumption, and sustained inquiry into nature and art. They include raw materials, processed material substances, and, in one case, finished products made from raw materials. Of interest to a range of early modern experts, including knowledgeable artisans, educated merchants, polite connoisseurs, and university-trained practitioners, they were quotidian things and at the same time scientific objects. They circulated between workshops, laboratories, marketplaces, coffeehouses, salons, hospitals, dispensaries, studies, and lecture halls. Following their circulation and their different uses allows social distributions of past commodities to be analyzed together with the systems of knowledge, social distinction, and power that conferred value and meaning upon them. Milk, for example, studied in Barbara Orland's chapter, was used by many different social groups and thus had different values and meanings in eighteenth-century France. Several chapters in this volume focus on the production of materials in workshops and laboratories and on experiments that explored the techniques of production, the quality of materials, or their chemical composition. By this means they seek to unravel the traces of materiality, the thickness of things, and the challenges emerging from materials in the course of physical intervention.[1] Other chapters illuminate the ways in which the manipulation of matter generates a common ground where production and consumption meet, or the rise of communities that cannot be easily fitted into frameworks dealing with production or consumption alone. Taken together, the essays in this book address a broad variety of early modern forms of knowledge about materials. These range from ineffable bodily skills and technical competence to articulated know-how and connoisseurship of materials, from methods of measuring, data gathering, tabulation, and classification all the way to analytical and theoretical knowledge. The old dichotomy between artisanal and scholarly knowledge, or hand and mind, is thus replaced by a differentiated continuum of forms of knowledge.

The forms of knowledge and practice of the hybrid experts explored in

1. The expression "thick things" was coined by the anthropologist Clifford Geertz. For a discussion of its relevance in the history of science, see Alder 1997, 13; Alder 2007; and Mauskopf's chapter in this volume.

this volume have traditionally fallen outside the scope of both the history of science, with its focus on theoretical knowledge, and the history of technology, with its emphasis on machines and the "big picture" of industrialization. In recent decades, extensive work on the rise of an eighteenth-century consumer society and the attendant transformations of commerce, production, and new modes of inquiry into nature has laid the foundations for a revision of this state of affairs; our volume is indebted to these new histories of consumption and commerce.[2] In this it follows several other recent publications in the history of science which have begun to pay serious attention to the question of how consumption, production and trade affected the development of the sciences.[3] We are particularly indebted to recent approaches that interconnect the history of science and technology, highlighting early modern practitioners or sites which permitted an erosion of the older dichotomies between hand and mind, scholar and artisan, learned and practical knowledge about the material world.[4]

One way of characterizing our volume might be as an inquiry into the interconnectedness of the sciences, technology, and society in the early modern period, through particular sorts of material objects and practices. However, both *science* (especially in the singular) and *technology* can have anachronistic meanings when applied to the early modern era. Neither the disciplinary structure of the modern sciences nor the technological networks of industrialized societies existed between 1500 and 1800. The expression "learned inquiry into nature" serves to describe a range of fields, such as mechanical and experimental philosophy, experimental history, natural history, anatomy, physiology, medical theory, alchemy, chemistry, astronomy and optics.[5] Similarly, the expression "arts and crafts" may be more appropriate for the period before the late eighteenth century than the term "technology."[6] Yet these terminological substitutes, while heuristically useful, do not in them-

2. See Berg 1991, 1994; Berg and Clifford 1999; Campbell 1987; Fox and Turner 1998; McKendrick, Brewer, and Plumb 1982.

3. See Cook 2007; Jacob 1997; Jacob and Stewart 2004; Jenner and Wallis 2007; Long 2001; Roberts, Schaffer, and Dear 2007; Schabas and De Marchi 2003; Schiebinger and Swan 2005; Smith and Findlen 2002; P. Smith 2004; Stewart 1992. Among the older publications which deal with similar themes are Grossmann 1987 (a translation of Grossman 1935); Hessen 1971; Merton 1970; Musson 1972; Musson and Robinson 1969; Rossi 1970; Zilsel 2000.

4. See, in particular, Roberts, Schaffer, and Dear 2007.

5. See Cunningham and Williams 1993; Park and Daston 2006.

6. It is both noteworthy and significant that the fine and mechanical arts were not distinct in early modern discourse, as Hilaire-Pérez remarks in considering the sixteenth-century origins of the notion of "génie" (2000, chap. 1). See also Smith 2004.

selves solve the problems tackled in this volume. They still leave us with the task of writing a historically more appropriate narrative of the events that eventually gave rise to the modern sciences and their close ties to technology and commerce.

In the early modern period, inquiries into the natural world came to privilege observational and hands-on knowledge about material bodies, machines, and instruments. In numerous domains, practices depended upon a blending of artisanal labor, experimental inquiry into the material world, and learned investigation of nature. Contemporaries also advertised the phenomenon, as in the claim by Thomas Sprat, secretary of the Royal Society, that Bacon's *New Atlantis* would be accomplished when "Mechanick Laborers shall have Philosophical heads; or the Philosophers shall have Mechanical hands."[7] The significance of this emphasis on experience and the material, as compared to textual exegesis and philosophical disputation, has long been a subject of debate among historians of science.[8] Some historians of science would insist that the formation of the observational and experimental sciences in the early modern period was rooted firmly in a long natural philosophical tradition, transmitted by texts, denying that the contemporary arts and crafts had any significant and constitutive impact on this historical process.[9] Several factors have contributed to the perpetuation of the sharp distinction between the history of science and of technology, which our volume questions. Moral hierarchies of long standing among learned elites have privileged intellectual over manual activity, and theory over practice. Similarly, early modern elite polemics portrayed the handworker as limited by ignorance and blind routine to purely mechanical action. Such asymmetries continue to color historians' accounts of the relationship between higher learning and the arts and crafts, mundane things and scientific objects.

The early modern natural philosophies and matter theories had, of course, historical roots that clearly differed from craft traditions. The contributions to this volume are not concerned with theories about the whole universe and the ultimate structure of matter; such theories were sometimes of interest to our experts, in particular those living before the eighteenth century, but it is questionable that they provided any heuristic means for their practices of observation and experimentation. Instead, this volume studies limited objects

7. Quoted in Stewart 1992, 101. This study illustrates the complexity of interactions between patrons, artisans, and natural philosophers in eighteenth-century England.

8. See, for example, Grafton and Siraisi 1999.

9. See, for example, Newman 2006. A lengthy historiographical tradition in history of science endorses this view, including eminent scholars such as Rupert Hall, Alexandre Koyré, and George Sarton. See, in particular, R. Hall 1969; Koyré 1943, 1968; Sarton 1931.

of inquiry, such as materials, which were uncontested parts of communal practices and explained in technical texts. In so doing, we show that, in effect, it is often impossible to make distinctions between objects of natural inquiry and of artisanal innovation. This is particularly obvious in the case of alchemy and eighteenth-century chemistry, which put materials at center stage.[10] Similarly, for the cases of early modern medicine and natural history, recent studies have explored the ways in which texts and practices of observation, experimentation, and material manipulation were interconnected.[11] These relations might be seen as akin to the interaction between the scholarly and commercial worlds with which this volume is concerned, especially in consideration of the fact that early modern naturalists often sought ways of improving the exploitation of natural resources within the arts and medicine.[12] It is this middle ground that persistently remains invisible if the history of science is separated from the history of technology and the history of medicine; for example, in claiming that "in many ways the advance of scientific knowledge had virtually no direct effect upon technology," one historian supports his argument using the examples of gunnery, metal-working, and mining, precisely the examples addressed in this volume by Mauskopf and Bartels.[13]

Arts like mining, metallurgy, printing, pottery, gunnery, pharmacy, distillation, dyeing, and calico printing underwent extraordinary technical transformations in the period from 1500 to 1800. In these arts the production and manipulation of materials required both high levels of technical competence and familiarity with systematic experiential knowledge. Many artisanal producers insisted upon a scientific understanding of the natural processes in which they intervened as a necessary prerequisite for good practice. These arts thus created and deployed new forms of representing knowledge, ranging from maps showing mineral deposits to pharmaceutical recipes to distillation treatises, from geometrical diagrams of bullets to theories of metal formation. By the late eighteenth century, the training of apothecaries, distillers, miners, assayers, and other skilled artisans often involved scientific lectures, demonstrations, and textbooks covering many learned disciplines, including chemistry, botany, hydrostatics, and mathematics. Numerous pro-

10. On the artisanal and scientific dimensions of materials in eighteenth-century chemistry, see Klein and Lefèvre 2007.

11. See W. B. Ashworth 1996; Smith and Findlen 2002; Park 1985; Siraisi 1990; Grafton, Shelford, and Siraisi (1992).

12. See Cook 2007; Cooper 2007; Spary 2003.

13. Mathias 1991, 36. Such statements rest upon histories of technology and science that include only famous inventors and scientific practitioners.

fessional schools of mining, pharmacy, surgery, artillery, engineering, and architecture were founded during the eighteenth century. Along with economic and philosophical societies, academies, journals, and public teaching, such institutions fostered the improvement of commercial production, the development of consumer markets, and the intensification of agricultural output, while simultaneously contributing to the scientific understanding of nature. Many new sciences of the early modern period, such as chemistry, botany, mineralogy, geology, and agronomics, were heavily indebted to, and intersected with, the arts and crafts. There was thus both extensive familiarity with learned knowledge-claims and language among certain types of artisanal experts, and extensive occupation with artisanal practice and production by university-educated men of science. Our volume addresses precisely this middle ground, where technical competence, connoisseurship, and learned natural knowledge were converging and from which hybrid experts emerged, borrowing skill, language, and explanations from both the artisanal and the scholarly worlds.

Many of the individuals considered in this volume, such as the academician and dyeing inspector Charles-François Du Fay, studied by Nieto-Galan, or the naturalist and ceramicist Bernard Palissy, studied by Shell, were skilled practitioners and learned investigators of nature at the same time. The production of materials such as metals, gunpowder, dyestuffs, or chemical remedies, which involved skill, expert judgment, and complex systems of organization and quality control, was an ongoing technical and epistemic challenge. Experts such as assayers, military comptrollers, or apothecary-chemists united the qualities required to meet such challenges, for it was they who often united an artisanal training with some form of higher education. Such hybrid experts frequently experimented upon materials in laboratories with a view to technical improvement and profit, but they also sought to generate new facts, classifications, analyses, and representations of material substances, which would hold up well in the scholarly and artisanal domains alike.

Laboratories, addressed in the chapters by Klein, Nieto-Galan, Mauskopf, and Spary, were sites where mixed artisanal and learned practices and forms of knowledge flourished in the early modern period.[14] They represented a strand of early modern expertise developing around the manufacture of materials, as well as chemical techniques such as distilling, smelting, or dissolving. Laboratories were the product of a long tradition, going back to late medieval alchemy, in which innovative forms of material production, technical exper-

14. On early modern artisanal and academic laboratories, see Klein 2008; P. Smith 2006b.

tise, and text-based understanding developed in parallel.[15] Well into the nineteenth century, they remained sites of *chemical* operations almost exclusively and served learned chemical inquiries into nature, technological innovation, and commercial production. "Laboratories" were thus established not only at academic institutions but also at apothecaries' shops; mines and foundries; mints; arsenals; dyeing, porcelain, and chemical manufactories; distilleries; and perfumeries. Like traditional family workshops, early modern laboratories were small-scale workplaces. But unlike such traditional workshops, they were also sites of technical ventures and systematic observation, which had a galvanizing effect on the development of the experimental sciences. Laboratories significantly contributed to an ongoing process in which the ancient separation between hand and mind was being surmounted. Around 1600 the humanist and physician Andreas Libavius, for example, had to defend himself against accusations that he had no authentic knowledge about chemistry because he did not possess a laboratory.[16] Although a large proportion of scholars still looked down upon manual labor, by the late eighteenth century, this group was counterbalanced by a powerful group of practitioners who relied upon experimental skill and craftsmanship, including chemists, experimental natural philosophers, and physiologists. This group, along with anatomists, cameralists, technologists (*Technologen*), engineers, mining officials, assayers, pharmacists, and other types of hybrid experts, emphatically argued for the amalgamation of hands-on experience with higher learning. In so doing, they were amply supported by administrators.

In what follows we will address several more specific questions that are central to this volume. Why materials and not, say, instruments? Can we gain new insight into the historical development of the observational and experimental sciences by focusing on material substances rather than instruments? Do such studies contribute to our understanding of the historical roots of the technosciences that we witness everywhere today? What do materials tell us about the mixed artisanal and learned practices of the past? How do materials differ from other kinds of objects discussed in recent historical studies of material culture? By addressing questions like these, we seek to clarify our approach and arguments still further through a discussion of certain recent approaches to the history of science and technology. These include the historical and philosophical studies of "scientific objects," along with their differentiation of scientific objects from quotidian things; historical studies of instruments; studies of materials within the history of technology; and stud-

15. Some alchemical laboratories were also sites of religious revelation; see P. Smith 2006b.

16. See Moran 2007, 54, 91.

ies of objects in museology and commodity history. Among the important questions raised by this volume are many which must remain the province of subsequent inquiries. This volume has its central emphasis in the eighteenth century, the period when mixed technoscientific practices involving materials were transformed into sustained social institutions. A detailed account of the historical process by which the union of artisanal and learned practices involving materials developed before that time, from the late medieval period onward, must remain the subject of future studies.

## Scientific Objects

Historians and philosophers of science have interested themselves for some time in the formation, investigation, and accreditation of the objects of scientific inquiry, or scientific objects.[17] Gaston Bachelard argued that the objects constituted by a "scientific mind" were segregated from ordinary objects through an "epistemic rupture." This separation was achieved, he suggested, by accuracy, precision instruments, mathematical representation, and a permanent rigorous criticism, which replaced ordinary "primary experience."[18] Drawing upon Bachelard, Lorraine Daston has recently demarcated quotidian objects—"the solid, obvious, sharply outlined, in-the-way things of quotidian experience"—from scientific objects. In contrast to quotidian objects, which "possess the self-evidence of a slap in the face," she points out, scientific objects are "elusive and hard-won."[19] Typical examples of such objects might be the atoms and electromagnetic forces of nineteenth-century chemistry and physics or the gene of twentieth-century genetics. These objects were, indeed, utterly unfamiliar to ordinary people at the time of their first experimental investigation. They were entities that were not accessible to the observer by means of the senses; instead, they were constituted in a highly mediated way, through the use of instruments, experimental intervention, measurement, and the work of representation. The task of measuring, individuating, and identifying them all fell to specialized scientists, and their very existence was often fiercely contested at first.[20]

In the early modern period, many university-based men of science were

17. See Daston 2000b; Hacking 2002; Latour 1999; Rheinberger 1997.

18. Bachelard 1996.

19. Daston 2000a, 2; see also Knorr-Cetina 1999, chap. 5; Latour and Woolgar 1979.

20. The question of the relations between representation and materiality is a complex one, about which we will not make any general claims. The philosophical problem of realism is lurking here, and antirealists would certainly question the applicability of the term *objects* to such entities. Among the extensive philosophical literature on scientific realism, see, for example,

also preoccupied with invisible entities, such as substantial forms, corpuscles, spirits, the celestial ether, or the vacuum. All of these were also highly controversial entities. Controversies concerned not only their identification and understanding, but even their very existence. For example, in early modern philosophical debates over the vacuum, it was unclear whether it should be understood as a real entity or as a philosophical speculation. Robert Boyle would use the air pump to prove its existence by experiment, but the credibility of experimentation and physical manipulation as a proof strategy was fiercely contested by many Aristotelian philosophers.[21]

By contrast, all of the materials studied in this volume were rooted firmly in the artisanal practices and market culture of the early modern period. The existence of gunpowder, dyestuffs, metals, clay and ceramics, ethers, vegetable materials, mineral waters, ink, milk, and liqueurs was thus never contested, though the ways of their identification as well as their meaning and values were subject to debate. Some of these materials were not readily available to ordinary people; others, like milk or mineral waters, were familiar to all but remained, both epistemologically and practically, problematic substances. All provoked new inquiries into nature by means of observation and experimentation.

Hence, in certain respects our materials resemble the early modern philosophical entities and modern scientific objects. They were curious and demanding objects, whose production, investigation, and use required advanced knowledge or expertise, and several of them were also hard-won, recalcitrant things, produced only by the most advanced of technical maneuvers and instrumentation. All materials studied in this volume were challenging things which provoked investigators to expand and refine their activities and understanding. Yet at the same time they were not elusive in the sense of the invisible entities of early modern experimental philosophy, of modern particle physics or molecular biology. All of them spoke irresistibly, and not only by interpretation and representation. All of them continually asserted their physical presence through color, odor, explosion, fermentation, or putrefaction, and they were maneuverable by actors' hands and tools.

Such objects were thus not constituted within a closed space of scholarly, or scientific, investigation of nature distinguished by an "epistemic rupture"

---

Hacking 1983; Pickering 1989. On the problems of distinction or individuation and identification in experimental research, see also Klein and Lefèvre 2007.

21. On Boyle's experiments with the air-pump, see, for example, Shapin and Schaffer 1985. Similar challenges were faced by Galileo in producing philosophical knowledge using a telescope; see Biagioli 1993, chap. 2.

from other forms of inquiry in the arts and crafts and everyday life. They were investigated using methods and concepts belonging to the scholarly world, but were never severed from the world of artisanal production, commercial circulation, and everyday consumption. Likewise, the practitioners who feature in this volume did not have the luxury that Karin Knorr-Cetina has attributed to modern laboratory science, of "the detachment of objects from their natural environment and their installation in a new phenomenal field defined by social agents."[22] Instead, their efforts were directed primarily at the object-in-the-world, and they endeavored to produce new natural knowledge and new social benefit and commercial profit at the same time. Even chemists' shift toward analysis was not necessarily a move toward ontological parsimony, in which the object was stripped of observable qualities to become a partial and metonymic version of itself. Instead, quite often, the products of chemical analysis themselves became new commodities, so that analysis was itself an enrichment of the quotidian world of commerce through the proliferation of new entities.[23]

For many areas of early modern inquiry, it is thus problematic to insist on a clear-cut distinction between learned discourse on the one hand, and the world of commerce on the other. The objects of interest in this volume were at once scientific objects and quotidian things, subjects of learned natural inquiry and products of labor, objects of understanding and sources of profit.[24] Compared to the early modern natural philosophers' hidden entities, these objects were uncontested parts of the material culture and practices of early modern societies. But we are not suggesting that the phenomena and challenges arising from a material substance were the inevitable results of a Kantian thing-in-itself; rather, they manifested themselves as one of the outcomes of practical manipulations of a material, and they were as contingent and multifarious as these practices. As a consequence, their meanings and values varied considerably among the experts discussed in the contributions to this volume. Differences in meaning and value of the actors' objects of inquiry also hinged on ontologies, which underwent dramatic changes in the period covered by this volume.[25] The sixteenth- and seventeenth-century practitioners studied by Bartels, Smith, Shell, and Johns manipulated materials in a world replete with astrological agencies, invisible spirits, occult forces, and

22. Knorr-Cetina 1999, 26–27; see also Pickstone 2000, chap. 4.

23. See Klein and Lefèvre 2007. See also the chapters by Eddy, Klein, and Orland in this volume.

24. For the concept of labor used here, see Lefèvre 2005.

25. On historical ontology, see Hacking 2002; Klein and Lefèvre 2007.

hidden chemical principles. A sixteenth-century alchemist ordered commodities such as metals into a web of signs, relating them to the planets and other celestial agencies that spurred the generation and transmutation of metals, as Smith's chapter shows. Vermilion, Smith points out, was simultaneously a pigment and a symbol of blood, generation, and life. Palissy's ceramics, studied in Shell's chapter, were precious commodities highly esteemed by wealthy patrons and at the same time simulacra of spontaneous generation and fossilization. Johns points out that ink bore many associations with magic. In the eighteenth century, such entities and relationships were fading away. As Klein's chapter shows, the early eighteenth-century concept of the ether was reminiscent of the ancient concept of subtle celestial spirits; but in the later eighteenth century, chemists redefined ethers as laboratory substances and the products of ordinary chemical reactions, governed by the laws of chemical affinity.

## Instruments

In the past two decades, historical studies of the material culture of early modern societies and of the relations between the experimental and observational sciences and the arts and crafts have placed instruments, bodily skill, and the question of the repeatability of experiments at the forefront of inquiry.[26] Such studies, as well as general accounts of the reconfiguration of learned knowledge and practice during the early modern period, have demonstrated the extent to which the generation of learned natural knowledge crucially depended on instruments and the skills of instrument makers and experimenters. Instruments and repertoires of bodily skills enabled experimenters to intervene physically in nature and to generate new phenomena. These two resources also shaped experimental programs, directed or constrained experimenters, and contributed to the definition of new experimental goals. Our volume shares the concern of scholars of early modern instruments and their makers with material culture and with the relationship between artisanal and learned knowledge. However, there are also significant differences between our approach and most studies of instruments, which affect our historical understanding of the differentiation and scope of the early modern

26. For examples of the growing literature on early modern instruments and skills, see Anderson, Bennett, and Ryan 1993; Bennett 1986, 1987, 2002, 2003; Bennett et al. 2003; Daumas 1953; Holmes and Levere 2000; Hunter and Schaffer 1989; Bertoloni Meli 2006; Morrison-Low 2007; Porter et al. 1985; Schaffer 1989, 1997b; Stewart 1992; A. Turner 1987; G. Turner 1990; Van Helden and Hankins 1994.

sciences as well as our view of the historical roots of modern technoscience. Moreover, most historical studies of instruments focus on epistemological and methodological questions; only a few of them study the historical actors' ontologies as well.

Among the instruments of the early modern period, historians of science have devoted most attention to mathematical, optical, and philosophical instruments.[27] Other instruments, such as those used in chemistry, anatomy, surgery, or natural history, have failed to achieve the central epistemological and methodological place accorded to mathematical, optical, and philosophical instruments in the historiography of early modern science and technology. This primacy reflects the general emphasis placed upon the mathematical and physical sciences of the early modern period: astronomy, optics, mechanics, and experimental philosophy, which were integrated and restructured as modern physics in the nineteenth century. These fields have received a disproportionate amount of historical attention, as historians of physics have put them at the forefront of a battle over the history of science-as-idea versus science-as-practice in the early modern period. Historical studies of controversies over experimental method, the accreditation of experimental facts, the social relations between learned and artisanal experimenters in the early modern period, and some more recent attempts to historicize our notions of observation and experimentation have encouraged the view that experimental philosophers and instrument makers were the main representative types of early modern scientific practitioners, and mechanical and experimental philosophy the characteristic forms of scientific inquiry. This volume suggests some revisions to this picture.

The "philosophical instruments" of central importance within seventeenth-century natural philosophy—most notably, the air pump and the electrical machine—had no commercial standing prior to the eighteenth century. Instruments of this type have the least in common with the materials studied in this volume.[28] Philosophical instruments were initially used by natural philosophers to discover the truths of nature. They thus served specific problems of proof and demonstration confronted by the natural philosophical community. These instruments rightly deserved the designation "philosophical," because their function was defined wholesale within the new experimental philosophy. Philosophical instruments intervened in

27. On the classification of early modern instruments, see Bennett 1989, 2003.

28. Bennett 1989, 2003; A. Turner 1987, 126–132, 149–153, 171–182. It should be noted, however, that in the eighteenth century the electrical machine and other philosophical instruments became more widely distributed; see Bennett 2002; L. Roberts 1999; A. Turner 1987, 171–182.

nature and thereby produced novel experimental effects that could not be observed otherwise. In so doing, they contributed to studies of hidden entities such as the vacuum or atoms, whose existence had long been a subject of controversy in scholarly debate. Furthermore, philosophical instruments were built by natural philosophers or by skilled instrument makers in close contact with natural philosophers. On occasion they were also used for experimentation by instrument makers themselves. But in all cases, their use and significance was not stabilized in artisanal practices outside and prior to experimental philosophy. Hence, with respect to the seventeenth-century philosophical instruments, the interaction between the artisan and the experimental philosopher clearly differed in type from the blending of artisanal labor and learned natural inquiry that evolved around the raw materials and processed material goods discussed in this volume.

The seventeenth-century philosophical instruments tell us little, or almost nothing, about the theme studied in this volume: the truly hybrid experts and mixed artisanal and learned practices in the early modern period. Instead, these devices entailed epistemological and methodological problems which were specific to certain particular domains of experimental philosophy but played no significant role outside of this context. The latter issue is discussed in detail in Steven Shapin and Simon Schaffer's seminal study *Leviathan and the Air-Pump* (1985). Shapin and Schaffer show that air-pumps in seventeenth-century England, Holland, and France all varied in design. Apart from Christiaan Huygens, who was present at Boyle's air-pump trials in 1661 and replicated Boyle's air-pump, no experimental philosopher at the time possessed an air-pump identical to Boyle's, even though some possessed full textual accounts of Boyle's pumps.[29] As a consequence, Boyle's experiments with the air-pump could not be perfectly replicated elsewhere. Shapin and Schaffer conclude that early modern experimenters achieved consensus about matters of fact only through local eye-witnessing of an experimental performance using a particular instrument.[30] But their insight into the "intensely problematic nature of replication" in early modern experimental disciplines is not necessarily applicable to all styles of experimentation and natural inquiry in the early modern period.[31] Philosophical instruments were not widely distributed in the seventeenth century, not least since their high

29. Ibid., 230.

30. Elsewhere, Shapin and Schaffer have also independently paid considerable attention to the problematic role of materials in early modern natural philosophy; see, for example, Schaffer 1989, 2004.

31. Shapin and Schaffer 1985, 226.

price put them beyond the reach of all but the wealthy and the instrument makers they patronized. These were elite instruments, which produced a specific methodological problem of the replicability of experiments that ought not to be generalized. Most of the material objects studied in this volume had never been the exclusive property of wealthy gentlemen, nor did their production depend on rare philosophical instruments owned by practitioners of experimental philosophy. For these reasons, some of the most intensely debated historical problems in the study of early modern experimentation—replicability and the public witnessing of rare experimental effects—do not play a significant role in the essays assembled in this volume.

Unlike such philosophical instruments, there were numerous mathematical instruments, such as rulers, sextants, quadrants, astrolabes, theodolites, compasses, plane tables, and circumferentors for surveying, which had long been used for navigation, mapping, astrological prediction, and similar activities.[32] Many new optical instruments of the seventeenth century, such as the telescope and the microscope, also had their origins in the world of commerce.[33] Like the materials studied in this volume, these instruments were both applied in commercial contexts and used for scientific observation and experimentation. Hence, studies of these mathematical instruments can provide insights into the interconnection of learned and artisanal practices of the early modern period that are similar to those afforded by the study of materials. Furthermore, certain more mundane mechanical devices and machines, which have been considered in an extensive literature, most recently by Domenico Bertoloni Meli, also have interesting features in common with the materials studied in this volume.[34] As Meli points out, many of the simple early modern experimental devices such as the beam and the pendulum had practical applications. The beam, for example, was of central importance to engineering, and the pendulum first attracted Galileo's attention as a time-measuring device.[35] Pendulums and beams—as well as cranes, pumps, and mills—were technical devices with widespread applications in the early modern period. Like gunpowder, dyestuffs, metals, or chemical remedies, they belonged to the material world of the craftsman and engineer and were commodities with an economic value and a productive capacity.

Such mechanical devices made the transition to the scholarly world not

32. On mathematical instruments, see Bennett 1986, 1987; Bryden 1993; G. Turner 1990.

33. See Bennett 2003; Clifton 1993; Dijksterhuis 2007; Simpson 1989; Schaffer 1989; Van Helden 1977; Wilson 1995.

34. Bertoloni Meli 2006. See also Grossmann 1987; Hessen 1971; Lefèvre 2000; Renn and Valleriani 2001; Valleriani 2007; Zilsel 2000.

35. Bertoloni Meli 2006, 5.

only as instruments but also as the objects of scientific inquiry in their own right. Models of pumps, mills, and other machines were used in demonstrations of natural philosophy—like the pendulum used by Galileo in investigating dynamics—in a manner analogous to the samples of gunpowder, pigments, or chemical remedies stored in eighteenth-century laboratories for experimental use.[36] In today's science museums, the elaborate, often beautifully decorated philosophical and mathematical instruments of the early modern period are accorded a prominent place. However, as Meli rightly observes, "museum holdings are not representative of the tools used by seventeenth-century scholars, especially those unadorned objects used for research rather than display."[37] This is a problem that the simple and mundane tools of the early modern mechanical arts have in common with the materials and material objects discussed in this volume. With the exception of Palissy's ceramics, none of the latter has ever had a prominent place in science museums. But such objects, which may never have existed in a state suitable for display, played a prominent role in the material culture of the emerging experimental disciplines.

However, mathematical and optical instruments as well as simpler mechanical devices differ from the materials studied here in one important aspect. Being primarily or exclusively involved in investigations that lent themselves to quantification, precision measurement, and mathematical treatment, they direct our attention to a specific domain of early modern expertise that has long been at center stage in the history of science, namely, astronomy, optics, and mechanics, three early modern sciences that were heavily dependent upon such instruments. Yet in alchemy and eighteenth-century chemistry, anatomy, botany, physiology, and mineralogy, instruments of this sort played a minor role or none at all.[38] The studies of materials in this volume thus seek to contribute to a more balanced picture of styles of observation and experimentation in the early modern period, which no longer privileges applied mathematics and natural philosophy.

Tools, instruments and machines, were commonly used to manipulate matter, to produce things and natural phenomena, and to investigate natural

36. On the use of instruments and models of machines in lecture demonstrations of early modern natural philosophy, see, for example, A. Turner 1987, 190–202.

37. Bertoloni Meli 2006, 5. See also our discussion below on objects in museology.

38. These latter disciplines gradually integrated some mathematical and optical instruments, such as the microscope in natural history or the camera obscura in studies of the physiology of the eye (see Wilson 1995; Dijksterhuis 2007; Lefèvre 2007). Likewise, alchemists used the balance, and eighteenth-century chemists continually expanded their toolbox of precision instruments (see Holmes and Levere 2000).

objects and processes. They were artifacts designed to function as a means to achieve certain goals. In the early modern period, the ontological status of these mechanical devices shifted. From being regarded as tools, they came to be considered worth studying in their own right. The Baconian program of experimental history and experimental philosophy was a significant moment in this development, because it redefined instruments and machines as objects embodying natural forces and laws. Instruments, machines, and other mechanical artifacts became legitimate targets of natural inquiry. Historical studies of these objects can thus also contribute to our historical understanding of ontologies of the past. Even so, these mechanical devices formed a small subsection of the natural objects studied in the early modern period and were only one type of material object involved in artisanal labor and learned inquiry into nature; the other was the raw materials that were the subjects of labor and inquiry and the processed substances made from them. The things studied in this volume belong to this second class. A study of early modern materials can thus extend and enrich our understanding of the early modern sciences and material culture beyond a narrow focus on mathematics, mechanics, and experimental philosophy.

## Materials in the History of Technology

Many historians of technology are virtually silent on the majority of materials addressed by our volume, even though few of the substances explored in these chapters were simple natural products.[39] Each was the end result of some processes of refinement and transformation through knowledgeable manipulations; sometimes these products demanded highly complex processing and considerable expertise. Many well-known studies in the history of technology fall into two main methodological difficulties, as far as the materials considered in this volume are concerned. They often take the term *technology* to mean a unitary invention, often a machine or tool, but sometimes also a practice. From this standpoint, the history of technology boils down to a history of the accumulation of mechanical devices or "inventions" and thus inevitably becomes a winner's account.[40] In addressing materials, our project shows up some of the shortcomings of the mainstream

39. Notable exceptions are discussed, in particular, by Mauskopf and Popplow in this volume.

40. The teleological tendency to identify "technology" with "successful invention" is evident, for example, in O'Brien 1991. For an interesting discussion of technology and technological systems, including the relationship between innovation and invention, see R. Adams 1996, 11–35. The editors are grateful to Bernhard Rieger for drawing this reference to their attention.

historiography of "technology." At the simplest, it should be taken as a call to break with such histories and to do justice to the full range of material objects and their uses in early modern cultures. Though none of the materials here presented constitutes a machine or mechanical device, in early modern terms such substances certainly fell within the province of "mechanics" as a discipline. In defining that branch of mixed mathematics, Bernardino Baldi, writing in 1589, would extend it to all "sensible subjects" and to the demonstration of the "several effects that occur in them."[41] The early modern mechanical arts addressed the production of useful effects from the world accessible to the senses, and so were very broad in scope, including activities that have since migrated to more refined domains of practice, such as painting and sculpture.

The "history of technology," taken literally, is not the history of machines and other artifacts but the history of the *study* of the arts. Definitions in French works such as the *Dictionnaire technologique* (1822–1826) show that, by the late eighteenth and early nineteenth centuries, "technology" was construed as a project both of describing and of rationalizing the arts, as well as a program for collecting facts.[42] Writing in the late eighteenth century, the German professor of economy Johann Beckmann, commonly seen as the founder of *Technologie* in Germany, had earlier defined technology (*Technologie*) as a science and a teaching discipline: it was "the science which teaches the processing of (*verarbeiten*) natural bodies (*Naturalien*), or knowledge about the arts and crafts."[43] He further pointed out that the "natural bodies" subjected to the labor process were designated "materials" (*Materialien*): "the raw, or partly processed, natural bodies (*Naturalien*) that are the subjects of labour in the arts and crafts are designated materials."[44] A century later, Karl Marx would adopt Beckmann's definition of materials for his famous definition of the labor process in the first volume of *Das Kapital*.[45] According to Marx, "the elementary factors of the labour process are 1, the personal activity of man, i.e., work itself, 2, the subject of that work, and 3, its instruments"; as

41. Quoted in Marr 2006, 154.

42. The earliest use of the word *technology* in many dictionaries, meaning simply the description of all technical terms proper to the sciences and arts, is Moscherosch 1656, written by a Strasbourg police official. But compare Francoeur 1822, x–xi, xxvi–xxxvii, for a typical later statement concerning the ways in which the arts were to be reformed through scientific knowledge. On the recent currency of the term *technology*, see also Oldenziel 1999, 14–15; on artisans as automata, see especially Schaffer 1999.

43. Beckmann 1777, xv.

44. Ibid., xi.

45. For Marx's studies of Beckmann, see Timm 1964, 60.

the most significant "subject of work," he highlighted "raw materials," which he further defined as "the subject of labour [that] has, so to say, been filtered through previous work."[46]

The essays in this volume question the adequacy of all models of the history of technology and science that assume a one-way flow of knowledge between the sciences and the arts, in either direction. Such models are problematic in several ways, not least in the imbalance of historiographical significance they accord to scientific or artistic knowledge, depending on the preference of the historian in question. In the domains of practice and knowledge studied in this volume, material objects and knowledge-claims moved back and forth between the marketplace and the laboratory, the workshop and the academy, the scenes of artistic and scientific practice. *Circulation, exchange,* and *appropriation* seem more appropriate terms to capture such complex and intertwined relations than the metaphor of flow recently deployed by the economic historian of technology Joel Mokyr.[47] A very different perspective results from Stewart's account of British natural philosophers and entrepreneurs collaborating on countless hydraulic, engineering, and urban reform projects. The sheer number of failed inventions and unfulfilled projects to which Stewart alludes underlines the fact that the history of technology cannot be told as a succession of triumphant, functional inventions. This volume instead heeds the call of historians of technology like Edgerton and Goody for attention to use, skill, appropriation, and materiality.[48]

Another point of comparison with our endeavor might be the history of "technics," relatively unfamiliar to an Anglophone audience.[49] In its concern with mundane materials and their makers and users, our history of technics nonetheless departs from much French *histoire des techniques,* which tend to be limited to reconstructions of past artisanal activities or machines.[50] Instead, it is closer to a historical approach that unites the history of the mechanical arts and invention with cultural history, and whose leading ex-

46. Marx and Engels 1975–2003, 35:188. In other words, the "raw materials" that are to be transformed by the labor process may already themselves have been transformed by previous work.

47. Mokyr 2002, 2. Critiques of this model of knowledge date back over thirty years; see, for example, Chartier 1987; Cooter and Pumfrey 1994; Fish 1980; Holub 2003; Layton 1974; Secord 2003; Suleiman and Crosman 1980.

48. Stewart 1992, 162ff, and part 3; Bijker 1985; Edgerton 2006, ix–xvii; Gooday 2000.

49. In using the term *technics,* our aim is not to ally our enterprise with well-known Anglophone uses of the term in philosophy or in the writings of Lewis Mumford; see Kaplan 2004; Mumford 1986, 2000.

50. Such as, for example, Amouretti and Comet 1998. See also Jacomy 1990.

ponents include scholars such as Maxine Berg, Liliane Hilaire-Pérez, and Natacha Coquery. The practices and materials studied in our volume were all extensively discussed in writing, often by different constituencies, including apprenticed experts, university-educated practitioners, and sometimes also users. This allows a unique insight into the continual trafficking between material manipulations, explanations, and uses, and into the various purposes served by made materials and claims to material expertise.

As we have argued above, the engagement with matter can never be understood to be simple or linear; it was always mediated by debates about method and meaning. Thus, the contributions to this volume remind us that the stakes underpinning different technical projects for transforming matter were plural and sometimes even in conflict. The manipulations of matter in the early modern period were closely connected to the social order as well as to commodification and wealth, as in the making of luxury fabrics or the marketing of new materia medica. They might also exemplify religious debates or political goals of various kinds. Since some seminal work by historians of technology in the 1980s, technological devices have come to be regarded as possessing meaning and functionality not in their own right, but rather within technological systems made up of heterogeneous elements, social, mechanical, economic, and otherwise.[51] Some recent studies in this vein are also paying increasing attention to materiality and the multidimensionality of material artifacts in ways that relate significantly to the approach taken by this volume.[52] At the same time, even this more recent turn in the history of technology continues to be predominantly concerned with high-tech and highly mechanized artifacts, rather than with the mundane processed materials studied in this volume. It remains to be seen whether approaches that present technology as a mode of disciplining bodies and of political agency will prove equally fruitful for the materials that are the subject of our volume.[53]

## Objects in Museology and Commodity History

Another possible approach upon which a history of materials might fruitfully draw originates outside the history of science and technology. Within archae-

51. Cowan 1983, 13; Hughes 1983, 1987; Layton 1974, 38–39; See also Cowan 1997, 2–3; Alder 1997, xii–xiii, 14–19; Bijker, Hughes, and Pinch 1987; Hecht and Allen 2001.

52. See, in particular, Alder 2007; Edgerton 2006; Gooday 2000. Hecht and Allen 2001 also express views similar to those we have discussed above concerning expert power as a heterogeneous thing, partly based around performative action, and concerning the central role played by the material nature of artifacts in their success.

53. For example, Hecht and Allen 2001; Hughes 1983.

ology, anthropology, and museology, objects have long held a prominent, indeed central, place as the subject of inquiry.[54] To some extent this tradition of interest in material culture can be traced back to the work of George Stocking.[55] In recent decades, it has been developed further within museological studies and, concomitantly, the history of natural history and instruments.[56] Material objects often play central roles in this literature, and outstanding work has focused upon the varied problems of preservation, presentation, and display. Indeed, a museological approach to scientific objects has done a great deal to encourage attention to the problems of making and the identities and activities of collectors and instrument makers.[57] The call of such scholars for greater attention to the material is thus a welcome and timely one. However, this approach must, of necessity, remain limited to certain particular categories of objects and excludes the vast majority of materials under consideration in the essays included in our volume. There are several reasons for this. In some cases the material involved is simply too ephemeral; the survival of foodstuffs or remedies since the early modern period is an exception. Orland's chapter, indeed, addresses a material whose most salient feature was its lability. In museological approaches to material culture, the essential property of the materials in question is often their immutability, which has been taken by more than one scholar as the foundation for their epistemological status.[58] But preservation was not a central attribute of the materials brought together here, and certainly not a common attribute. Nor, in general, did they possess other characteristics that might have rendered them suitable subjects for display, such as visual beauty (Palissy's ceramics, discussed by Shell, are the exception here).

Our historiography of the material must adopt a different course than studies of objects in museology, and one that also diverges from another

54. The aim of such accounts is often to generate a big picture of material culture, whereas our focus is more narrowly conceived. While anthropologists tend to treat a given material as univocal in a given culture, our materials often possess multiple significances. Myers (2001) calls for a new approach to material culture that acknowledges the multiple and mutable regimes of value to which a given object may be subject within or between cultures. His call for attention to objects as part of "global as well as local circuits of exchange, display, and storage" is exemplified in several of the contributions to our volume.

55. Stocking 1985; see also Thomas 1991.

56. For museological approaches, see Hooper-Greenhill 1992; Pearce 1989, 1994, 1990; Elsner and Cardinal 1994. For histories of early modern collecting that have attended to materials, see, in particular, Findlen 1994; te Heesen 2002.

57. The work of Jim Bennett, in particular, moves between questions of display and questions of making; see Bennett 1987.

58. Latour 1986; Pomian 1987.

group of studies with some affinity to the aims and concerns of our project. In recent years, the materials of pharmacy and natural history have attracted increasing attention in their guise as commodities. Interest in the drug and specimen trades has grown concomitantly. During the early modern period, many European countries became increasingly implicated in global networks of trade and exchange, beginning with the Portuguese in the fifteenth century. By the eighteenth century, Portuguese supremacy in world trade was contested by the Genoans, Venetians, English, Dutch, and French; many other European nations undertook long-distance trade to a greater or lesser degree.[59] Arguably, the rise in early modern collecting was made possible by this increasing European access to distant parts of the world, as well as by changes in trading patterns. The largest literature on this subject addresses the transformations these events brought about in the "world of goods," to borrow the phrase of Mary Douglas and Baron Isherwood. Over the period covered by this book, well-to-do Europeans found exotic and luxury goods like sugar, coffee, tea, chocolate, silk fabrics, and Asian ceramics increasingly within their purview.[60] It was precisely in these two centuries that everyday life for large sectors of European society underwent extensive change, and that change took the form of possessions. Many of the made objects and materials purchased by European consumers served purposes of self-fashioning in different ways—for the Dutch in the late seventeenth century, the display of learning and virtue was mediated by natural history collections. The consumption of exotic drugs in remedies produced by apothecaries played an important role in constructions of sickness for early modern medical clients. Moral, financial, social, and epistemological purposes became intertwined in the purchase, manipulation, and deployment of such materials.[61]

Increasing attention to materials as commodities has begun to change our map of the world of learning in the early modern period, in part by forcing attention to the dual function objects might play at once as a commodity and a subject of learned inquiry. Yet at the same time, the full significance of the global movement of objects as a *commercial* problem has not always been explored. At the different points where objects changed hands and were reworked or used, different systems of valuation intersected. Under such circumstances, those who could make prescriptive statements about the au-

59. Chaudhuri 1985; Cook 2007. On the movement of commodities as a historical issue, see Appadurai 1986.

60. Douglas and Isherwood 1996. See also Hobhouse 1985; Levere and Turner 2002; Mintz 1985; Schivelbusch 1993; W. Smith 2002; Walvin 1997.

61. Cook 2007; Hochstrasser 2007; W. Smith 2002; Schiebinger and Swan 2005.

thenticity, rarity, and value of individual materials performed an important function as mediators between consumers and their goods. The practitioners who are the actors in our volume could thus be seen as gatekeepers of change in both directions. In their hands, matter was both transformed and transformative: it left their workshops and laboratories in a new and different condition, but also went on to alter the condition (social, physical, moral) of consumers and clients, from courts to commoners.[62] Both producers and consumers—as well as the relationship between them—were thus reconfigured through their engagement with the new materials that flooded their societies, whether obtained from long-distance trade or from local manufacturing. This is a major reason why we would side with those historians of consumption who scrutinize the changes in everyday life that preceded and fostered the rise of large-scale or mechanized production in the Industrial Revolution.[63] From such a perspective, it is possible to see the practitioners addressed by this volume not as isolated and peculiar figures, but rather as key parts of a wholesale transformation of the economic world.

Not only are the historiographical trends currently manifest in commodity history and the history of consumption generating an increasingly interesting and rich revisionist picture of the Industrial Revolution, they are also helping to confer a new standing upon the material as the subject of historical inquiry. In the words of Daniel Roche: "Some contest or even still deny the right of Enlightenment historians to place on a par phenomena which derive from a clearly definable intellectuality (in their eyes) and those which derive from material culture, which is relegated to a lower register of analysis. Some go so far as to find it amusing, in the name of an idealist philosophy open to challenges and of a purblind epistemology, that questions of the following sort should be posed: what role do water, light or food play in Enlightenment culture?"[64]

Roche's implication, that the history of knowledge cannot be divorced from the history of the material world, is one that we believe to have fruitful implications for our volume. In particular, the attention of the contributors to this volume turns in a reflexive direction, by addressing the types of accounts that could be given of the nature and powers of made materials, both by makers and by users. In this sense, we seek to bring the agendas of

62. Poulot 1997, 350–51.

63. See Berg and Eger 2003b; Berg and Clifford 1999; Pennell 1999, 558; Reddy 1984; Roche 2000a; Shammas 1990.

64. Roche 2000b, 11.

historians and sociologists of science and technology together with those of historians of material culture, commodities, and consumption.

## Primary Sources

Secondary sources can be found in the cumulative bibliography at the end of the book.

### PUBLISHED SOURCES

Beckmann, Johann. 1777. *Anleitung zur Technologie, oder zur Kenntnis der Handwerke, Fabriken und Manufacturen, vornehmlich derer, die mit der Landwirtschaft, Polizey und Cameralwissenschaft in nächster Beziehung stehn. Nebst Beiträgen zur Kunstgeschichte.* Göttingen: Vandenhoeck.

Dictionnaire technologique. 1822–1826. *Dictionnaire technologique, ou Nouveau dictionnaire universel des arts et métiers, et de l'économie industrielle et commerciale.* 22 vols. Paris: Huzard-Courcier.

Francoeur, Louis-Benjamin. 1822. Discours préliminaire. In *Dictionnaire technologique, ou Nouveau dictionnaire universel des arts et métiers, et de l'économie industrielle et commerciale.* Paris: Huzard-Courcier, 1:vii–lxvii.

Marx, Karl, and Frederick Engels. 1975–2003. *Collected Works.* 49 vols. London: Lawrence & Wishart.

Moscherosch, Johann Michael. 1656. Technologie Allemand & Françoise, das ist, Kunst-übliche Wort-Lehre, teutsch vnd frantzösiche. Strasbourg: Iosias Städeln.

PART ONE

# The Production of Materials

## Introduction to Part 1

The chapters in this first part of our book are concerned with early modern expertise involved in the production of materials, including vermilion, ceramics, metals, ink, and ethers. They explore obstacles and incentives emerging in commercial production or experimental preparations of materials, and they follow the various ways in which practitioners met such challenges. Furthermore, they seek to come to grips with materiality by studying the obstinacy and resistance of materials as well as surprising effects occurring during their transformation processes. What surprises and obstacles emerged in the production of, say, silver in the Harz region of the sixteenth century? What forms of useful knowledge were developed in response to these obstacles? What was, in this local context, the specific role of the hybrid experts who are at the center of this volume? How did these experts combine geometrical, mineralogical, geographical, chemical, and other forms of learned knowledge with local hands-on experience? But not only immediately useful knowledge is at stake here. The chapters are also concerned with the historical actors' understanding of their materials, the meanings and values they attributed to them, and the orders of things they established. They thus also provide a glimpse of the experts' ontology and of ontological shifts during the period under investigation. Pamela H. Smith's and Hanna Rose Shell's essays, in particular, show that the sixteenth-century practices of making materials were replete with invisible agents and signs, which faded away in the centuries to follow.

In her chapter about vermilion and metalworking, Pamela H. Smith studies in fine detail late-medieval and early modern artisans' ways of making, measuring, and naming materials along with their ontology, so strange today, and their impact on the development of alchemical theory. The production of the deep red pigment vermilion was a complicated and dangerous technical process that had been explored and described continually in written recipes since the early middle ages. By the early eighteenth century, numerous accounts of innovations in vermilion making appeared in print, which reduced the labor, length, and danger of the process. Such written accounts on the production of materials, Smith argues, contributed to the articulation of alchemical matter theories. Furthermore, the web of signs and correspondences addressed in these accounts—such as the correspondence between the red color of vermilion and blood, generation, and life—manifests a "vernacular science" that differed profoundly from modern understandings of nature. "While the origins of many techniques to investigate, observe, and analyze the material world can be found in the early modern period," Smith points out, "the meanings of matter and the practices of manipulating matter were rooted in a different understanding of the world."

The sixteenth-century ceramicist, natural historian, philosophizing experimenter, and teacher Bernard Palissy, studied by Hanna Rose Shell, is a compelling example of a hybrid expert. Trained as a craftsman, he produced and sold beautiful ceramics and at the same time sought to explore nature by molding objects out of clay. As Shell points out, Palissy wrestled almost endlessly with his raw material, clay, to transform it into simulacra of plants, animals, coloration, generation, and fossilization. Apart from his ceramics, Palissy is today also known for his writings on geology and agriculture, especially his *Discours Admirables* (1580). His texts, Shell argues, can be best understood in direct relation to his trials with clay and creation of ceramic artifacts. Clay was, for Palissy, a vital medium for inquiry into terrestrial and organic processes. Ideas about nature were thus not only expressed in, but also generated by, the process of casting nature in clay.

Christoph Bartels presents a detailed account of innovations in mining and the production of metals, especially silver, copper, and lead, in the German Harz region from the late medieval until the early modern period. His chapter illuminates the metallurgical ventures of many ingenious experts who are obliterated in the history of science, despite the fact that they contributed to the development of early modern sciences such as chemistry, geology, and mineralogy. To this group of experts belonged, in addition to the quite well-known sixteenth-century assayer and controller of the mint Lazarus Ercker, Heinrich Albert von dem Busch, an expert in ore prospecting and stratigraphy, who also introduced new and efficient methods in mining administration; Claus von Gotha, an engineer who solved the urgent problem of drainage in flooded mines by creating new hydraulic machinery; Daniel Flach and Caspar Illing, who introduced new methods of land and mine surveying; the Saxon miner Carl Zumbe, inventor of a new mining technique in connection with rock blasting with gunpowder, which he tested systematically; the administrator Christoph Sander, who significantly contributed to the standardization and rationalization of local smelting processes by introducing new methods in calculating input and output factors; and the young artillery officer Georg Winterschmidt, who participated in the search for a more efficient motor than waterwheels and developed the water-pressure engine around 1750. Early modern mining and metallurgy, Bartels argues, relied significantly on land surveying, stratigraphy, ore prospecting, assaying (the chemical analysis of ores and metals), data collection, and mathematical data processing as well as the writing of technical instructions and treatises.

Adrian Johns's chapter turns to ink, its astonishing varieties and ways of production from the medieval period until the beginning of industrialization. As ink recipes presented in books of secret and of natural magic demonstrate, the preparation of inks always required practical knacks. It meant long, arduous, and dangerous labor, which stimulated continual trials and adjustments to new techniques of application; Johns provides compelling examples from the ways of preparing printers' ink. Like the producers of chemical remedies and dyestuffs, addressed by Klein and Nieto-Galan in this volume, the makers of different varieties of ink had to find ways to ensure the quality of their products. They developed clear criteria for judging their quality, discerning good ink from poor ink, and for avoiding adulteration. As Johns points out, adulteration was a permanent concern, which "reflected the dispersal of knowledge and skills" involved in ink making. In addition to its practical use as a material, ink also bore associations with natural magic and alchemy in the sixteenth and seventeenth centuries. In the second half of the late eighteenth century, it further became an object of entrepreneurial competition and extensive experimentation. The history of ink, Johns concludes, is thus a history of an entire system that includes the material substance, paper, instruments, techniques, and places as well as people, skills, and attitudes.

Ursula Klein takes studies of the production of materials—in this case a particular group of chemical remedies, the ethers—as a route to explore late eighteenth-century apothecaries' shifts from commercial production to natural inquiry and vice versa. In the course of the second half of the eighteenth century, apothecaries performed endless trials to facilitate techniques for producing ethers and to improve methods of identifying them. The search for unambiguous ways of identifying ethers here was embedded both in attempts to standardize medicines and avoid adulteration as well as in the writing of experimental histories of material substances. As Klein's chapter demonstrates, there were also apothecaries who smoothly carried their explorations of production techniques and methods of identification further to chemical analysis and theoretical explanation. Such shifts of inquiry contributed significantly to the recognition of apothecaries as learned chemists. Although they clearly manifest apothecaries' aspiration for higher learning, they often hinged on professional ethos and commercial interests as well. The early modern apothecary-chemists were truly hybrid experts participating both in the world of merchants and artisans and that of university professors and academicians. And it was the hybrid experts themselves in this case, rather than university-educated scholars interested in the arts and crafts, who integrated artisanal expertise into the academic system and thus contributed to its deep social and intellectual reorganization.

2

# Vermilion, Mercury, Blood, and Lizards: Matter and Meaning in Metalworking

PAMELA H. SMITH

Historians have debated the relationship of science and technology for many years and, although no consensus has been reached, most would now view science and technology as occupying a kind of continuum, rather than existing in entirely different spheres or possessing completely separate ends. This chapter strives toward such a reconsideration of the relationship between science and technology by examining the ways in which matter and its manipulation—making—are related to deductive and propositional knowledge—knowing—in early modern Europe. Such a focus, I think, leads to interesting consequences. It undermines conventional ideas about matter and natural materials, for it shows that even natural materials cannot be viewed as transhistorical entities. Moreover, it makes clear that historians must pair their discussions of primary texts in the history of science with an account of the practices that accompanied the texts, for these often seem to have formed parts of a continuum, particularly in the texts we now see as part of the history of chemistry.

To begin such an examination, we can draw three important essays to our assistance: First, Helen Watson-Verran and David Turnbull, in *Science and Other Indigenous Knowledge Systems,* note that scientific knowledge is heterogeneous; there is no term that "captures the amalgam of place, bodies, voices, skills, practices, technical devices, theories, social strategies and collective work that together constitute technoscientific knowledge/practices."[1] For this reason, they use *assemblage.* Such technoscientific assemblages can

1. Watson-Verran and Turnbull 1995, 117. This view has been helpfully developed in two special issues (Technoscientific Productivity) of *Perspectives on Science* 13, numbers 1 and 2 (2005), including essays collected by Ursula Klein.

bridge "making" and "knowing," that is, can make the transition from local to general knowledge if they develop a means of transmitting knowledge. This transition was accomplished, they argue, in the templates used by medieval cathedral masons; in the calendars that the Anasazi built into their edifices; in the control of large amounts of information by means of the ceque and quipa by the Inca; and in the memorized map of the heavens and the concepts embedded in navigational practices used by Polynesian navigators. All meld local practices into stable assemblages that connect local to general forms of knowledge.[2]

Second, Roger Chartier, in *Culture as Appropriation: Popular Culture Uses in Early Modern France,* considers readers, printers, and the contents of chapbooks. These inexpensive vernacular books have always been considered a cultural form that was solidly "popular," indeed not just as a component of popular culture, but as its very marker. Against this, Chartier argues that the notion of separate, culturally pure "cultural sets" must be replaced with a "point of view that recognizes each cultural form as a mixture, whose constituent elements meld together indissolubly." Chartier demonstrates that no single cultural boundary between popular and elite existed in early modern France, but rather, shifting boundaries undergoing constant repositioning was the norm. In his analysis of chapbooks, the redrawing of cultural boundaries was closely related to the publishing strategies of printers. One could argue that the proclaiming of the new experimental philosophy in the seventeenth centuries was a moment of such redrawing of cultural boundaries.[3]

Finally, the work on "everyday technology" of Jean Lave brings us to reconsider the relationship of thinking and doing and of knowing and making.[4] As we all know, schemas of knowledge making and knowledge makers are hierarchical, almost without exception placing knowing above doing and making. Tim Ingold, in arguing for the primacy of doing over thinking, that is, of experiential knowledge over propositional knowledge, as well as for the embodied and situated nature of *all* knowledge, gives us a new starting point: "We do not have to think the world in order to live in it, but we do have to live in the world in order to think it."[5] As I will argue in what follows, we need to think about lived experience and the ways in which principles might be

2. We might differ from them in the necessity of an "inscription device" for making the transition from local to general, because many of the chapters in this volume show that knowledge can be disseminated and theorized when contained in embodied techniques, folk songs, routinized practices, and other nonwritten forms.

3. Chartier 1984.

4. Lave 1988.

5. Ingold 2000, 418.

embodied and lived, rather than articulated and externalized. I believe that if we take seriously the attempt to see making and knowing, elite and popular, and science and technology as parts of a continuum, we can write a new history of science, one in which phenomena that used to be seen as local can be integrated into an overarching narrative of the making of knowledge. The history of science might in the broadest conception be written as the history of material life and the human engagement with nature.

## Matter and Meaning

The production of materials in early modern Europe was intimately related to the investigation of nature, both at the time this production was carried out in artisanal workshops and as the investigation of nature developed into what we consider to be the modern empirical practices of natural science. In what follows, I will touch briefly on several ways in which the production of materials and the practices by which artisans manipulated natural materials were related to what would become modern science. First, in recent work, historians of chemistry have pointed to the importance of practices in early modern metalworking—for example, in assaying, smelting, alchemy, and mining—for the development of new empiricist practices in the investigation of nature (see also Bartels, this volume). Quantitative measurement was one of these practices,[6] and there is no doubt that precise quantitative measurements were critical in metalworking—the first enclosed balance was used by assayers, while measurements of heat were crucial for the smelting and casting of metals and for the making of steel. While quantitative measures are just one of the marks of modern science that can be discerned in early modern artisanal practices, it is important not to lose sight of the meaning of matter to early modern people. While the origins of many techniques to investigate, observe, and analyze the material world can be found in the early modern period, this chapter will attempt to provide examples of some of the ways that the meanings of matter and the practices of manipulating matter were rooted in a different understanding of the world.[7]

Let us begin, then, with quantitative measurements. A sampling of measurements used in medieval and early modern artisanal recipes gives us a glimpse of the difference between our world and that of the early modern artisan. For example, volumes are expressed as "a walnut shell" of liquid, or pea-sized, or "a hazel-nut of gum *armoniacum*," "a pea of Armenian bole,"

6. Newman and Principe 2002.

7. See also Shell's contribution to this volume.

"a small pot of honey," "four drops of spittle,"[8] or "as much of the mash as would equal a goose egg,"[9] two-fingers wide, as well as the measures, more conventional to us, *Maß*, *Lot*, and *Unzen*. Numbers of recited pater nosters measured the passage of time, as did, in one case, the length of a long summer day and a night.[10] These appear to us as very open-ended measurements. This may well be deceptive, however, for if one considers the techniques associated with achieving the correct heat in the furnace—one of the most important aspects of metallurgy—things look rather different. One treatise instructs the metalworker to throw a lock of female hair into the mass and when it catches fire, he will know to take the mixture off the fire.[11] A sixteenth-century metalsmith's treatise states: "To know the right heat, dip a little piece of paper into your metal. If it blackens without burning, you have got the right heat. But if the paper bursts into flame and burns, it means that your metal is too hot."[12] The early twelfth-century metalworker Theophilus writes in relation to producing a mold: "When it is completely red-hot inside, take it off the fire and let it lie until it is cool enough to be held for a short time in your hand."[13] Even in the twentieth century, such measurements were used, for example, by potters who lit adjacent kilns at the correct temperature by covering over the window opening between the kilns with a sheet of newspaper. When the paper reached a high enough temperature, it caught fire, lighting the adjacent kiln at the required temperature.[14]

Artisanal recipes and treatises employ many measurements that make use of all five senses. These measurements were often expressed in a language which, because of its sometimes animate idiom, can sound to us imprecise. For example, taste was often employed: vitriol could be identified by its biting, sharp to the taste, pungent to the tongue, astringent nature, while rock alum had "a bitter taste with a certain unctuous saltiness."[15] Other measure-

8. *The Strassburg Manuscript* 1966, 67. This collection of recipes is written in Old Middle German, but the date of its composition is uncertain.

9. *Ibid.*, 39.

10. Kärtzenmacher 1538, xiii verso.

11. *Rechter Gebrauch* 1531, IX verso; this recipe is for making silver from tin from which to produce cups.

12. From a recipe for casting in a cuttlebone mold from an anonymous late sixteenth-century collection of goldsmith's recipes and techniques; Bibliothèque nationale de France (BnF), Paris, Ms. Fr. 640, n.d., f. 145r.

13. Theophilus 1979, 181. Theophilus was in all probability the monk Roger of Helmarshausen.

14. Reese Rawlings, 2003, Course in Conservation, Victoria and Albert Museum, lecture at Imperial College.

15. Biringuccio 1540, 95 (on vitriol), 98 (on rock alum).

ments relied upon hearing: "Put your cuttlefish bone very close to the fire. If you hear little cries, it means that your bone is dry enough."[16] In a process for hardening mercury, the material in the crucible makes a loud bang to sign that it has had enough of the fire.[17] And, in another, "If the tin cries very much, it means you added enough lead and not too much; if the tin cries softly, it means you added too much lead."[18] The sixteenth-century Italian metalworker Vannoccio Biringuccio writes of tin that it is purer when it shows itself granular like steel inside. Also, when bent or squeezed by the teeth, it makes cracking sounds "like that which water makes when it is frozen by cold." Furthermore, good iron ore was also indicated by the presence of bole and another red, soft, fat earth that made no crackling noise when squeezed between the teeth.[19] In his dramatic account of casting bells, Theophilus advises the caster to "lie down close to the mouth of the mold" as the metal is poured into the bell mold, "and listen carefully to find out how things are progressing inside. If you hear a light thunderlike rumbling, tell them to stop for a moment and then pour again; and have it done like this, now stopping, now pouring, so that the bell metal settles evenly, until that pot is empty."[20] An anonymous goldsmith's treatise advises the assayer to place the gold to be assayed into the acid and let it work until the water is full of dissolved silver. In order to make sure that it has dissolved all available silver, the assayer should listen carefully: "and to have sure knowledg thereof, laye your eare unto the said glass, and if it be full laden and chardged with silver it will sound in this wise, bott, bott, bott; Then lett your said glass coole."[21]

The terms of these measurements sound unfamiliar, as they attempt to put into words the experiential and sensory knowledge of artisans that was almost always left unarticulated. Such observation was subjective but also learnable and replicable. Their unfamiliarity should not mislead us as to either their precision or their sophistication.

More than any other type of knowledge, except perhaps medicine, artisanal knowledge was based on observation and experiment, and it is clear from manuals that artisans undertook constant experimentation. In the first half of the sixteenth century, the writer on metalworking, Vannoccio Biringuccio, advised constant trial: "It is necessary to find the true method by doing it again and again, always varying the procedure and then stopping at the

16. BnF, Ms. Fr. 640, f. 145v.

17. *Rechter Gebrauch* 1531, III.

18. For casting with lead, see BnF, Ms. Fr. 640, f. 131v.

19. Biringuccio 1540, 67.

20. Theophilus 1979, 173.

21. *The Goldsmith's Storehouse* [ca. 1604], chap. 12 (Assaying of Gold), fols. 22v–24r.

best."[22] "For this reason it is necessary to have a superabundance of tests, and to test and try enough to find the desired aid [to smelting], not only by using ordinary things but also by varying the quantities, adding now half the quantity of the ore and now an equal portion, now twice and now three times, so that the virtue that the ore contains may better defend itself from the fire and from the evilness of its companions."[23]

An anonymous sixteenth-century metalsmith's manuscript collection of recipes and accounts of his work testifies to many experiments over the years, particularly with sands of differing compositions for casting. To take one example from its numerous accounts of "experiments": "I tried four kinds of sands for use with lead and tin: chalk, crushed glass, tripoly and burned cloth." At another point, he explicitly refers to the experimental nature of his work: "Since my last experiences, I molded with burned bone, clinker and burned felt."[24] His manuscript also contains numerous observations on the behavior of animals he has caught and kept for life casting. For example,

> Keep your snake in a barrel full of bran, or, better, in a barrel full of earth in a cool place, or in a glass bottle. Give your snake some live frogs or other live animals, because snakes do not eat them dead. Also I've noticed that when snakes want to eat something or to bite, they do not strike straight on, on the contrary they attack sideways as do Satan and his henchmen. Snakes have small heads, but very large bodies, they can abstain from eating for 7 or 8 days, but they can swallow 3 or 4 frogs, one after the other. Snakes do not digest food all at once, but rather little by little. . . . If you worry and shake your snake, it will bring up digested and fresh food at the same time. Sometimes 2 or 3 hours after swallowing a frog, it can vomit it alive.[25]

Another goldsmith's manual from around 1604 makes clear the firsthand observational and empirical underpinning of assayers' practices: In all, he wrote, assaying metals "asketh a good Judgment, gotten rather by years & experience, then by speculation & dispute." Besides a "grounded experience in this Science or mysterie," the goldsmith "should have a perfect Eye to vewe, & a stedye hand to waye [weigh] for other mens senses cannot serve him."[26]

The precision, eyewitness observation, and experimentation that underlay practices of metalworking would be incorporated into the new experimental

22. Biringuccio 1540, XVI.

23. Ibid., 143–44. He goes on: this is a "treatment" for ore—the "ores must be so tormented that the obstinacy of their hardness is overcome" (144).

24. BnF, Ms. Fr. 640, f. 86v.

25. Ibid., f. 109r.

26. *Goldsmith's Storehouse* [ca. 1604], f. 6v.

philosophy in the seventeenth and eighteenth centuries. But this is just one way in which artisanal practices intersected with the development of a new science. Another is illustrated by an example from mining: According to the beliefs of German miners, spirits—*Bergmännlein*—made their homes underground. These spirits also appeared in woodlands and around springs, but were especially prevalent in the mines. While the presence of these gnomes was a sign of the productivity of the mines, humans had to show respect for these guardians of the earth's treasure. Any harm done to the spirits could cause them to transmute noble metals into worthless dirt. This was especially true of one group of these sprites, called *Kobolden,* who produced a corrosive ore, named by the miners *Koboldertz,* which corrupted valuable metal ores. The medical reformer Theophrastus Bombastus von Hohenheim, called Paracelsus, took up this terminology for what today is known as a cobalt-containing ore. Via his writings, this terminology flowed seamlessly from miners' beliefs to chemistry textbooks, and today Paracelsus is viewed in the history of chemistry as the discoverer of cobalt.[27] This is yet another example of the way in which the making and producing of materials intersected with and formed the basis of what have been regarded as theoretically driven enterprises, such as the emergence of modern science.[28]

Another intersection between artisans' practices and scholars' theory in the making of materials can be found in the production of the deep red pigment vermilion from sulfur and mercury. The remainder of this chapter will examine the intersections between pigment making and metallic theory and argue that, while the development of theoretical concepts out of making practices can be put in the familiar terms of theory and practice, we should also pay attention to the more alien dimensions of the worldview that underlay the practices of pigment making and metalworking.

## Vermilion

In the last half of the seventeenth century, the Amsterdam paint seller Willem Pekstok (1635–1691) and his wife Katalina Saragon (d. 1681) recorded a number of recipes, many of them for sealing waxes, pigments, dyes, borax, and other materials of their trade; but they also noted eye remedies, medicines for fever, bladder stone, and other ailments; as well as brandy wine distillations and the making of hippocras, a spiced wine. The longest portion of their

27. Webster 1982, 18.

28. Other examples are numerous, such as Isaac Beeckman and Descartes, as recounted by van Berkel 1983.

recipe collection consisted of detailed instructions for vermilion production on a large scale.[29] Vermilion, created by combining mercury and sulfur, was employed as a pigment to produce a deep saturated red. It was a synthetically produced form of a naturally occurring mercury ore known as cinnabar (HgS). Ground cinnabar, which had been widely used both in the East and the West, was replaced by the artificially manufactured cinnabar as the technique spread, probably from China to the Islamic World and thence to Europe.[30] Pekstok's instructions are unusual because of their thoroughness, beginning with a description of the shed needed for the operation (with its twelve-foot-high roof, four-foot-deep hearth, and forty-foot chimney), and moving on to the construction of the furnaces, pots, trivets, luting mixture for sealing the pots (including even the proportions of potter's clay, sandy clay, wood ashes, horse manure, crushed glass, and so on of which the luting substance was mixed), and finally the process of vermilion manufacture itself. Each section included measurements for all materials employed, as well as illustrations of the pots and rings that held them in place. The shed, furnace, pots, and ironwork are described in such a way that it appears the Pekstoks constructed and produced all pots and ironwork on the furnace site.

The process itself is clearly a spectacular one, and the Pekstoks were pursuing it on a large scale. Amsterdam was the center of vermilion making in Europe in the seventeenth century; only China produced finer pigment.[31] Their recipe calls for eighty livres of quicksilver and twenty-five livres of refined Italian sulfur.[32] The sulfur was melted in an iron pot, taken off the fire, and mixed with forty livres of quicksilver. This mixture was stirred with an iron shovel and allowed to rest until covered with a thin film, at which point forty more livres of quicksilver was added and stirred well until the two ingredients were entirely mixed and began to harden. The contents of the pot were dumped onto a wet iron slab and spread to a thickness of about two fingers. The worker had to be alert with his water and wet brush in case any flame broke out, and he should have eaten a "thick piece of bread and butter" before be-

29. The recipes as well as the last wills of Willem and Katalina were probably copied by their son Pieter, who inherited his father's business in 1691. The Pekstok papers are held in the Amsterdam Gemeente Archiv # N 90.23 (Pekstok papers n.d.), and the description of vermilion making has been published by van Schendel (1972).

30. Gettens, Feller, and Chase hazard this account of the spread of vermilion making (1994, 160). Wallert believes the technique predated the translations from Arabic to Latin (1990, 155).

31. Harley 2001, 127.

32. The following description and quotations are from van Schendel's transcription and translation of the process in Schendel 1972, 71–82.

ginning the mixing process to protect himself from the fumes.[33] The mixture would then turn dark blue on the outside and an even silver color on the inside without any individual drops of quicksilver being visible. When the molten slab of metals had reached this dramatic stage, it was to be pounded with a pestle in a large wooden bowl and divided between eight to ten small pots.

A large pot was then hung in the furnace on its trivets, piled all around with bricks, and heated above the flame. The firing had to be even and carefully controlled so that the pot did not burst. When the pot and its surrounding bricks were heated to a glowing red (about five hours), it had to be sounded with an iron bar to ascertain that it had not cracked and the bricks removed to halfway down the pot. Ash was then laid upon the top layers of bricks to keep the flame from reaching above the midpoint, the fire was stoked, and three of the small pots full of the mercury-sulfur mixture were added to the red-hot vessel. After a short time, a flame about three to four yards high shot out of the pot. When this began to die down, three more pots were added. This was repeated until about one hundred livres of the mixture had been added. The fire, which had been kept hot during this stage, was allowed to die down slightly and the pot was covered with an iron lid. One watched the pot for any sign of a "greasy mixture" rising to the top, a sign that the sulfur was not sufficiently burned to begin the sublimation process. If this occurred, the fire had to be stoked again until the "greasy inflammability" burned up. The next stage of the process reached a culmination in the subliming process that produced the pigment:

> When a thick clear flame appears out of the blueish fire all full of slate-like glittering, this is the correct flame and indicates that the substance sublimes. Place your lid so that it is lighted up by the subliming flame. Now it is not greasy and goes out immediately. When a smooth dry red is shown on the lid shining out of the brown, and when this flame is present, then the pot has to remain covered because, if it remains open, the vermilion escapes constantly. But one has to take care that the lid remains constantly loose.
>
> When the flame comes out forcefully between the pot and the lid one must reduce the fire under the pot. By doing this the flame in the pot also diminishes and slowly starts to burn against the lid.

If the fire is not hot enough, "the flame in the pot reacts accordingly and turns into black smoke." This signified a complete loss of the quicksilver because it had evaporated without subliming. If all was going well, the lid had

33. For an exploration of the relationship of butter to mercury, see P. Smith 2005.

to be removed in order to watch the process. If the lid became stuck to the top of the pot by the subliming vermilion, one had to "slam the pot so hard" that the lid might fly off, but the instructions admonish, "Do not be afraid, but put it back on again," and "Stoke the fire the right way (because when you allow the lid to bake too fast to the pot, which happened to me once, and when I knocked off the lid, the pot, the vermilion, and the lid flew about my ears in ten thousand pieces; that is why you may not be sleepy, but always be awake and careful)."

The sublimation process took about two hours, at which point the lid was removed and the sublimed cake stirred loose from the sides of the pot with an iron bar. A row of bricks was removed in order that sublimation would continue lower down in the pot. After firing for "seven, eight, or nine hours, depending on the number of little pots, each one taking about an hour, and on how well the fire reacted, one feels inside the pot with a broomstick. If it is empty, another ninety livres of the sulfur-mercury mixture should be added and the whole process repeated at least twice. If one wanted to make the cake really heavy, one could knock the top of the pot open with a pointed hammer ("rather dangerous," as the writer records) and put in a last load of forty to sixty livres of the mixture.

At the end of the last burning, the fire was to be put out and the covered pot allowed to cool and then set upside down on a slab. The pot was then chipped away from the cake of vermilion and the two to three hundred-livre bell-shaped cake was bound with a hoop to keep it from disintegrating. The Pekstok instructions end with notes on how to deal with pots cracked while firing or subliming and which kinds of fuel (driftwood, turf) might be employed. "In conclusion, this is the perfect manner in which to make large cakes of vermilion from two hundred to three hundred [livres] in weight. You will learn the use and preservation of it. People now make cakes of four hundred, five hundred, and five hundred and fifty [livres] from the year 1650 until 1671."

The survival of such clear and detailed notes on vermilion making is unusual, although many less elaborate recipes put into writing in Europe since the eighth century are still extant today. Vermilion had been made in Europe since the early middle ages; the first recipe for it is contained in an eighth-century collection, a recipe that continued to form a basis of written recipes throughout the middle ages. Daniel V. Thompson has noted that although the techniques do not differ much from one recipe collection to another, the proportions of mercury to sulfur vary with some frequency, ranging from 2:1 to 4:1, as does the notation of whether to add sal ammoniac or not.[34] As Thomp-

34. D. Thompson 1933–1934, 62–70.

son records, recipes become less frequent by the fourteenth century precisely because it was such a familiar process. Indeed, vermilion was already almost universally used as a red pigment by manuscript illuminators by the twelfth century.[35] By the time of the painter Cennino d'Andrea Cennini's painting manual, *The Book of the Art* (late fourteenth century), it could be bought from the apothecary, as is evident from Cennino's comment that vermilion "is made by alchemy, prepared in a retort," but "I am leaving out the system for this, because it would be too tedious to set forth in my discussion all the methods and receipts. Because, if you want to take the trouble, you will find plenty of receipts for it, and especially by asking of the friars. But I advise you rather to get some of that which you find at the druggists' for your money, so as not to lose time in the many variations of procedure."[36] By the sixteenth century, in Germany and no doubt elsewhere in Europe, vermilion was a midpriced commodity sold in apothecary shops. Price lists from Munich shops show that it stood at around half a pfennig per gram in lists of prices between 1488 and 1553 that ranged from .07 pfennig/gram for Armenian Bole to 3.20 pfennigs/gram for cochineal (or kermes?).[37] From the early eighteenth century, numerous accounts of vermilion making appeared in print, in part due to the invention in Germany in the 1680s of a wet process for vermilion manufacture that considerably reduced the labor, length, and danger of the process.[38]

## Mercury

Cennino's comment that vermilion is made by alchemy indicates not just the red pigment's artificial nature and its fiery production, but also the degree to which artists' practices overlapped with the textual tradition of alchemy. Indeed, vermilion was of particular interest to scholars, such as Albertus Magnus, throughout the Middle Ages because it was produced by the combination of the two principles of all metals, sulfur and mercury. The view that sulfur and mercury were the basic components of all metals first entered Europe textually in the work of Jabir Ibn Hayyan, when it was translated from Arabic to Latin, and continued to form the cornerstone of alchemical theory up through the seventeenth century.[39] The principles of sulfur and mercury

35. Ibid., 70.

36. Quoted in ibid., 69.

37. Burmester and Krekel 1998.

38. See, for example, the German literature: Schießl 1989, 175–76. On the wet process, see Gettens, Feller, and Chase 1994, 2:163.

39. See the work of Allen G. Debus, B. J. T. Dobbs, and, briefly, Holmyard 1990; Stillman 1960. In the sixteenth century, Paracelsus added a third principle—salt—to sulfur and mer-

were not identical with the familiar material forms of these two elements, but rather these "principles" drew upon the physical characteristics of material sulfur and mercury. In a pure form in nature, sulfur, on heating, turns a dark red color, and then, when cooled rapidly, forms a glassy red substance. Native mercury, on the other hand, is liquid at room temperature and possesses a silver glittering quality. In alchemical theory, these qualities were the essential components of the principles of mercury and sulfur, and, as such, they accounted for the behavior of metals. In alchemical texts, sulfur was viewed as the hot, fiery, male principle, representing the qualities of fire and air and giving metals their combustibility. It combined with the wet, cold, female principle of mercury that possessed the properties of earth and water. Mercury accounted for the liquidity of metals when heated. But mercury also possessed a solid state that gave way to its silvery, fluid state on heating, eventually vaporizing. It thus appeared capable of uniting the qualities of matter and spirit.[40] Because of mercury's unique capability to embody all metallic states, alchemical authors believed it played a central part in the transmutation of metals, and it became one of the cornerstones of the theory of matter that was articulated in alchemy. Indeed, transmutation was described as a process in which a base metal was transformed into a noble one by undergoing a process of putrefaction and regeneration, going through stages in which the mass first turned black (putrefaction) and then underwent a series of color changes, finally turning red before transmuting into gold. The philosopher's stone, the substance that could instantly turn the mass from putrefying black to gold was often described as a red powder. Thus, pigment making and alchemical theory appear to have been intimately related in more ways than one. Not only were the two principles of metals in (al)chemical theory, sulfur and mercury, also the ingredients of vermilion production, but the outward manifestations of both processes of transformation bore strong resemblances.

As the art historian Arie Wallert has noted, when alchemical theory arrived in Europe in the twelfth century in textual form, craftspeople had already been combining mercury and sulfur to produce a red powder for at least two centuries.[41] Indeed, the practices of vermilion production always predated the articulation in texts of a theory of metals as well as antedat-

---

cury, probably reflecting the importance of the production of salts beginning in the fourteenth century. Gunpowder, saltpeter (from which fertilizer was produced), and the preservative properties of common salt came to have increasing importance in the fourteenth and fifteenth centuries.

40. For an excellent discussion of the significance of mercury in alchemical theory, see Figala 1998. See also Dobbs 1975, 35–37, for the matter-spirit problem in alchemy.

41. Wallert 1990, 155.

ing the description of alchemical transmutation as a process in which metals composed of sulfur and mercury undergo a transformation involving the same color changes observed in the process of vermillion making. This may indicate that the sulfur-mercury theory of metals actually arose from the practice of making vermilion, in other words, from the work of craftspeople and their practices. Thus, one of the most pervasive and enduring theories of matter and its transformation in metallurgy—the sulfur-mercury theory of metals—may have emerged from the making of materials.

## Red

The two preceding sections have illustrated some of the various ways in which artisanal practices and the development of early modern science intersected. In the case of the theory of matter just discussed, the relationship between the practices of pigment making and the two-principle theory of metals appear to fit a pattern familiar to historians and philosophers of science. Theory and practice in this case were related and probably reciprocally and continuously informed each other. It would be misleading to conclude from the familiarity of this relationship between alchemical theory and artisanal practice, however, that the artisanal conceptions of the world and material processes, what we might call the "vernacular science," that underlay these practices overlapped so tidily with our conceptions of nature and the relationship between practice and theory. In what follows, I attempt to map out a system of associations—the "vernacular science" that I believe underpinned the practices of pigment making and metalworking—a web of correspondences among red, blood, and gold that can be teased out of artisanal recipes and practices.[42]

As just discussed, the red powder of vermilion, produced by combining mercury and sulfur, manifested in material form a theory of metallic transformation, but, just as significantly, the red color of vermilion possessed potent significance, being associated with generation and life, especially in relation to blood and gold. Blood was the carrier of life heat, and gold could stimulate heat and life when prepared as potable gold or even when worn on the body.[43] Red substances were associated with blood and regeneration. For example,

42. I discuss the meaning of "vernacular science" in a somewhat different way in P. Smith 2004. Some of the most successful efforts at getting at the parameters of an artisanal worldview are contained in Cole 2002; Bucklow 1999, 2000, 2001; and, with some reservations, Prown and Miller 1996.

43. Albertus Magnus 1958, 19.

coral was used against bleeding: "And it has been found by experience that it is good against any sort of bleeding. It is even said that, worn around the neck, it is good against epilepsy and the action of menstruation, and against storms, lightning, and hail. And if it is powdered and sprinkled with water on herbs and trees, it is reported to multiply their fruits. They also say that it speeds the beginning and end of any business."[44]

Red components, such as the pigment vermilion, were often added to processes related to gold,[45] even when they seem not to have had any practical effect on the chemical process. This would indicate that red was seen as an essential ingredient in processes that sought to generate or transform.[46] The components of vermillion also often appear in recipes for gold pigments, such as that for mosaic gold (tin, or stannic, sulfide, $SnS_2$), a sparkling golden pigment that imitated pure gold, which call for tin, sulfur, mercury and sal ammoniac. Cennino Cennini lists one, which calls for "sal ammoniac, tin, sulphur, quicksilver, in equal parts; except less of the quicksilver."[47] Conservators have determined that the quicksilver (or mercury) is unnecessary to produce mosaic gold and appears to refer back to the principle that sulfur and mercury must be combined in order to form gold.[48] It is perhaps also possible to understand the mosaic gold recipe as a transmutation of tin by means of the red powder of vermilion.[49]

## Blood

Red was associated naturally with blood and in particular with the blood of Christ, which had purified mortals of their sins, just as gold purified their bodies and just as ores were purified by smelting and refining. A close relationship existed between religious devotion and metallurgical labor. For example, in indicating colors to be employed in illuminating manuscripts in the Middle Ages, scribes and illuminators often used a cross to indicate where the red pigment vermilion was to be used.[50] Cennino Cennini seems

44. Albertus Magnus (thirteenth century) 1967, 81.

45. For example, *Goldsmith's Storehouse* [ca. 1604], f. 55r: chap. 25: "To make saltpeter with Vermilion for water that will part gold from silver as well as aquafortis. 4 oz of vermilion added to the lye, boil etc until it become saltpeter."

46. Bucklow cites pigment recipes in which red and gold are associated; for example, Bucklow 1999, 145–47.

47. Cennini 1960, 101–2, "Mosaic gold".

48. Wallert 1990, 158–59.

49. Bucklow 1999, 146.

50. Gage 1998, 39.

to have equated vermilion with blood and with the life force that blood carried. In "How to Paint Wounds," Cennino specified that a painter must "take straight vermilion; get it laid in wherever you want to do blood."[51] In a long passage, Cennino describes precisely how one is to lay in the flesh tones of living individuals in a fresco. He specifies that this flesh tone is never to be used on dead faces. Where this color in fresco is to be made from red ochre pigment, on panel, vermilion is used. Cennino called this color *incarnazione* and clearly regarded its use as akin to the incarnation of life in a body.[52] This giving life to (or "incarnating") an image was for Cennino a straightforward artisanal technique by which the abstract principle and profound miracle of the incarnation of God and the Word in human flesh could be imitated.[53] No stronger link between the material or the spiritual and between devotion and the making of materials may be found. This simultaneously material and spiritual understanding of the production of materials surely was an important component of the "theory" that underlay artisanal practices, although it was a lived, rather than theorized, reality.

Blood was also regarded as an extremely powerful agent. It was often cited as the only way to soften or cut hard gemstones such as diamonds. Most such recipes called for goat's blood, but one recipe noted that human blood was also good for making gold and silver.[54] It is a feature of these recipes that they include details that would appear to indicate actual use:

> If you want to carve a piece of rock crystal, take a two- or three-year-old goat and bind its feet together and cut a hole between its breast and stomach, in the place where the heart is, and put the crystal in there, so that it lies in its blood until it is hot. At once take it out and engrave whatever you want on it, while this heat lasts. When it begins to cool and become hard, put it back in the goat's blood, take it out again when it is hot, and engrave it. Keep on doing so until you finish the carving. Finally, heat it again, take it out and rub it with a woolen cloth so that you may render it brilliant with the same blood.[55]

51. Cennini 1960, 95.

52. Kruse 2000.

53. Through such a practice, Cennino acknowledged the transformative power of art and artisan.

54. Theophilus 1979, 189–90. See also *Rechter Gebrauch* 1531 (f. III v), on making precious stones capable of cutting and of casting. Kärtzenmacher 1538 (xxix verso) states that philosophers have hidden the fact that human blood is good for making silver and gold, and that all should know that menstrual blood, or "sanguis rubei collerici," is the best for the art of making gold and silver.

55. Theophilus 1979, 189–90. This recipe endured for at least four hundred years, which probably indicates its power in organizing an understanding of the world. Four hundred years after Theophilus's treatise, the anonymous German author of a printed treatise on metalwork-

The significance of blood itself for artisans can be seen in Benvenuto Cellini's (1500–1571) account of casting his statue *Perseus Beheading Medusa* in 1545–1551. In the midst of the casting, Cellini was abruptly attacked by a fierce fever which increased until he believed himself to be dying. A wraith entered his bedroom to proclaim the ruin of his work, at which Cellini sprang from his bed to find the metal "curdled" in the furnace. He proclaimed to his assistants, "from my knowledge of the craft I can bring to life what you have given up for dead, if only the sickness that is upon me shall not crush out my body's vigour."[56] He piled young oak onto the fire and the metal began to "glow and sparkle." As what he called the "corpse" of the metal came back to life, Cellini recovered from his fever. The furnace cracked from the heat and Cellini opened the mouths of the mold to let the metal flow in, but still the metal would not flow, so Cellini ordered all his pewter dishes thrown into the molten mass. He fell to his knees in prayer, calling out, "O God, who by infinite power raised Yourself from the dead and ascended into heaven!" On this, the mold filled in an instant.[57] Michael Cole has pointed out the manifold connections in this account between vivifying blood, the infusion of matter with spirit,[58] and Cellini's own successful employment of a vivifying force that had sent life coursing back through the dead metal just as it had reentered his own veins. Through his art, Cellini was able to capture the spirits that had resuscitated the dead metal to flowing, corporeal vitality.

Similarly, the Dutch apothecary's son and sculptor Adriaen de Vries (1556–1626) displayed a similar understanding of the flow of metal and the flow of blood. De Vries left evidence of his working processes and knowledge of natural materials not in writing, like Cellini, but instead in his sculptures themselves. For example, instead of clipping off the channels by which the

---

ing offered a strikingly similar formula for cutting and engraving crystal and precious stones: "Take goose and goat's blood and dry it until it is hard. Take crystal or any stone. Pound the blood to powder, pour ashes on it. Let it mix well in a container. Mix in strong vinegar, then lay the stone in it, warm it a little. And the stone allows you to cut or form it as you want. Throw it in cold water and it will become hard again in an hour." *Rechter Gebrauch* 1531, III verso. The anonymous author of the *Goldsmith's Storehouse* writes, however, "some Aucthors doe wryte that the Dyamon cannot be broken, butt with the new warme bludd of a goate, but it is not soe, for Dyamon Cutters have dayly experience to the Contrary, who doe continually use the powder of Dyamons" (*Goldsmith's Storehouse* [ca. 1604], 62v).

56. Cellini 1967, 123. In recounting the casting of another work, Cellini again emphasized his ability to resuscitate the dead: "owing to my thorough knowledge of the art, I was here again able to bring a dead thing (*un morto*) to life" (Cellini 1967, 125).

57. Cellini 1956, 343–48.

58. Perhaps for Cellini the process even mimicked the preparation of an elixir of life (Cole 1999, 222–25).

metal entered the mold (the sprues) on some of his most difficult finished statues, he left small sprue ends to indicate that the piece had been created by a single casting. In his last statue, a large depiction of Hercules (*Hercules Pomarius,* 1626–1627), de Vries left the sprues (concealed as vines) feeding directly into the figure's veins, thereby preserving, as Francesca Bewer has put it, "the very channels through which the master metalsmith infuse[d] the figure with life."[59]

## Lizards

This correspondence between blood, red, and gold is also important in a puzzling set of recipes such as that set down by the twelfth-century metalworker Theophilus for Spanish gold, concocted from "red copper, basilisk powder, human blood, and vinegar." In order to produce the basilisk powder, two twelve- to fifteen-year-old cocks were put into a cage, walled like a dungeon with stones all around. These cocks are to be well fed until they copulated and laid eggs, at which point toads then should replace the cocks to hatch the eggs, being fed bread throughout their confinement. Male chickens eventually emerged from the eggs, but after seven days they grew serpent tales. They were to be prevented from burrowing into the floor of their cage by the stones, and, to further reduce the possibility of escape, they were to be put into brass vessels "of great size, perforated all over and with narrow mouths." These were closed up with copper lids and buried in the ground. The serpent-chickens, or basilisks, fed on the fine soil that fell through the perforations for six months, at which time, the vessels were to be uncovered and a fire lit under them to completely burn up the basilisks. Their ashes were finely ground and added to a third part of the dried and ground blood of a red-headed man, which was then tempered with sharp vinegar. Red copper was to be repeatedly smeared with this composition, heated until red-hot, then quenched in the same mixture until the composition ate through the copper. It thereby "acquire[d] the weight and color of gold" and was "suitable for all kinds of work."[60]

This recipe has excited much comment among historians, being viewed as a garbled set of instructions for making brass or for the chemical process of cementation in which gold is purified.[61] The curator and conservator

59. Bewer 2001.

60. Theophilus 1979, 119–20.

61. Halleux (1996, 887–88) argues that it is based on Arabic recipes, perhaps part of the Jabirian corpus.

Arie Wallert has interpreted it as an alchemical recipe, with blood forming an alchemical "cover name" (a secret term known only to the adepti) for sulfur, and basilisk ash for mercury.[62] Sulfur and mercury were the essential components of the philosophers' stone to turn base metals—in this case copper—into gold. In his fascinating study of the relationship between red and yellow pigment recipes, Spike Bucklow views this recipe as evidence for the centrality of the vermilion-making process as providing the model for other processes of metallic transformation.[63]

Where Theophilus calls for basilisks, a later set of recipes calls for lizards. In a 1531 text that includes pigment-making and metalworking recipes, entitled the *Rechter Gebrauch der Alchimei,* there are several recipes for making noble metals through a process of catching, feeding, and burning lizards. As in the instructions for softening hard stones by means of goat's blood, this recipe opens with quite precise instructions on how to catch these lizards. It directs the reader to move very quietly in "felt slippers," to quickly snatch the lizards before they give off their poison, and then to immediately plunge them into a pot of human blood. A recipe for making lizard-rib gold follows, which calls for two pounds of filed brass and a quart of goat's milk, and continues: In a pot wide at the bottom and narrow at the top, with a cover that has air holes in it, place nine lizards in the milk, put the cover on, and bury it in damp earth. Make sure the lizards have air so that they do not die. Let it stand until the seventh day in the afternoon. The lizards will have eaten the brass from hunger, and their strong poison will have compelled the brass to "transform itself to gold." Heat the pot at a low enough temperature to burn the lizards to ash but not to melt the brass. Cool the mixture, then pour the brass into a vessel, rinse it with water, then put it in a linen cloth and hang it in the smoke of sal ammoniac. Once it is washed and dried again, it will yield a "good calx solis," or powdered form of gold.[64] This recipe may employ lizard ash as a cover name for mercury, or the recipe may aim to produce the painter's pigment mosaic gold.

The association of lizards with mercury occurs again in a book of secrets ascribed to Albertus Magnus, written no later than the fourteenth century. Among many secrets for lighting a house, one calls for cutting off the tail of a lizard and collecting the liquid that bleeds from it, "for it is like Quicksilver," and when it is put on a wick in a new lamp, "the house shall seem bright and

62. Wallert 1990, 161.

63. Bucklow 1999, 145.

64. *Rechter Gebrauch* 1531, f. XIII.

white, or gilded with silver."[65] In this recipe, lizards, mercury, and the noble metal, silver, all were associated.

Lizards were associated with processes of putrefaction and generation more generally,[66] just as blood and mercury were associated with the generation of substances, both in pigments and metals. Animal and mineral generation might even be combined, as they are in a recipe that claims to yield a gold pigment that is produced by mixing mercury with a fresh hen's egg then putting it back under the hen for three weeks.[67] A similar identification of mercury with processes of generation appears in the 1540 *Pirotechnia* by the metalworker Vannoccio Biringuccio, who noted that the prospector for mercury should be on the lookout for verdant mountainsides, for all places where mercury is engendered "have abundant water and trees, and the grasses are very green, because it has a moist coolness in it and does not give off dry vapors as sulphur [does]."[68]

## Conclusion

Discussion of metalworkers' techniques and their practices of precision and experimentation have dominated historians' accounts of the relationship of metalworking to the development of modern chemistry. In contrast, I have attempted in this paper to show how we might take account of these intersections between artisanal techniques and the development of modern ways of investigating nature, while at the same time delineating a less familiar worldview or "vernacular science" of materials and nature that appears to have underpinned and informed artisanal practices in pigment making and metalworking. Not surprisingly, this worldview does not overlap neatly with a modern scientific understanding of the world. While I have done no more in this chapter than suggest the outlines of the web of correspondences, I think it is important to acknowledge that, while the phrase "the working of materials" today conjures up the manipulation of inert matter with clear productive and economic meanings, in the early modern period, it had an entirely different meaning. Metalworking in the sixteenth century was part of a web that included vermilion, the color red, blood, mercury, gold, and lizards, and it gave access to the powers of nature, transformation, and gen-

65. Albertus Magnus 1973, 104.

66. See P. Smith 2004, 117–23.

67. *Kunstbüchlein* 1538, f. 19v.

68. Biringuccio 1540, 83.

eration. The manipulation of metals in early modern Europe was not simply about the handling and transformation of inert materials, but rather allowed the artisan to investigate and engage in life forces and in the relationship of matter to spirit, and even to imitate the most profound mysteries, such as the incarnation. On the one hand, this was a mundane and hard-headed practice that produced useful goods, but on the other, these artisanal techniques gave access to the greater powers of the universe. These practices, moreover, were neither rote nor random nor untheorized. Rather, they were tied to a kind of lived theory, one not necessarily systematized and articulated in words, and they reveal an underlying set of principles. This is an assemblage of knowing and making, and a means for struggling with matter, a way of living in the world, in Tim Ingold's terms.

There is, then, no doubt that the making of materials can be regarded as related to the emergence of modern science, that is, it can be seen as investigative. It was informed by (and informed) particular theories of matter, and it involved a set of empiricist techniques that were assimilated into the new experimental philosophy in the seventeenth century. But such characteristics—as important as they might be for modern science—could never be the meaning that making and matter held for its early modern practitioners.

## Primary Sources

Secondary sources can be found in the cumulative bibliography at the end of the book.

### UNPUBLISHED SOURCES

Bibliothèque nationale de France, Paris. Ms. Fr. 640. n.d.

*The Goldsmith's Storehouse.* [ca. 1604]. Washington, DC. MS V.a.179. Folger Shakespeare Library.

The Pekstok papers. n.d. Amsterdam. Call-Nr. N 90.23. Gemeente Archiv.

### PUBLISHED SOURCES

Albertus Magnus. 1967. *Book of Minerals.* Trans. D. Wyckoff. Oxford: Clarendon.

———. 1973. *The Book of Secrets of Albertus Magnus of the Virtues of Herbs, Stones and Certain Beasts. Also A Book of the Marvels of the World.* 1530 English translation, ed. M. R. Best and F. H. Brightman. Oxford: Clarendon.

Biringuccio, Vannoccio. 1540. *Pirotechnia.* Trans. C. S. Smith and M. T. Gnudi. New York: Basic Books, 1943.

Cellini, Benvenuto. 1956. *Autobiography.* Trans. G. Bull. London: Penguin.

———. 1967. *The Two Treatises on Goldsmithing and Sculpture.* Trans. C. R. Ashbee. New York: Dover.

Cennini, Cennino D'Andrea. 1960. *Il libro dell'Arte (The Craftsman's Handbook).* Trans. D. V. Thompson, Jr. New York: Dover.

Kärtzenmacher, Petrus. 1538. *Alchimia. Wie mann alle farben/wasser/olea/salia und alumina damit mann alle corpora/spiritus und calces preparirt/sublimirt und fixirt machen sol. Und wie mann diese ding nutze-auff das Sol und Luna werden mög. Auch von solviren uund schaidung aller metal/Polirung aller handt edel gestain/ fürtrefflichen wassern zum etzen/schaiden und solviren/undt zletst wie die gifftige dämpf züverhüten ein kurtzer begrif.* Strassburg: Jacob Cammerlander.

*Kunstbüchlein, Auff mancherley weyß Dinten und allerhandt farben zu bereiten. Auch Gold unnd Silver/sampt allen Metallen auß der Federn zu schreiben/Mit viel anderen nützlichen künstlin. Schreybfedern unnd Pergamen mit allerley Farben zu ferben. Auch wie man Schrifft und gemälde auff stähes lene/ Eysene Waffen/und dergleichen etzen soll. Etliche zugesetzte Kunststüclin/vormals im druck nye außgangen. Allen Schreybern/Brieffmalern/sampt andern solcher Künsten Liebhabern/gantz lustig und fruchtbar zu wissen.* 1538. Augsburg: Michael Manger.

*Rechter Gebrauch der Alchimei/ Mitt vil bisher verborgenen uund lustigen Künstien/Nit allein den fürwitzigen Alchmisten/sonder allen kunstbaren Wercklеutten/in und ausserhalb feurs. Auch sunst aller menglichen inn vil wege zugebrauchen.* 1531. Frankfurt am Main: Egenolff.

*The Strassburg Manuscript: A Medieval Painters' Handbook.* 1966. Trans. V. Borradaile and R. Borradaile. London: Alec Tiranti.

Theophilus. 1979. *On Divers Arts.* Trans. J. G. Hawthorne and C. S. Smith. New York: Dover.

# 3

# Ceramic Nature

HANNA ROSE SHELL

Clay consists of fine shards of rocks (usually granite and feldspar) that have decomposed. Ceramics, human artifacts made of baked clay, cannot exist without people to produce them. As such, they always have been, and always will be, integrated into more or less complex aesthetic and economic systems and scientific worldviews.[1] The sixteenth-century natural historian and ceramicist Bernard Palissy, studied in this chapter, channeled philosophies of nature into clay commodities. Best known today for his biomorphic earthenware, Palissy (1510–1590) has been celebrated as geological theorizer and religious martyr. Yet above all, he was a craftsman. He was a polymath potter of forms, words, and natural knowledge, and his novel ceramic basins can be understood as simultaneously artworks, natural formations, and teaching tools (see figure 3.1).

Palissy had a productive—and tumultuous—career as ceramicist, geologist, and public lecturer. Born into a French artisanal family in 1510, the young Palissy trained as draftsman and glass-painter, receiving a few years of vernacular education. From 1528 until about 1535, he traveled around France, working as painter, land surveyor, and mapmaker while pursuing interests in natural history, geometry, geothermy, and alchemy. After settling down near Bordeaux, he began to experiment with ceramic practice. His labors brought forth grottoes and fountains, as well as his popular rustic basins—the sculpted ponds encrusted with animals and gilded by glazes. In about 1540, Palissy set up his first pottery studio. He built a kiln and collected a

I thank James Bono, Paula Findlen, Katherine Park, Lorraine Daston, Denise Phillips, Mike Reed, and Pamela Smith, all of whom read and commented on drafts of the original paper on which this chapter is based. An earlier formulation of my argument appears as Shell 2004.

1. Prown 1982; Coutts 2001, 2.

FIGURE 3.1. Etching of a life-size bronze statue of Bernard Palissy. Palissy holds a rustic platter with fossilized seashells and crystals. There is a kiln at his feet, and he wears a potter's apron over gentleman's dress (Burty 1886, 9).

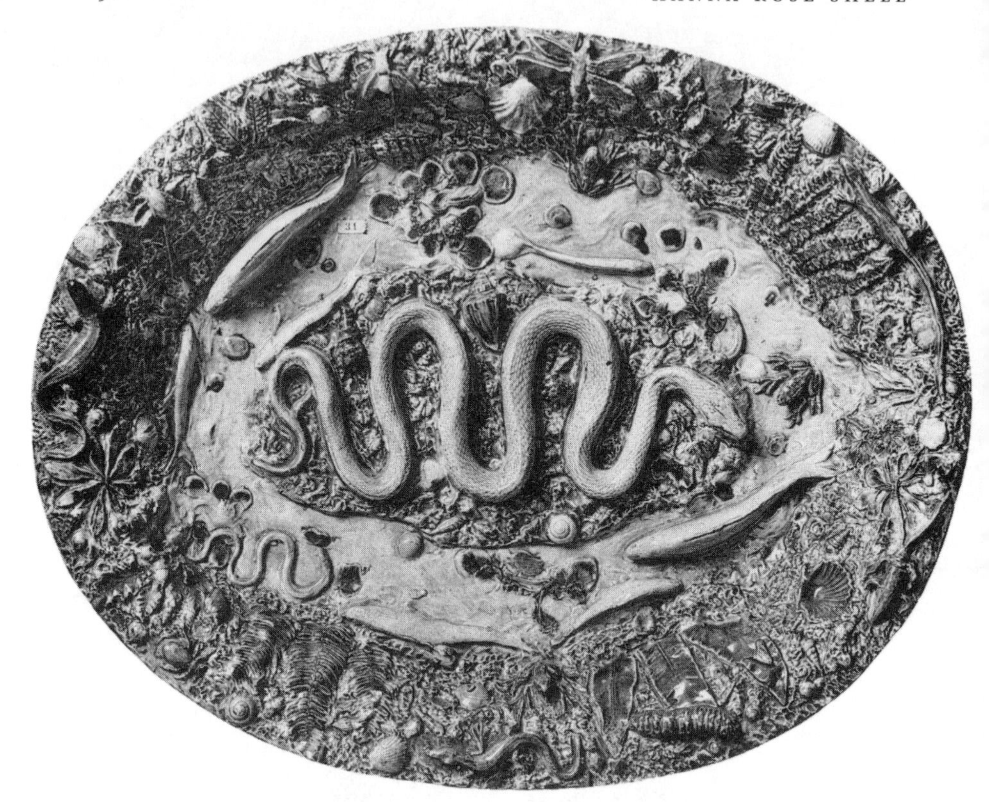

FIGURE 3.2. Typical rustic basin by Bernard Palissy, mid-sixteenth century (card 29 in the set of photogravures published by Editions Albert Morance in 1923 in connection with Ballot 1924).

wide range of viscous clays from deposits around the Bordeaux region. When mixed with water, clay particles (largely aluminum silicate) form a tenacious paste;[2] Palissy wrestled almost endlessly with it. He molded, dried, and fired the clay, creating durable earthenware pieces. Within a couple of years, he had established an urban workshop, wherein he employed members of his immediate family, casting nature into ceramics (see figure 3.2).

"It is impossible to imitate anything whatsoever in Nature without first having studied her effects, taking her as both pattern and example."[3] In his ceramic work, Palissy represented the natural environments most central to

2. Clay consists mainly of aluminum silicate and is derived mostly from the decomposition of feldspathic rocks. Particular kinds of clay are known as brick, fatty, fire, plastic, porcelain, and potter's clay.

3. Palissy 1996, 53.

his linked theories of coloration, generation, and fossilization—freshwater ponds. In his earthenware, Palissy incorporated multicolored glazed ceramic figurines cast from shells, plants, and live amphibian and marine specimens to create teeming representations of murky ponds. Ideas about nature were both expressed in and generated by the process of fashioning these decorated earthenware vessels.[4] In his plateware, Palissy explored and explicated terrestrial processes as variations on an environmental theme. He aimed to understand the life (and death) of ponds—and the generation that occurred therein—through their re-creation in clay. The plates' commodity status was entangled with the very quandaries about natural processes that both motivated and facilitated their fabrication. Palissy discovered and enacted theories about the generation of forms, colors, and fossils in nature in the production and contemplation of his biomorphic pottery.

Palissy's rustic earthenware plates, esteemed by museums and collectors today and widely imitated in the nineteenth century, became marketable commodities in the sixteenth century. Patrons included Henri II, Ann de Montmorency, and the Royal Palace of Fontainebleau. Catherine de Medici was so enamored of his wares that she became the Protestant potter's protector. In 1563 she appointed him royal "Earthworker and Inventor of Rustic Ware" (*Ouvrier de Terre et Inventeur des Figulines Rustiques*), a title he held until his death in 1590.[5] But Palissy's ceramic wares can be understood in direct relation to his writings on geology and agriculture, particularly the extended series of dialogues published as *Discours Admirables* (1580). Inhering in these plates were not just style and social capital, but also theories about natural processes. His earthenware productions served as expressive embodiments—both heuristic and illustrative—of his innovative, and sometimes controversial, theories about natural history. Their visual and material contours bespoke his ideas about three phenomena—coloration, generation, and fossilization in nature. This chapter examines how Palissy's work with clay related to his lifelong study of these phenomena. I begin with a discussion of Palissy's life-casting technique as a strategy for "harnessing" nature. Sections on theories of coloration, generation, and fossilization, and their objectification in clay, follow. In these sections, I describe how Palissy aimed to make tangible in clay his theories about these phenomena. Finally, I offer an account of his particular attitude toward the exhibition of ceramic and natural artifacts. Palissy's ceramics came to exist as sculptural renditions

4. Hanschmann 1903.
5. Dimier 1934, 17.

of the natural world, displayed both in the residences of his noble patrons and the on the shelves of his personal gallery.[6]

## Casting Life: Artisanal Limnology

From 1545 until 1590, Palissy relied on the same method to cast life into clay. Its goal was the artificial petrifaction of aquatic and amphibian creatures, their literal transformation into earthenware.[7] As has been explored by several scholars, a host of sixteenth-century artisans in Germany and Italy used life-casting techniques in their production of elaborate metallic art.[8] Among these life casters, Palissy was unusual both for his use of earthenware (as opposed to metal) and for the sophistication of his expressions of theories about organic processes; the latter he did with both earth and words.[9] Throughout his life, Palissy imitated, through life casting and ceramic posing, exactly those habitats and specific species that surrounded him. He gravitated toward caves, ponds, and small lakes for natural inspiration. Insects, frogs, snakes, crayfish, turtles, and small plants attracted Palissy's aesthetic and scientific eye.

Renaissance life casters stored their life models in bottles and casks containing various combinations of bran, peat, and damp earth. In these workshop menageries, as in nature, one animal's life might be sacrificed at any point for the sake of another's survival. Frogs might provide their bodies as nourishment to hungry snakes at the bequest of the life caster.[10] A perfectly orchestrated wresting of life from the captive animals ensured clay immortality in the mold. Ideally, the life casts would be used for many years of production and reproduction.[11] The life caster nearly killed the specimens just before casting them, through brief immersion in jars of urine or vinegar. He then coated the dying specimens in a greasy substance before immersing them in flattened clay. After the plaster set, the life caster would make a clay impression of the plaster relief, rendering the animal body as a mimetic three-dimensional representation. The natural life of the specimen was thereby channeled into the artifactual life of the mold.[12]

6. Kris 1926.

7. Planiscig 1927, 450–64.

8. P. Smith 2006a, 74–80, 83–100; P. Smith 2004, 59–95; Forssman 1956, 13; Kemp 1999, 72–78; Daston and Park 1998, 282.

9. Shell 2004.

10. Amico 1996, 86.

11. Kris 1926, 151.

12. Paris, Ms. Fr. 640 (Microfilm 1965: folios 122, 123, 124, 140, 141 recto and verso), Bibliothèque Nationale de France.

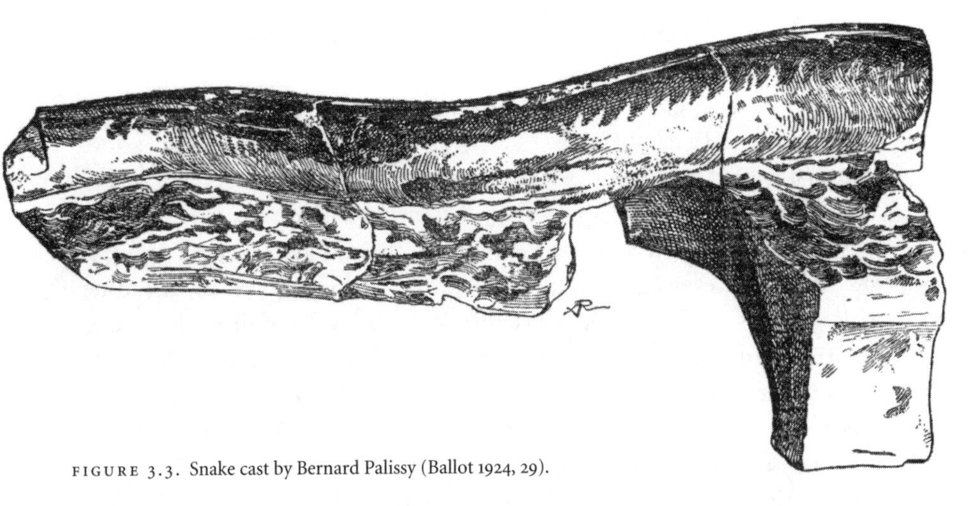

FIGURE 3.3. Snake cast by Bernard Palissy (Ballot 1924, 29).

## Nature's Coloring Book

The contemplation and re-creation of color—in both the field and the workshop—were integral to Palissy's sense of observing and making nature. The production of ceramic glazes was about knowing as much as seeing, natural history as much as aesthetics. Color formation has rarely been considered in relation to ideas about scientific or natural knowledge. And yet, comparable to the making and using of dyes and inks, producing artistic color and studying natural form were related enterprises in the sixteenth century.[13] Art historians have understood patterns of coloration in Italian Renaissance art in temporal and spatial terms and have shown that attention to color formation was not unusual among Renaissance artists.[14] And in Palissy's case, ceramic coloration through the production and application of glazes and enamels may be understood in terms of his theories about natural color formation.

In *Discours Admirables,* Palissy links color and life in his descriptions of both nature and the workshop. In nature, according to Palissy, the ability of an organism to generate its own colors bespeaks its dynamism. In particular, color transformations provide evidence of organismic growth and development. In one section of the *Discours,* "Pratique" alerts his interlocutor, "Theorie," that in nature "you can see that none of these things [animals or plants]

13. See the contributions by Johns and Smith in this volume. In several aspects this practice was continued in the eighteenth century; see the chapter by Nieto-Galan in this volume.

14. Hills 1999, 337–50; see also Gage 1999.

retains its original color. But rather in their growth, things change color; and in one thing there come to be many colors."[15]

"Growth changes color" is a key concept that emerges repeatedly in Palissy's writings. In all plants, he asserts, "fruits change colors in their growth and maturity." In the same text, Palissy comments on the marvel of organic coloration: "Observe the seeds when they are cast into the earth: they have but a single color: coming into growth and maturity they then take on various colors . . . even in a single flower. Similarly you will find snakes, caterpillars and butterflies of many beautiful colors. . . . All of these things draw nourishment from the earth, just as their color also comes from the earth." Later in the same essay, he describes reptiles in nature gilded with "marvelous colors, by a labour that no painter nor embroiderer could imitate in their fine works."[16] If neither painter nor embroiderer, perhaps a glazier? In his pottery work, Palissy chose the latter occupation, striving to impart clay with living color.

In glazing the cast pond plants and animals, Palissy imparted precisely these "marvelous colors" and lively (i.e., lifelike) adornment. Near the end of the *Discours,* he notes similarities between color change in plants and rocks. "Just as fruits of all kinds change color as they ripen . . . similarly rocks, metals and other minerals, even clays, change color during their decoction."[17] His hope was to imitate such transformations by harnessing the earth's own techniques. To this end, he tried to decipher the causes of color transformation in nature.

Palissy struggled to create particular shades of glazes—frog green, periwinkle blue, viper gray. He experimented with minerals and water mixed with a lead-based tincture in varying proportions. Particular colors resulted from the specific minerals used and the oxygen content present in the kiln during and after firing. Palissy developed glazes using formulations derived from tin, iron, steel, antimony, cobalt, copper, sand, and lead, as well as the ashes derived from burned wine lees.[18] Experience hands-on with nature was crucial to discovering the secrets of glazes. As an example, Palissy notes: "There are a great many materials that cause the colors of stones, and many of these are unknown to men; Nonetheless experience, which has ever been the mistress of the arts, has vouchsafed to me that iron, lead,

15. Palissy 1996, 139.
16. Ibid., 110, 137, 151.
17. Ibid., 381.
18. Coutts 2001, 40.

silver and antimony can produce no color other than yellow."[19] Observation and physical experience are essential here. To know glazes, one must know nature, and vice versa. As in the utopian Salomon's Hose, described by Francis Bacon's *New Atlantis,* knowledge of nature comes through imitating and demonstrating the phenomena in question.[20] The glazier will be able to understand the colors of stones after imitating them. When a glazed pot is kilned, single tones may be transformed into a whole rainbow of naturalistic hues, each with its own properties of generation and destruction. As he recalls in *On the Art of the Earth*: "Some of my enamels turned out fine and well melted, others poorly melted, others were burned, because they were made of different materials which were fusible to different degrees; the lizards' green burnt before the colors of the serpents had melted, likewise the color of the serpents, crayfish, turtles and crabs had melted before the white had attained any beauty."[21] The oven's heat generates color shifts—for better or worse. The heated plate thereby becomes a microcosm presenting animate color as well as terrestrial structures—in either perfected or degraded form.

Palissy used a variety of tools to heighten the illusionistic effects of color on clay. Through coloration, he returned his cast animals, plants, and rocks to a more naturalistic (or "true to life") appearance. According to Palissy, color in both craft and nature comes from the earth. And yet in both cases, humans and other organisms can channel coloration for their own devices. He expostulates on the ability of fish and sea creatures to craft their own shell-houses. As he writes: "I have often admired the colors of these same shells, and have not been able to understand their cause."[22] Palissy exalts mollusks as natural craftsmen who channel rainbows and earth deposits in order to paint rainbows on their calcified shells. As he notes in *Recepte Véritable,* snails—like the artisan in the studio—fuse their bodily labor with their productions; whereas for Palissy it is ceramic ware, for snails it is their colorful, calcified shells. Along these lines, he actually proposes that humans should try to build human domiciles along the conceptual and architectural terms of snail shells writ large, a proposal that later found favor with Gaston Bachelard.[23]

19. Palissy 1996, 253.
20. Bacon 1626, 481.
21. Palissy 1996, 305.
22. Ibid., 127.
23. Kemp 1999.

## Pond Generations

Palissy used ceramics to explore not just individual animal forms, but the generation of the environments in which they lived: freshwater ponds. He chose to produce pond scenes because these environments exemplified his own theories of natural generation. Terrestrial processes would be laid out as variations on an environmental theme. Palissy's plate production was a means for his exploration of generation within the natural environments they represented.

Palissy's first task as rustic potter was the manufacture of earthenware rocks. These fabricated rocks would serve as molds; in their inverted form, they would take on the shape of basins. The basin, once cast from the rock, became the stage on which he would direct a cast of clay-rendered aquatic and amphibious creatures. With animal and plant molds in hand, and before glazing, Palissy composed his plate scenes. Sculpting the finished piece involved choosing and positioning the animal actors. Model frogs, snakes, and other animals were placed onto a rough-hewn basin, cast from the rock mold. To this end, Palissy fired a slab of various clays and soils that had been crafted into an appropriate shape. His ceramic mound came to evoke the rocky shores of a fertile pond.

Palissy employed needles, palette knives, and other tools to integrate the ceramic interstices of earth and animal, water and vegetable.[24] But what kind of synthesis was occurring here? What did Palissy's ceramic microcosm represent, the articulation of which motivated his labors decade after decade? Close examination of a representative basin provides the beginning of an answer (see figure 3.2). This basin exemplifies Palissy's ceramic style, depicting a pond of some depth, in the middle of which is a simulated island. The whole object is approximately eighteen inches long, twelve inches wide, and five inches deep. As is standard in the naturalistic pottery, a spotted, slate-gray viper writhes back and forth along the oval island protrusion. In this three-dimensional composition, the viper takes center stage; it is both the centerpiece of Palissy's work and the main locus of danger within the habitat represented therein.

A menagerie of plants and animals surrounds the serpentine centerpiece. Cast frogs, fish, and crayfish scatter both in the pond water and on its rocky shores. Frogs are oriented away from the viper's mouth, as if struggling to defy a possible death sentence. Meanwhile, three fish circulate around the pond's perimeter. Swimming near the border between pond and shore, they

24. Amico 1996, 93.

seem to occupy a transitional zone between water and earth. The pond itself is figured as a donut or wheel; here, the fish's unidirectional motion implies the cyclicality of the life of both organisms and habitats. The shores, figured around the lips of the basin, display an array of cast shells and barnacles. These animal remains represent mollusks' discarded calcified packaging.[25] Like fossils, they are traces of life impressed in earth. Here, as throughout Palissy's ceramic ware, the basin captures an organic scene in which violence and aggression provide the focus for a theater of life, death, and transformation.

Over the course of decades, Palissy staged this ceramic theater again and again. Indeed, perusal of Palissy's rustic oeuvre presents a startling consistency of subject matter. A coiled or writhing snake attacks a water hole, scaring away a cluster of animals and causing the fish to thrash about in circles. Reptiles, amphibians, and fish gather in simulated bodies of water—all forever petrified in clay. At several points in *Discours Admirables,* Palissy describes ponds as related to the generation of life. In one such discussion, he depicts the particular natures of stagnant ponds. In Palissy's account, putrefaction, generation, congelation, and petrifaction are the staples of pond activity. The danger of pond environments lies exactly in these transformative capabilities. As Palissy reports:

> There are a great many kinds of pools (*mares*) both natural and artificial, several call them *claunes.* . . . The air and the sun warm all these and by this means generate and produce several kinds of animals. And inasmuch as there is always a great quantity of frogs, serpents, asps and vipers always gather around these *claunes* in order to feed on the frogs. There are also commonly leeches in them, so that if the oxen and cows remain in these pools for some time, they cannot fail but be stung by the leeches. I have several times seen serpents lying curled up at the bottom of such ponds.[26]

Palissy's description brings to mind the composition of his rustic plates; he evokes the same curled vipers waiting to pounce on an array of small, medium, and large frogs. His explanation for the production of life in the ponds is mimetically enacted in the process through which he figuratively brings his

25. Bachelard 1958, 105–36.

26. Palissy 1996, 29–30. The term *mare,* relatively common in the sixteenth century, referred to a freshwater pool or a pond, either natural or constructed. Emile Littré defined *mare* as "a small extent of still water, natural or artificial" (Littré 1882). According to Olivier de Serres, "A *mare* is a large trough, hollowed out with gentle slopes on all sides, such that livestock can easily descend; it deepens in the centre, though moderately, where spring and rain water collect." (de Serres 1600, 784). *Claune* referred to a shallow pond that was either naturally or artificially coated with any of a number of clay-based residues.

ceramic ware "to life." Palissy steered toward these kinds of places because he believed them to be sites of generation through putrefaction of organic bodies, followed by congelation of various salts, waters, and minerals.[27]

Congelation and putrefaction are both recurring themes in Palissy's writing. *Putrefaction* referred then, as now, to decomposition of animal and vegetable substances, generally accompanied by a putrid smell or appearance. Putrefaction has consistently tended to evoke danger and corruption. In the sixteenth century, *congelation* referred to the solidification of a liquefied substance through a variety of processes, the application of heat usually among them. Putrefaction and congelation were part of the often-shared discourses of alchemy and artisanal culture in the period.[28] According to Palissy, congelation was the seat of generation among the lower, aquatic animals. Frogs especially, he felt certain, clustered in locations where salts turned "congelative water" to mud, to amphibious life, and back again through processes of generation and putrefaction. As he describes his understanding of pond life, if "there is only green slime, it is the sign of putrefaction, and the beginning of the generation of something." And furthermore, "there is no part of the earth that is not full of some kind of salt, that causes the generation of many things."[29] Such evocative, generative, and often congelative, environments inspired his theories of the earth, as well as providing animal, plant, and mineral models for incorporation into his constructions.

Palissy's attraction to these pond environments—reptiles and amphibians in particular—was by no means exceptional. He worked within a context of Renaissance fascination with inhabitants of the aqueous and mineral worlds, and in frogs, salamanders, and snakes specifically.[30] Two French authors, Pierre Belon (1517–1574) and Guillaume Rondelet (1507–1566), indeed, two of the few authors whom Palissy is thought to have read over the course of his lifetime, considered pond reptiles at length. Both Belon and Rondelet wrote on the generative origins of pond dwellers, including three of Palissy's favorite subjects for capture and casting—frogs, fish, and snakes. In the illustrated volume *La Nature et Diversité des Poissons,* first published in French in 1555, Pierre Belon discusses amphibian generation at length. Notably, for Belon, amphibians included almost all animals that lived underwater (including fish). One section of his text focuses on the mechanisms through which frogs both spawn and degenerate. According to Belon, frogs are apparently

27. Schnapper 1988, 18.

28. Newman 2004.

29. Palissy 1996, 35, 39.

30. Schnapper 1988, 69–71. See also Smith's contribution to this volume.

generated through two related processes: putrefaction and egg deposit. Annual generation and degeneration work alongside a regular spawning process; all the while, "their habitat and food source is all the many vermin that are engendered from excrement and from the corruptions of the dirty waters in which they dwell." Frogs and their food sources merge and converge. In Belon's section titled "Grenouille," he writes: "At the end of six months, the frog turns back into silt. But when the spring comes, they fuse together, resolving again to form frogs. . . . These frogs, nonetheless, also breed among each other; they spawn by laying eggs, thereby making little ones that begin as only a set of big eyes and a tiny head." Belon concludes, "that which I find most admirable about the frog is that at the end of about six months it turns back into silt. And when spring arrives, they come together again: nonetheless they also breed and make eggs and little ones."[31]

Frogs kept on popping up. Guillaume Rondelet's *L'Histoire Entière des Poissons* appeared in French in 1558.[32] Rondelet's massive compendium covered many more underwater species than had ever been described before; Palissy had almost certainly read the translated volume by 1560.[33] Within, Rondelet devotes an entire volume of his work specifically to "Les Animaux Palustres," defined as those organisms that live in stagnant marshes. Like Belon, Rondelet is nervous about how to classify frogs and warns against their ingestion.[34] He also expresses concerns about frog mating and cannibalism practices. And finally, the great Italian polymath Cardano considered frogs to be at the center of processes of spontaneous generation. (Spontaneous generation here refers to the development of living organisms without the agency of preexisting living matter.) As he wrote in an essay on spontaneous generation, almost certainly read by Palissy in translation by the 1560s, "Frogs are born of impure waters and sometimes of rain: it is believed, however, that a certain number of imperfect animals are born, without seed, from corruption."[35]

Palissy's descriptions of the *mares,* and the life energies lurking within, evoke his description of his own ceramic workshop in *On the Art of the Earth.* The presence of heat in both deepens the analogy between ceramic studio and freshwater pond. In both the pond and the studio, intense heat

31. Belon 1555, 48–49.

32. Rondelet 1558. This is a vernacular translation of Rondelet's original Latin treatise on aquatic organisms: Rondelet 1554–1555.

33. Fragonard 1996, xxxix.

34. Rondelet (1558, 168) distinguishes between aquatic and terrestrial snakes and frogs, water-dwelling creatures tending to be much more colorful than the land-dwelling varieties.

35. Cardano 1556, n.p.

FIGURE 3.4. Pond in clay. Rustic basin by Bernard Palissy (Burty 1886, 31).

activates forms of life. According to Palissy, in the pond environment, the sun produces animals; it breathes life into water and earth by activating the various congelative salts lodged within both. He notes that cool ponds will not "generate" until heated; when "it is cooler it can produce no animal, since no generation, either of animate or vegetative things, ever occurs in the absence of a heated humor."[36] In the absence of heat, there will be no life. By contrast, when aerated and heated, the pond springs to life. Aeration and heat are similarly responsible for animation in Palissy's workshop. Therein, the kiln literally "fires" the form into action; heat activates the glazes and solidifies the bodies. The basin lives as naturalistic pond only once fired into a solid, multicolored plate.

Danger—risk to both human and animal health—marks the life cycle of both the natural and the ceramic pond. Palissy warns his readers that, because

36. Palissy 1996, 30.

of intense processes of putrefaction, warm ponds are hazardous. As he attests: "In truth such waters cannot be good, neither for man nor beast." For, he continues: "these waters, thus aerated and heated, cannot be good; and very often oxen, cows and other livestock die, which might have taken their diseases from such infected watering-holes."[37] Not only water, but also the pond creatures themselves, might be dangerous. Indeed, in the sixteenth century, frogs and serpents in particular—two fixtures of Palissy's compositions—were associated with mysterious illnesses.[38] According to Rondelet, water frogs, snakes, and salamanders tend to be venomous as well as cannibalistic.[39]

As in the generative pond, so in the ceramic workshop; danger went along with vast generative potential. In the *Discours,* Palissy presents the kiln as a similarly treacherous habitat, simultaneously ripe with generative potential. Palissy describes slaving away in a full-body sweat, burning his family's clothes and furniture to feed the hungry generative fire. In the oven, pieces of pottery are "exploding and making a fusillade among themselves like a great many musket shots and cannonades . . . [leaving] some of them with split heads, others with broken arms and legs."[40] Palissy's characterization of his own work habits recalls the Paracelsian idea of bodily fusion of the flesh of the creator with the material of his creation. Paracelsus, the sixteenth-century physician who exalted artisanal handiwork, claimed that knowledge inhered in objects themselves. Palissy, however, perhaps seeing alchemy as a threat to his own artisanal practices, was ambivalent about Paracelsian applications of elemental transmutation; he distanced himself from the alchemical traditions in his published discourses while remaining deeply indebted to them.[41] Water, heat, danger, and theories of nature and its manipulation fused (and at times shattered) in Palissy's kilns of knowledge.

## Fossils and Fabrications

For Palissy, ponds were sites not only for the generation of technicolored life, but also for the memorialization (through putrefaction and petrifaction) of death.[42] In the first place, life casts were—in the most literal sense—earthen impressions of animal and plant bodies. As molds, they remained uncolored (without glazing), thereby looking highly reminiscent of fossils collected in

37. Ibid., 29, 30.
38. Morley 1853, 61.
39. Rondelet 1558, 160–63.
40. Palissy 1996, 279.
41. Debus 1991, 32.
42. Camporesi 1988, 67–90.

the field (see figure 3.3). Thus, craft and nature fused in the constructed fossil.[43] Once fired in the craftsperson's kiln, these objects served as both record and mold.

According to Palissy, just as pond water eventually turns into earth, pond animals will turn into fossilized remains. In *Discours Admirables,* Palissy notes that pond habitats are prime spots for the collection of figured objects, such as shells and rocks.[44] In the sixteenth century, any object dug out of the ground was considered to be a fossil (*fossile,* coming from the Latin *fodere,* meaning "to dig up"). In this period, "figured stones" referred to the specific subset of fossils that resembled organic (animal and plant) bodies. Figured stones often beguiled the viewer because they seemed to cross between ostensibly different realms of nature.[45] Of particular concern was the presence of figured stones representing animals that no longer lived in the area in which their fossilized representations were found. The puzzle was a cause of great concern, because, according to biblical scripture, there could never have been a species that either appeared or disappeared at any point after the initial creation.

Palissy's plate production materialized an antediluvian explanation for fossil formation in nature. His constant molding and firing of rustic plates in the kiln enacted the transformation from water to earth to animal to rocky impression in the heated pond. From the High Middle Ages onward, most European naturalists tended to see fossils as only incidental in their resemblance to life forms. By contrast, Palissy and some of his Renaissance peers saw fossils as petrified remnants of once living beings.[46] Palissy argued for the organic origins of fossils, contending that they bore the impressions of once-living representatives of extinct or extant species. As he ultimately concluded: "After I had observed all the shapes of the stones very closely, I found that none of them could take on the form of a shell, nor of any other animal, unless the animal itself had formed its own shape."[47]

Historians here place Palissy within a community of Renaissance fossil theorizers; but they generally depict Palissy as less discriminating than his better-educated, nonartisanal contemporaries.[48] These contemporaries—the

43. Findlen 1990, 293–97.

44. Schnapper 1988, 19–20.

45. I use *fossil* in the modern sense—that is, to refer to mineral forms shaped by organic plant and animal bodies.

46. Duhem 1909, 281–82.

47. Palissy 1888, 166.

48. Earth scientists themselves, however, have often drawn inspiration from Palissy's geological theories, expressed both in his textual and ceramic productions. In addition to Georges Cuvier, eighteenth-century naturalists including Buffon, Fontanelle, and Voltaire engaged with

German Agricola (1546), the Swiss Gesner (1565), and the Italian Cardano (1550)—worked from Pliny and other ancient authors, developing explanations and categorization schemes for fossils. German physician and mineralogist Georgius Agricola published the first "soundly based" classification of what he called "Inanimate Subterranean Bodies" in 1546. Agricola believed that, in general, fluids circulating in the earth would enact processes of petrifaction, leading to the formation of both minerals and fossils.[49] Meanwhile Gesner drew from Cardano's compendium of the universe, *De Subtilitate,* to develop a classification based on Aristotelian elements—geometrical figures, heavenly bodies, terrestrial bodies, and crafted objects.[50]

Palissy interpreted fossils by combining his own observations and activities as both naturalist and ceramicist.[51] Rather than distinguishing between fossils that resemble nature and those that resemble constructed objects (as Gesner had done), Palissy actually used human craft to elucidate natural process.[52] As he articulated in *Discours Admirables,* Palissy believed that fossil formation resulted from the interaction of several natural processes and generative substances.[53] "Generative salt" and "congelative water" fused into matter through petrifaction and putrefaction, "the generative and preservative glue and putty of all [fossil-like] things."[54] Further, in his opinion, salt is "an unknown and invisible body, like a spirit and yet replacing and supporting the thing in which it is enclosed."[55] Additionally, such fossilization always required the application of heat or humidity. As Palissy reports: if salt "never felt humidity, many things in which it [salt] exists would last forever . . . for generation can never take place without a humor warmed by putrefaction."[56]

Such theories are written all over Palissy's plates. Every basin depicts a rough-hewn pond in which water, mud, plant, and animal have been fused into a congealed aqueous community. As he describes fossilized shellfish,

---

Palissy's work on fossils (Rossi 1984, 93–94). In an article written for the Académie Royale des Sciences, Antoine-René Ferchault de Réaumur argued that Palissy had been the first to realize that the specific layering of fossils in the earth outside Paris was caused by a prehistoric sea, rather than by a biblical flood (Amico 1996, 188–90).

49. F. Adams 1938, 93.

50. Gesner 1565. Gesner classified "fossil objects," including both geological and human artifacts, into fifteen groups based on what he saw as varying degrees of simplicity or complexity (Torrens 1985, 205).

51. Daston and Park 1998, 255–303.

52. Rudwick 1972, 26.

53. Amico 1996, 44.

54. Palissy 1580, 178.

55. Palissy 1996, 199.

56. Ibid., 199.

fish, and amphibians found in the earth's crust: "I maintain that shellfish . . . were generated on the very spot, while the rocks were but water and mud, which since have been petrified along with these fishes."[57] Later on, he reiterates: "I have, as explained above, worked to make you understand above that these fishes were generated in the very place where they changed in nature, while keeping the same shape that they had when they were once alive. . . . The fishes and the water have petrified at the same time."[58]

This sense of drying up into fossil formation evokes the environment of his own studio. He himself confirms the correlation between ceramic practice (in the workshop) and terrestrial process (in the world) when he notes, "Just as all the many kinds of metals and other fusible matter adopt the form of the molds and shapes in which they are thrown or placed, even when placed into the earth itself . . . so the materials of all kinds of rocks in nature take on the shape of those things around which rock has congealed."[59] Palissy re-created his field sites in the rustic pond ceramics, compressing nature and time.[60] On the one hand, he represents water and live animals (the pond environment). On the other hand, by rendering them all in clay, he represents the animals' eventual transformation via death, putrefaction, and petrifaction into fossilized monuments of their former living selves.

## Knowledge through Artifacts

Palissy encouraged his texts' readers to develop knowledge through sensory experience with three-dimensional commodities. Significantly, Palissy ends the *Discours Admirables* with a transcript of the label text for a small "teaching exhibit" he established near his workshop in Paris, and which he had insisted his audience visit after each of his Paris lectures.[61] Unlike the famous *Wunderkammern* and *cabinets de curiosité*, established during the Renaissance largely to celebrate the strange and multifarious wonders of nature, Palissy's museum had a foundation in a specific theory, a narrative (or cluster

57. Ibid., 238–39. Palissy claimed fossils resembling marine organisms found inland had originated there, in ancient lakes or ponds that had since congealed into dry land (216–18, 230).

58. Ibid., 244–45.

59. Ibid., 361.

60. Michel Jeanneret (2001, 50–81) has argued that da Vinci's creative process—sketching rock outcroppings and cataclysmic rains—played a crucial role in the development of his complex geological theories. The connection between da Vinci's artistic and intellectual development is also discussed in Duhem 1906 and Perrig 1980.

61. Kemp 2001, 19.

of origin stories) about generation.[62] Interaction with the material culture of the museum—either collected objects or produced rustic-ware—developed knowledge of nature.[63] Thus did Palissy advise his readers: "I have formed a cabinet in which I have placed many admirable and monstrous things, which I have drawn from the bowels of the earth, and which provide certain witness to what I say, and no man will be found who is not constrained to confess these things to be true, after he has seen the things that I have prepared in my cabinet, in order to convince all those who would not otherwise want to trust my writings."[64]

It was not unusual for a Renaissance collection to contain fossil objects; many contained "petrifactions," from shark's teeth to gallstones.[65] Indeed, a host of seventeenth-century catalogs describe petrified things, including various people, animals, and household objects mysteriously turned to stone.[66] And yet, apart from collections such as Palissy's, such objects were generally mixed in with a range of other items, until well into the eighteenth century.[67] Meanwhile, his description of a teaching museum presages, and may even have directly influenced, Francis Bacon's depiction of the galleries of Salomon's House, wherein natural resources and artificial imitations thereof would be exhibited side by side, available for demonstration through imitation. There, for example, were "fossils and imperfect minerals . . . and other rare stones, both natural and artificial" as well as "great and spacious houses where we imitate and demonstrate meteors.[68] In fact, Bacon seems to have visited Palissy's workshop in Paris and probably attended several of his lectures.[69]

62. Findlen 2002.

63. Pérez-Ramos 1988, 3–31; Allbutt 1913, 234–47.

64. Palissy 1996, 12.

65. Schnapper 1988, 18–19.

66. Another documented sixteenth-century fossil collection belonged to the German physician Johann Kenntman (1518–1574); his inventory included over sixteen hundred specimens. In his 1565 treatise, Kenntman included a woodcut illustration of his mineral cabinet (Torrens 1985, 206; see also Oakley 1965, 9–16, 117–25).

67. Natural historians remained puzzled about fossils long after Palissy's time. Were they rocks *sui generis,* games of nature (*lusus naturae*), or did they take their appearance from animals and plants from which they were derived? As the scholar Lhuyd wrote to John Ray in 1690, to answer such questions, "nothing would conduce more than a very copious collection of shells, of the skeletons of fish, of coral, pori etc., and of those supposed petrifactions" (Scheicher 1985, 35). See also Pearce and Arnold 2000; Grote 1994.

68. Bacon 1626, 487. There were also "great and spacious houses where we imitate and demonstrate meteors; as snow, hail, rain, some artificial rains of bodies and not of water, thunders, lightnings; also generations of bodies in air; as frogs, flies and divers others" (481).

69. Farrington 1979, 14.

Palissy's teaching exhibit consisted not only of found objects but also of fabricated objects. Next to petrified rocks, he placed pieces of illustrative plateware: "I have put before your eyes this great pottery piece, crafted into the shape of a vase: but when it was touched by fire, it turned to liquid and collapsed. Its form was lost completely. . . . Don't you now believe that there is some unseen metal material in their earth from which comparable natural vessels are made? For otherwise it should have broken instead of having bent."[70] Comparison of nature and craft is purposeful. To highlight the processual aspects of the displayed natural formations, Palissy placed them beside the products of his own bodily and mental exertions—parallel enactments of the art of the earth.

The *Copy of the Labels* provides an "exhibit guide" to Palissy's museum. It consists of transcripts of the labels he placed beside his natural objects in the gallery. These labels are, in his words, "placed below the wonderful things that the author of this very book has prepared and put in order in his cabinet, in order to prove all the things contained in this book." In the labels, he is full of encouraging words, prompting viewers to experience the exhibit to its fullest potential. Palissy peppers his prose with imperatives: he impels individuals to see! touch! listen! and smell! The exhibit creator implores: "look . . . look . . . look at the slates! . . . Don't you see?" And then: "I have put this rock before your eyes." Each time, it is important that viewers recognize not only the object in front of them, but also geological or artisanal processes to which the object's current existence bears witness. "See all these kinds of fishes that I have put before your eyes, you will see a number whose seed has been lost and even, we do not at present know what they should be called: but that cannot alter the fact that it is evident to all, that the form of the same gives us clear knowledge that they were once alive." Natural purposes are thereby divulged: "You see clearly that all these shapes of shells reduced to stones, were formerly living fishes."[71]

Palissy promotes direct experience by readers, listeners, and visitors with exhibited objects presented in the collection. He finally characterizes the displayed objects as: "Marvelous things that are set as a witness and proof of my writings, arranged in rows or in layers with labels underneath; so that anyone may instruct himself." The objects in the museum—both natural and constructed—could, in their very materiality, prove the validity of his words. Here, in his opinion, it was not enough to simply look passively at the objects. They must be experienced through as many senses as possible. As he

70. Palissy 1996, 370.

71. Palissy 1996, 361, 365–66, 369, 365–66, 369.

concludes: "In proving my written reasons, I satisfy sight, hearing and touch: for this reason, defamers will have no place in my domain: as you will see when you come to see me in my little Academy."[72] Palissy praises the power of sensory experience—as either maker or museum visitor—in relation to three-dimensional objects.

Understanding Palissy's oeuvre as experiential interplay is all the more plausible because he recorded his ideas into dialogue form. Palissy repeated his aforementioned lecture series on natural history from 1576 to 1584.[73] In 1580, he published the *Discours,* wherein he compiled a series of short dialogues—all inspired by his lectures—staged as imaginary conversations between "Theorie" and "Pratique." Palissy employed the dialogue form, popular in the period, to harness his experiences as performer and educator. Palissy's approach to dialogic expression was anything but a throwback to the ancients. Rather, his dialogues participated in a movement to the vernacular, interactive realm of drama. In the back and forth of the dialogue structure, text could preserve the effective breaks of speech.[74] Palissy's philosophies of nature thus are expressed as a forceful engagement between two voices, evoking the dramatic realm of orality and aurality rather than textuality.

## Conclusion

Palissy cast, wrote, and spoke his mind on matters of craft, nature, and knowledge. His ceramic wares, as well as the exhibition strategy he laid out in *Copy of the Labels*, exemplify a model for learning through sensing—as well as knowing through making—in three dimensions. In Palissy's models, mental reflection, physical exertion, and organic materials fused into three cohesive dimensions. He compressed animal, vegetable, and mineral entities, as well as natural philosophies into his clay pond renditions. Which is to say, Palissy's ceramic commodities impressed both nature and time into clay.

Other sixteenth-century naturalists interested in the processes of generation and fossilization represented specimens with two-dimensional images.[75] But Palissy went farther—deeper into the earth itself—to materialize his

72. Palissy 1996, 16, 17.

73. Among Palissy's regular listeners was the "Alençon" group of alchemical physicians (Fragonard 1996, xv).

74. Jean Céard has analyzed Palissy's employment of the dialogue form in terms of the larger literary culture of the Renaissance (Céard 1987, 1991). Ernest Dupuy alluded to a possible interpretation of Palissy's texts along these lines (Dupuy 1902, 212–17; see also Kushner 1981, 149–51).

75. Pyle 2000, 69–75.

ideas about its living inhabitants. Simultaneously objects of decorative art, models of geological theory, and products of both artisanal and intellectual labor, his basins articulated a new artisanal philosophy of knowledge. His rustic ceramic-wares should be seen as novel and enduring, articulations of early modern natural philosophy, embedded in a commodity culture in which ceramics were just beginning to become highly prized aesthetic objects. A specifically sculptural, and ceramic, form of scientific experience and natural knowledge was in the making.

## Primary Sources

Secondary sources can be found in the cumulative bibliography at the end of the book.

### UNPUBLISHED SOURCES

Paris. Ms. Fr. 640. (Microfilm 1965, folios 122, 123, 124, 140, 141 recto and verso) "How to Cast Little Lizards," "How to Cast a Shrimp," "Vipers and Serpents," "How to Cast Turtles," "Molding a Turtle," and "Frogs." Manuscrits Occidentaux du Bibliothèque Nationale de France.

### PUBLISHED SOURCES

Agricola, Georg. 1546. *De Ortu et Causis Subterraneorum.* Per H. Frobenivm: Basileae.

Bacon, Francis. 1626. *The New Atlantis.* In *A Critical Edition of the Major Work,* ed. Brian Vickers, 457–89. Oxford: Oxford University Press, 1996.

Belon, Pierre. 1555. *La Nature et Diversité des Poissons.* Paris: Charles Estienne.

Cardano, Girolamo. 1556. Des bestes nées de la putrefaction. In *Les livres de Hierome Cardanv intitvles De la subtilité, & subtiles inuentions, ensemble les causes occultes, & raisons d'icelles, traduis de latin en francois, par Richard le Blanc.* Paris: Guillaume le Noir.

Cuvier, Georges. 1817. *Essay on the Theory of the Earth, with mineralogical notes, and an account of Cuvier's geological discoveries by Professor Jameson.* Reprint, New York: Arno Press, 1978.

Gesner, Konrad. 1565. *De Rerum Fossilium Lapidum et Gemmarum.* Tiguri: Jacobus Gesnerus.

Palissy, Bernard. 1580. *Discours Admirables.* Paris: M. Le Jeune.

———. 1888 (1563, 1580). *Œuvres Complètes.* Ed. Benjamin Fillon. Niort, France: L. Clouzot.

———. 1996 (1563, 1580). *Œuvres Complètes.* Ed. M.-M. Fragonard, K. Cameron, J. Céard, M.-D. Legrand, F. Lestringant, and G. Shrenck. Paris: Editions InterUniversitaires.

Rondelet, Guillaume. 1554–1555. *Libri de Piscibus Marinis in quibis verae piscium effigies expressae sunt.* Lugduni Batavorum: Bonhomme.

———. 1558. *L'histoire entière des poissons, composè premierement en latin par maistre Guilaume Rondelet maintenant traduites en françois sans auoir rien omis estant necessaire à l'intelligence d'icelle.* Lyon: M. Bonhome.

Serres, Olivier de. 1600. *Le Théâtre d'Agriculture et Mesnage des Champs.* Paris: Imprimeur Ordinaire du Roy.

4

# The Production of Silver, Copper, and Lead in the Harz Mountains from Late Medieval Times to the Onset of Industrialization

CHRISTOPH BARTELS

Extensive residues from early mining and metallurgy as well as the preserved high-quality metal products prove that metalworkers possessed reliable knowledge and instruments for directing production processes from antiquity onward. This knowledge has since been widely supplanted by modern scientific instruments, and much older knowledge about the creation of metal products, which often required a high degree of skill, has been lost. Nevertheless, many of the material products have survived and, together with ancient texts on mining and mineralogy, attest to the competence of their producers.[1] A comparison of the products resulting from today's knowledge with those made between 1450 and 1750 reveals that the abilities of the producers must have been grounded in reliable experiential knowledge.

In debates over the making of materials nowadays, it is axiomatic that a single product often can be made in very different ways. Bread, for example, may be baked by hand, using traditional equipment and recipes; or it may be produced in baking plants, where production is underpinned by analyses in high-tech laboratories and methods of food design. The second method is able to generate large quantities of bread, characterized by a reliable quality and a low price. But is that bread of the highest possible quality? Could the "science-based" bread win out in a competition with that produced by the baker round the corner? Scientific analysis of the human metabolism might demonstrate that factory-made bread is just as nourishing as its traditional artisanal equivalent. But is it as tasty as the crafted product? Modern standards offer one mode of evaluation, but the individual character of products

1. For examples, see Yalçin, Pulak, and Slotta 2005; for general aspects of historical metallurgy, see Rickard 1932.

and the points of view of consumers offer another. Each mode points to different interests, expectations, and aspects of knowledge which make the product what it is. Similar questions might also be asked with respect to earlier periods and other kinds of materials. In this chapter I will examine historical sources that convey an idea of the late medieval and early modern forms of knowledge involved in mining and the production of metals. I will concentrate on a specific region, namely the Harz Mountains in Germany, which were rich in copper, silver, and lead. At the center of my historical *tour d'horizon* are practitioners who possessed hands-on knowledge in mining and metallurgy, complemented by knowledge of mathematics, geography, natural history (especially mineralogy and stratigraphy), alchemy, and data-collection methods in keeping with their administrative tasks. This chapter thus targets the same type of experts studied in other contributions to this volume, who worked and lived at the intersection of commercial production, administration, and a sphere of knowledge which may be understood as a "root" of the natural sciences.

The first part of this chapter presents an overview on the local site I am concerned with and the social and political context of mining and metallurgy in the Harz region from the late medieval until the early modern period. Metals are commonplace in everyday life, but are often produced in remote regions, rather distant from the societies such producers supplied.[2] Furthermore, precious metals have long been closely related to aspects of secular and religious power and representation.[3] They were indissolubly linked to artistic labor, which transformed metal into objects of adoration and representation. As my overview will show, advanced metalwork was formerly located close to centers of power, and the most sophisticated techniques of metalworking were often developed at the direct behest of political and economic leaders.[4] The second part continues this overview and further addresses some characteristic features of the technical development of mining and metal production in the Harz region in the period from 1440 until 1750.

The third and main part studies in finer detail forms of advanced expertise involved in mining and metallurgy. Metal production not only engendered groups of experts, it also developed into a field of advanced knowledge in its own right.[5] Metal products, especially those made from precious metals or semiprecious copper alloys (bronze and brass), had long exhibited the most

2. The Harz is a good example for this aspect. See Bartels 1992b.

3. Eliade 1980. See also the chapters by Shell and Smith in this volume.

4. Lücke and Dräger 2004.

5. For details, see Craddock 1995; Bachmann 1993; Weisgerber 1987; Eliade 1980.

advanced technical abilities and craft skills available in metal production.[6] This process of manipulating elements of the material world generated increasing knowledge about the structures of this world. It was not a blind ability to use and manipulate materials, lacking insight into the causal factors affecting materials and processes. From the very beginning, observation of the material world and experimentation were vital elements in the production of metals, without which they could not have been made. Consequently, the production of copper, silver, and lead generated a field of advanced written knowledge, as well as a group of experts, long before the beginning of the period covered by this chapter. As the archives at Goslar prove, literate persons began to be involved in the administration of mining and metal production as early as the eleventh century.[7] Around 1125, Theophilus Presbyter was already writing his *Schedula on diverse arts* (*diversarum artium schedula*).[8] The outlines of a written *scientia de Mineralibus* (science of minerals) existed in Albertus Magnus's *Book of Minerals* of circa 1250, well-known in the scholarly world. Albertus not only used the term *scientia,* but also outlined his use of that term: "It is not the purpose of a scientist (*physicus*) to discuss the origins of the material world (*materiae*). His starting point is the fact that this material world exists. Any science (*scientia*) accepts the existence of its basic subjects without asking how they came into the world." The task of the philosopher-scientist "was to inquire about what is possible in nature, based on natural powers and natural reasons."[9] Numerous manuscript copies of this work existed by the fifteenth century, and a printed version appeared in 1569.[10] None of Albertus's other manuscripts survive in as many copies as *De Mineralibus.*[11] Our central question, therefore, is not whether advanced knowledge or a "science" of minerals really existed between 1450 and 1750. It had already existed some two hundred years earlier, but it differed from many other sectors of advanced knowledge in the Middle Ages and the early

6. For basic features, see Bachmann 1993. See also Stöllner, Slotta, and Vatandoust 2004; Yalçin, Pulak, and Slotta 2005.

7. For details, see Bartels et al. 2007, chap. 2.2.4.

8. Wyckoff 1967; for details and references, see Bartels et al. 2007, chap. 3.1.6; for theological aspects, see Reudenbach 2003.

9. "(Propter hoc videtur dicendum, quod,) quamvis Physici non sit determinare de exitu materiae in esse, sed potius relinquere materiam esse, eo quod ipsa sit principium physicum, sicut qualibet scientia relinquit esse sua principia et non quaerit, qualiter in esse exiverint" (Fries 1981, 164). "In naturalibus habemus inquirere . . . quid in rebus naturalibus secundum causas naturae insitas naturaliter fieri possit" (6).

10. Albertus Magnus 1569.

11. For details and references, see Bartels 2002; Bartels et al. 2007, chap. 3.2.5; Riddle and Mulholland 1980; Weisheipl 1980.

modern period in being a genuinely medieval development, based on local experience rather than a revival of classical texts and knowledge.[12] The questions, then, are how influential and vital this *scientia* of minerals and metal production was from the late Middle Ages onward and how it was further developed in the period covered by this volume.

Three basic features are commonly conceived to be characteristic of science and technology: observation of the material world, causal inquiry as the basic mode of approach, and the definition of logical systems in which the character of causal chains can be described. These are highly abstract constructions. Studying characteristic examples of experts, I will try to flesh out what it meant in the late medieval and early modern period to possess advanced expertise in mining and metallurgy. Science and technology "map" natural processes to find out the rules and laws they follow. To produce (and not simply to pick up) materials means to reproduce and steer chemical and physical (or biological) natural processes. The ability to do so proves that this mapping had been performed to a certain extent and had produced reliable results. The knowledge of how to follow the rules and laws of nature represents the ability to "read the map."

## Overview: Policy, Mining, and Metallurgy in the Harz Mining Region

The Harz Mountains, with numerous ore deposits, are the northernmost of the German highland areas. The northwestern part of the Harz was in the hands of the Welfe dynasty, a house with a tendency to subdivide its territories repeatedly over the centuries. Up to 1495, territories at Braunschweig (with ore deposits at Bad Grund, Zellerfeld, Lautenthal, and Hahnenklee), an area north of the adjoining principality of Grubenhagen (with deposits at Clausthal, Altenau, and St. Andreasberg), and Goslar (with the Rammelsberg deposit) were the main political units in the development of metal production. In 1495, the territory of the house of Welfe was divided, and the dukes of Brunswick-Wolfenbuttel ("Mittleres Haus Braunschweig," 1495–1634) became the most important branch of the dynasty with respect to mining and metal production, rivaling the town of Goslar, until the town's mining and metal production rights were ceded to Duke Heinrich the Younger (b. 1489, ruled 1514–1568) by the treaty of Riechenberg. The principality of Grubenhagen, with significant ore deposits, existed until 1596, when Duke Philipp of

12. An important contribution was the development of a printed literature on metal production. This field has been studied in particular by M. Koch (1963). See, in addition, Slotta and Bartels 1990, 141–74.

Braunschweig-Grubenhagen died without issue. The territory then became the subject of conflict until the last member of the house of Brunswick-Wolfenbuttel died in 1634. This resulted in a new division of the Welfe lands. Along the old boundary between Brunswick or Brunswick-Wolfenbuttel and Grubenhagen, the Harz was now divided into two parts. The so-called Einseitiger Harz, the former Grubenhagen area, was ruled by the house of Celle-Calenberg. The Kommunion Harz remained a domain of the Welfe dynasty, corresponding to the former Brunswick-Wolfenbuttel. This arrangement lasted until 1788, when the house of Hannover absorbed the Einseitiger Harz.[13]

The council and inhabitants of the old imperial town of Goslar and the dukes of Braunschweig were involved in metal production in the northwestern Harz region around 1450, a time at which there were no similar activities in Grubenhagen.[14] Metal production in the northwestern Harz can be traced back to the time of the Roman Empire; archaeological finds suggest that copper production there probably began during the Bronze Age. It has been shown that the deposits were exploited more intensively during the seventh and eighth centuries, before the region was conquered by Charlemagne's armies in the late eighth century and incorporated into the Holy Roman Empire.[15] Some two hundred years later, the northwestern Harz was a property of the German kings. It subsequently passed to influential abbeys, various clerical or secular organizations, and noble houses, including the Welfe dynasty, a leading power in the north of Germany. Noble families had great influence in Goslar, but during the late thirteenth and early fourteenth centuries, their lands and influence were assumed by a new stratum of wealthy town patricians. The Welfe dynasty were landlords in the northern and western parts of the Harz Mountains and surrounding regions. Like many ruling houses, they were almost always in need of money, and mortgaged mining and forestal areas to Goslar's town council in the late thirteenth century. The most important of these mining districts was Rammelsberg, which the Welfe dynasty retained the right to take back once they had repaid the capital borrowed from the town.

Following a crisis at the beginning of the fourteenth century, mining and metal production in the northwestern Harz region, which had flourished

13. For details of the history of the territories, see Jarck and Schild 2000, especially the contributions by U. Schwarz and S. Brüdermann.

14. For details and references, see Bartels et al. 2007, chap. 3.

15. For the archaeological exploration of the Harz region in general, see Segers-Glocke 2000.

from the tenth to the thirteenth century, almost completely ceased for several reasons around 1360.[16] Only after 1410 did mining and metal production gradually begin to revive, both at Goslar and around Grund and Gittelde on the western borders of the Harz.[17] Centuries of medieval mining had exhausted rich deposits near the surface, and down to a depth of over 150 meters below the surface in some areas. It was difficult and costly to reorganize the old mines and to open up fresh ore reserves.

The district offered two types of nonferrous metal deposits. Rammelsberg contained a massive stratiform ore body of sedimentary origin, generated by submarine exhalations from mineral-rich hot springs on the Devonian sea floor. Lens-shaped ore bodies had developed on the seabed and were subsequently covered by thick sediments, which were folded as the Variscian Mountains formed during the Carboniferous. The ore deposit now consisted of two bodies, both plunging downward at an angle of around 45 degrees. Subsequently, during the Cretaceous period, the Variscian massif was eroded, flattened, and covered by sea before parts of the old mountain region were lifted up again, developing into the Harz Mountains during subsequent geological periods. One part of the Rammelsberg deposit (the *Altes Lager*, or Old Deposit) was partly eroded and thus exposed on the surface. The second part, situated nearby, did not reach the surface of the mountain and was not discovered until 1859. The massive, extremely dense and small-grained ore consisted mainly of galena, with a low content of silver, pyrite, sulfidic copper and zinc ores, and barium sulfate, with some sterile sediment in the substrate. The minerals were arranged in zones enriched by the main components. Up to the mid-nineteenth century, the useful components in metal production were copper, lead, silver, and later zinc.[18]

The second type of ore deposit was present in the northwestern part of the mountains, some ten to twenty kilometers south of Goslar. These deposits consisted of ore veins in underground faults, containing minerals generally similar to those of Rammelsberg, but with a very different distribution. Here, massive ore bodies of galena, zinc ores, and some copper ores were concentrated. The upper parts of the deposits, in particular, contained veins and pocketlike concentrations of rich silver ore, and the galena had a much higher silver content in some areas than the Rammelsberg ores. The ore veins of the Upper Harz formed zones with thick, massive ore bodies, extending almost vertically downward in most cases. When they eventually closed in

16. Bartels 1997b, 28–56.

17. Bartels 2004b, 65–76.

18. For more details and references on geology and ore deposits, see Bartels 1988, 9–11.

the twentieth century, the mines had reached a maximum depth of over one kilometer.[19] These deposits were the subject of productive activity from early medieval times until the 1360s or so. Activity stopped at the same time in this district as in the Rammelsberg deposit, though no precise information remains as to why. Nevertheless, when mining activities began again in the sixteenth century, it became clear that medieval activity had been extensive and had opened up the ore veins to an extent that may have made the operations too costly during a general economic depression, such as occurred in the decades after the Black Death.[20]

## Technical Developments in Mining and Metal Production in the Northwestern Harz from 1450 to 1750

After a boom in mining and metal production in the Harz Mountains during the Middle Ages, followed by a crisis in the fourteenth century, the two metalliferous districts of the region developed differently prior to industrialization.[21] Led by the town council of Goslar and the wealthy furnace-masters (*Hüttenmeister*), repeated attempts to revive production from the Rammelsberg ore deposit were undertaken from 1360 onward. These attempts gave rise to limited, but increasingly fruitful, activities from 1410 on, until the situation rapidly changed in response to increasing demand for lead after 1450, which drove prices up.

### SMELTING INNOVATIONS IN THE FIFTEENTH AND SIXTEENTH CENTURIES

Large quantities of lead were used in the separation of silver from copper, mined on the southern boundary of the Harz Mountains, near Eisleben, Sangerhausen, and Mansfeld. A new and productive technology for separating silver and copper (the *Seigerhüttenprozeß*) had developed during the fifteenth century, causing a boom in metal production after about 1460, which was driven by rapidly increasing demand for silver and copper in the European economy.[22] Silver, as the raw material for coin, was the basis of finan-

19. For details and references on geology and ore deposits, see Bartels 1992b, 23–31.

20. For details, see Bartels 2004a.

21. The following overview is based on several monographs and papers by the author. See Bartels 1988, 1989, 1990, 1992a, 1992b, 1994a, 1994b, 1997a, 1997b, 2000, 2002, 2004a, 2004b, 2004c, 2006; also Bartels, Bingener, and Slotta 2006, with references and broad discussions of the sources.

22. A detailed study is in Blanchard 1994.

cial systems. Copper was especially important in military equipment, luxury products, and all manner of durable household vessels. Lead was used both in the production of these two metals and in buildings, in pottery glazes, and—somewhat later—for military purposes as well. Some regions experienced a sort of gold rush. New deposits were discovered and often proved to be enormously rich. New towns were founded in remote regions with ore deposits, like Schneeberg in Saxony or Joachimsthal in Bohemia, developing in just a few years with the support of territorial rulers.[23] To this day, the town of Goslar bears the marks of this fifteenth- and sixteenth-century boom.[24] The size of the town's main entrance (*Breite Pforte*) and other fortifications, the town hall, several guildhalls, and palacelike buildings from the fifteenth and sixteenth centuries resulted from the explosion of wealth between 1460 and 1600.

This period can be divided into two phases of political development. The Welfe princes and their counselors were aware of the new boom and knew also that they held rights to the Rammelsberg deposit which could not be regained until they had repaid the loans their ancestors had accepted from the town.[25] In 1526/27, Duke Heinrich the Younger of Brunswick-Wolfenbuttel was able to pay back the money, but the town council rejected his demand for the return of all the mines, smelting works, and large forest areas. Both parties applied to the imperial court. The duke responded to the court's decision to favor Goslar's position with military action. In reply, the town council made military incursions into ducal lands. The conflict embroiled the emperor and neighboring rulers for twenty-five years. Initially, matters developed in favor of the town, but political developments ultimately enabled Duke Heinrich to realize his demands, though not without a certain degree of compromise with the town council. The signing of the treaty of Riechenberg in 1552 gave the duke and his counselors a leading position in the mining and metal production of Goslar.[26] The administration of mining and smelting by Heinrich and his son Julius (b. 1528, r. 1568–1589), a man of extraordinary economic and administrative skills, developed the metal production of Rammelsberg and the Upper Harz in general into a most remarkable and successful network of early modern enterprises.

23. On Schneeberg, see Wagenbreth and Wächtler 1990; for Joachimsthal, see Bartels 1997b, 56–69.

24. The relationship between mining and the development of the town of Goslar is discussed in detail and with an appendix of sources in Bartels 2004c.

25. A. Westermann 2005a, 2005b.

26. On the treaty of Riechenberg, see Rammelsberger Bergbaumuseum Goslar 2004, 189–203.

The Welfe dukes were similarly active with regard to the ore deposits within their dominion. Initially, they increased iron production, which had been revived around 1450, though it only took place to a limited extent before 1520. Once again, it was Duke Heinrich the Younger and his administration who successfully increased production. The duke had decided to make his territory into a modern state and to make the potential of metal production an economic basis of his policies. A high-risk player, he did not hesitate to use military force to realize his aims. As a politician, he was as talented as he was strong, and he was successful in achieving his goals. His son Julius also successfully increased production, though he was quite a different character, interested primarily in culture, science, and economy. He was a supporter of the principles of "good government" (*gute Herrschaft*), making the peaceful exploitation of his territory, economic welfare, and his position as the "father" of his subjects the primary goals of his reign.[27] He established an able mining and smelting administration, and continued his father's policy of recruiting foreign mining, hydraulic, surveying, and smelting experts. Particularly in the domain of metal production from the ores of the Rammelsberg deposit, one can see a fundamental change during the second half of the sixteenth century. Transformations in the production of silver, copper, and lead after the Middle Ages began at the end of the productive chain, that is, in roasting and smelting processes.

The Harz Mountains are just one example of a European development that gained in importance in the later fifteenth and sixteenth centuries.[28] Duke Julius's reign saw a boom in mining and metal production in the Harz Mountains.[29] Like his father, Julius used silver, copper, and lead production as a major resource for his state. The prince himself was the head entrepreneur, and parts of his administration were devoted to the management of the mining and smelting enterprises. Besides the duke, private shareholders also engaged in mining, but their influence was limited. All organizational, financial, and technical decisions in the field of metal production were in the hands of the ducal administration, while the metal trade was contracted out by the duke. This "directorial system" (*Direktionssystem*), as it was known, was a typical absolutist approach to metal production and prevailed in many European mining districts, though not in England.[30] From the 1560s onward,

27. Kraschewski 1978, 1989.

28. For overviews concerning the more important mining districts in Europe, see Slotta and Bartels 1990, 58–139.

29. For data on production and results of the mining operations, see Bartels 1992b, 490 and table 29 (pp. 726–31).

30. Henschke 1974.

smelting operations were run exclusively by the state; up to then, some private furnace-masters had been involved in metal production from Rammelsberg ore. From the revival of production at Goslar after 1410 up to around 1530, these private smelting enterprises had formed the basis of lead and silver production using Rammelsberg ores. Several attempts to produce larger quantities of high-quality copper had failed before 1580.[31]

From 1524 onward, Duke Heinrich the Younger systematically set about reviving mining and metal production in the Upper Harz. He proclaimed freedoms and privileges for miners willing to come to his territory and recruited experts from Bohemia. Within a few years, rich silver ore was being mined at Grund, Wildemann, and Zellerfeld, though it was quite difficult to extract from the main deposit between Wildemann and Zellerfeld in particular. Long adits had to be driven, and those dating from medieval times had to be restored and extended. At Wildemann, it took twelve years before work started to yield significant quantities. A major portion of the rich silver ore had been extracted during exploitation in earlier centuries, leaving behind masses of galena that was poor in silver, as well as ore sections that could not be extracted using existing methods. Though mining technology had developed rapidly with regard to mine drainage, shaft haulage, and ventilation, ore extraction techniques had not changed since the fourteenth century.[32]

As discussed above, silver was present in the deposits in two different qualities. There were rich silver ores of primary and secondary mineral composition, concentrated in the upper parts of the deposit, and there was a certain overall silver content in the lead-ore galena. The rich ores might contain 10 percent or more silver, while the galena had an average of between 0.03 percent and 0.07 percent, though some zones were richer. Up to the end of the sixteenth century, mining activity targeted first the rich silver ores and then the galena with relatively high silver content. The mines flourished from the 1550s onward, but rich ores were found only in limited quantities, and after some thirty years these resources began to be exhausted in several places. Thus, a new crisis began to develop toward the end of the sixteenth century, as production was considerably reduced, especially of silver, the most valuable product.[33] This depletion of rich silver and copper ores affected all of Europe's leading ore districts, beginning around 1550 in some regions like Tyrol, and as late as 1580 in others.[34] Across the region, the upper levels of

31. Bartels 1997b, 33–39.

32. For details, see Bartels, Bingener, and Slotta 2006.

33. Henschke 1974.

34. Bartels, Bingener, and Slotta 2006.

the deposits had already been opened up; the silver content reduced as the deposits descended, with a concomitant increase in the technical difficulty and, hence, cost of extraction. Significant portions of the ore veins extended beneath swampy flatlands and could be reached only by extending galleries into these areas over distances of five kilometers or more. Because the adits had not yet reached that part of the mountains, the presence of deposits there was a matter of speculation, and exploratory forays were very costly.

## A SEVENTEENTH-CENTURY INNOVATION: GUNPOWDER BLASTING

Thus, the mining and metal-production sector in the Harz Mountains had reached a difficult juncture when the Holy Roman Empire and adjacent states were caught up in the Thirty Years' War (1618–1648). The Harz region was affected particularly severely around 1626, when parts of the mining settlements were completely destroyed.[35] The Thirty Years' War devastated economic and social structures for decades in large parts of Europe. Many mining districts would not recover before the eighteenth century. The Harz Mountains experienced a different fate. In 1633, a new technique of extracting ore was introduced to the region, involving blasting rock with gunpowder.[36] This technique, whose European origin is unknown, is first documented at Le Thillot in Lorraine between 1617 and 1627. By around 1630, it had spread throughout European mining districts. The structure and mineral composition of the Upper Harz ore veins proved to be especially amenable to exploitation using the new technique. Major sections of these veins consisted of galena, often combined with nonmetallic minerals. Parts of the ore veins were several meters thick, the mineral content being mainly galena with a silver content of under 0.1 percent. Blasting made it possible to extract these parts of the deposits at low cost and in much higher quantities than by traditional methods of ore mining by hand. In addition, it allowed the extraction of parts of the deposits that had already been opened and were known to possess a high silver content, but whose ores had been too difficult to extract by hand. Seventeenth-century records from the Harz mining administration document the rapidity with which the new method of ore extraction was adopted and the systematic development of the technique. Just as progress in smelting had caused a revolution in this area of the sequence of operations during the latter half of the sixteenth century (and somewhat earlier in other

35. Bartels 1992a, 18; for general aspects, see Buchanan 1996a.
36. Bartels 1992b, 170–85.

mining districts), the blasting technique caused a revolution in mining during the seventeenth century.[37]

The Upper Harz is a striking illustration. The new technique made it possible to produce much higher quantities of raw ore, so that extraction operations became considerably accelerated, allowing the mines to offset the shortage of qualified miners resulting from the production crisis and the war. When blasting techniques began, efforts to reorganize mining to accommodate the effects of the war had begun to yield their first positive results. In the main deposit, near Clausthal, two mines returned to profitability in 1625 and 1629. Five other important mines followed suit between 1637 and 1649, and two more in 1660 and 1661. Profits flowed continuously for periods ranging between 25 and 130 years.[38]

## HYDRAULIC SYSTEMS

Detailed studies have revealed the great importance of blasting, particularly in the development of mines in the so-called Burgstätter range of ore veins near Clausthal, the most important mining subdistrict in the Upper Harz from the seventeenth to the nineteenth centuries. The extraction of larger quantities of ore entailed a rapid increase of transportation in galleries and shafts, as well as a rapid extension of underground structures, especially in depth. This extension generated an increasing demand for power. Mine machinery driven by waterwheels drained the workings and winched up ore and sterile rock. Huge pumping systems had been installed since the revival of mining activities in the fifteenth century at Rammelsberg and from 1524 onward in the Upper Harz. These systems were vital to the project of extending of the mines below the level of the deepest existing galleries, because the region was fairly wet, necessitating the drainage of groundwater.

During the sixteenth century (or even before, as in the case of Rammelsberg), the mines had become irreversibly dependent upon pumping engines. The pumping system required wheels both on the surface and underground, arranged in cascadelike groupings and housed in dozens of huge chambers, most of which were ten to twelve meters in height and length. In addition, the water had to be conducted to the shafts and organized in underground systems. Wherever possible, water-driven bull wheels were used to haul ore and rock through the shafts; many of the wheels were a century old or more, having been designed and built for the purpose of extracting comparatively small

37. Bartels 2000.

38. Bartels 2000, 188.

quantities of rich silver ores. Because these shafts were narrow, built to serve small hand winches, they had to be widened to make space for pumps, the rod systems that drove them, and larger haulage equipment. Furthermore, many new shafts had to be sunk. All these operations produced a need for additional pumping, haulage, and transportation on the surface. Most of the increasing mass of ore was processed in batteries of stamp mills, which crushed the minerals to sand, allowing ore to be separated from other components by gravity as it passed through networks of canals, basins, and washing-hearths. Here the crushed ore was washed in the water current, and the heavy particles were captured while the lighter, nonmetallic components washed away. The processed ore was then roasted, smelted, and refined in batteries of furnaces, each served by bellows driven by waterwheels.

The increasing need for hydraulic power transformed the Upper Harz region into an enormous hydraulic engineering site during the decades after the introduction of blasting. Between 1640 and 1660, three new artificial ponds were constructed in addition to some eleven to thirteen older ones, and the capacity of water reservoirs was thus enlarged from around 1.15 to around 1.95 million cubic meters. From 1660 to 1680, twenty-five new artificial ponds and lakes were constructed, with an overall capacity of 3.15 million cubic meters. The network of canals connecting the systems was extended to over 200 kilometers in length. Hydraulic systems, which had evolved over some 130 years between 1530 and 1660, were expanded by almost 150 percent in the next 20 years.[39] These developments demanded enormous investment, which was only possible thanks to rapidly increasing production and profit.[40]

## Advanced Knowledge and Experts in the Production of Silver, Copper, and Lead

The new technical developments deeply affected social structures, community life, and the social distribution of knowledge in mining. The proud mining craftsmen of the sixteenth century, who had highly developed abilities in identifying valuable minerals, extraction techniques which relied on handheld tools, and skills in sinking and timbering shafts or heading galleries, were replaced by unskilled miners, often recruited from farming families in the war decades. Their daily work was very different from that of the earlier craftsmen, for most underground workers spent their time boring blast

39. Bartels 2000, 186–205; similar developments are known from the Saxon and Bohemian Ore Mountains, Hungary, Slovakia, Sweden, and the United Kingdom; see Reynolds 1983.

40. For data, see Bartels 1992b, 1994b; for European data, see E. Westermann 1986.

holes, a dull, labor-intensive occupation. It is not surprising, given such radical change, that social conflicts arose during the period of these sweeping enterprises in hydraulic engineering. Such problems were exacerbated by the fact that new, well-paid groups of experts were ever more important and influential in mining.

Most of these new experts became shareholders in the mines, a phenomenon that had official support. In keeping with the mercantilist principles of mine management introduced in the sixteenth century (and to a certain extent during the fifteenth by Goslar's town council), the Welfe administrators developed mining and metal production into state enterprises with the involvement of private capital. From the 1680s onward, all the expert-officers involved in opening up a new mine were rewarded by the authorities with shares whose value depended upon the rank of the officer in question. Within a comparatively short time, the mines' administrators owned a considerable portion of the shares as private property, because their functions allowed them to direct all operations. On the one hand, such private shares stimulated efforts to run the mines profitably, to open up new deposits and mines, and to improve knowledge involved in the mining business. On the other, they stimulated all kinds of cost-cutting activities, including deskilling, lowering wages, and extending working hours. The economic interests of mining administrators thus led to a reorganization of the entire social system of knowledge as well as to social conflicts in the last years of the seventeenth century, which ended by generating a major opposition between the workforce on the one hand and the shareholders and expert-officers (and thus the state itself) on the other.[41] This situation was encouraged by the fact that, after successful exploitation during the seventeenth century, resources eventually began to run out, so that opening up further ore veins was vital to maintaining or, better still, extending mining operations.

## HEINRICH ALBERT VON DEM BUSCH: EXPERT IN ORE PROSPECTING AND STRATIGRAPHY

In 1689 Heinrich Albert von dem Busch entered the Hanoverian mining administration as an investor and expert-officer. He was to become a central figure in the field over the next forty years. Born in 1664, he came from a local noble family with numerous contacts to the Hanoverian nobility and administration, especially mining circles and leading shareholders. Only three

41. For details, see Bartels 1992b, chaps. 6 and 9; Bartels 1994b.

years after beginning work in 1689 as an assistant of the Mining Administration (*Oberbergamt*) Clausthal, aged 28, he was named vice president of this organization, then became president (*Berghauptmann*) in October 1695. In 1713 he was appointed to a ministerial post in Hanover, but his move to this new position did not entail abandoning his position as president of the Mining Administration, a post which he would hold for thirty-six years until his death in 1731. Even after moving from Clausthal to Hanover in 1709, he kept abreast of progress in mining and metal production in his district, as demonstrated by thousands of comments in the vast corpus of documents he studied. Simultaneously, Busch became the most important private shareholder in Harz mining.[42] He was not only a brilliant administrator and organizer, but also became an expert in the structural geology and behavior of the deposits. It was Busch who initiated ore prospecting, and when he realized that the administrative and economic structures of mining and producing metals were insufficient for his planning needs, he did not hesitate to change them. He became aware that there was a lack of investment in ore prospecting. Having tried in vain to get government funding for this purpose, he founded a banklike institution, the Clausthaler Bergbaukasse, funded by a new tax on beer and brandy. The Bergbaukasse was founded with a view to acquiring shares in unprofitable mines, so as to finance both the technical support of profitable mining and systematic ore prospecting.

At the time, no reliable prospecting method was available beyond following structures underground with exploratory galleries, or sinking shafts. This process had long relied upon private capital, but from 1693 onward profits had begun to fall sharply, and in consequence many shareholders had withdrawn from mining investment. Their shares were restored to the mining administration, but there was no further private interest in purchasing them. Busch had very clear ideas as to where to prospect, and after founding the Bergbaukasse in 1703 he immediately set about realizing his plans. He was able to do so because the Oberbergamt doubled as the administration of the new financial organization.

After just three years, the search was successful. In 1708, a two-meter thick vein was opened up with a silver content four times higher than usual. Beneath the swampy area east of Clausthal, an intact ore body with both rich silver ores and galena had been opened up, forming the basis for new mines. From 1708 until the end of the Hanoverian state in 1866, these new mines produced enormous profits. Shareholders in the Einseitiger Harz received an

42. Hoffmann 1978; Bartels 1992b, 285–306.

average profit of 398 talers per share in 1710, of 800 talers in 1718 (and thus the largest sum since the revival of mining in postmedieval times), and of even 1,068 talers on average (a total of 138,840 talers) in 1725. This was equivalent to four tons of pure silver or 140,000 weeks of work by miners, who were paid around 1 taler a week. The shareholders' profits were therefore considerably higher than the total income from mine work in the Clausthal district. To this sum of private profits should be added the fiscal profits of the state from mining, the metal trade and minting.[43]

This spectacular success would not have been possible without Busch's expertise in ore prospecting and new and efficient methods in mining and mine administration.[44] These were the greatest, but by no means the only successes in Upper Harz mining and metal production after 1700. In general, the years between 1710 and 1735 were the most profitable period in the postmedieval development of this region, but the high profits came at the expense of the pauperization of workers in mining, ore processing, and smelting. Social tension grew along with the profits, and in 1738 the situation ignited into open conflict.[45] Interestingly, the workers succeeded in bringing about a considerable change in their situation through their actions. Despite declining production of silver, copper and lead in the Harz Mountains after the big boom of the early eighteenth century, miners achieved a respected position as a social group with a canon of material and immaterial rights in a paternalistic society.

The years from 1750 onward are characterized by growth which resulted in metal production taking place on an industrial scale in the Harz region. By around 1800 the district was considered to be one of the leading European centers of mining, as well as of mining techniques and associated sciences such as mineralogy, geology, surveying, engineering, metallurgy, and chemistry. This reputation was forged not so much by the amount of production or the profitability of the district as by its scholarly, scientific, and technical reputation.[46]

43. Bartels 1992b, 295–300.

44. For details of the use of science-based technology, see Bartels 1992b, 318–22.

45. See Bartels 1992b, chap. 9; 1994a.

46. Bornhardt 1931; Roseneck 2001. Because of the structure and nature of the Rammelsberg deposit at Goslar, the production of metal from this ore body could not make use of the new developments that led to the revolution in Upper Harz metal production. Mining and metal production continued at the technological and metallurgical levels attained around 1600. Consequently, production continued, but on a lower level than during the years between 1460 and 1580, until the nineteenth century brought about new successes on an industrial basis. For questions of industrialization, see Burt 1995; Bartels et al. 2007, chap. 12.

## ASSAYERS AND ALCHEMY

The analysis of materials, assaying, was vital to all stages of metal production. The metal content of minerals was often far from obvious, as was the content of semiproducts and alloys. How was the quality (and thus the value) of a metal product to be controlled? At the start of our period, it was well-known, for example, that precious metals could be debased in various ways that were not easily detected. Fraud was a permanent threat. The silver content of coins is a prominent example. Up to the time of Frederick the Great of Prussia, sovereigns frequently undertook what they had officially prohibited: the blending of precious metals with other substances, with a view to generating coin to finance various enterprises. During wartime, in particular, such fraud tended to be the rule, rather than the exception.[47] Alchemy provided powerful tools for analyzing the content of metals in ores and for analyzing and controlling the content of coins and other metal goods.[48]

The historian of technology Lothar Suhling has reconstructed silver and copper production processes from minerals containing both metals as described in early sixteenth-century manuscripts. He has shown that one of the methods was developed circa 1480 by Peter Rumel and described by Leonard Härrer around 1500.[49] It becomes clear from Härrer's text and Suhling's reconstructions that assaying was involved in every stage of the production process. Smelting ore does not merely involve changing the aggregation of the material treated; the process generates several different resulting products. For example, melting copper ore with a silver content can either allow the silver to be separated from the copper or result in the loss of the precious metal into the slag heap. There are different possibilities for separating copper and silver, all of which involve the use of lead but rely on different reactions. Suhling's reconstruction of smelting and metal separation (see figure 4.1), based on these sources, does not cover the whole process from raw ore to the metals, but only those steps needed to separate silver from copper in the processing sequence. A reconstruction of all the steps involved in the production of refined copper and silver would be even more complicated.[50] Two

47. The Thirty Years' War was such a time, known as the "Kipper and Wipper" period in the Harz region; see Henschke 1974; Küpper-Eichas and Löning 1994.

48. On alchemy, see Newman 1994; Newman and Principe 2002; Smith's chapter in this volume.

49. Suhling 1976; see also Suhling 2000, 2004.

50. Suhling 2000, 198–99 (Fugger smelting work at Grasstein, 1540); Suhling 2004, 231 (smelting of lead-ores in Schneeberg and Gossensass, ca. 1500), with schemata of the processing of materials.

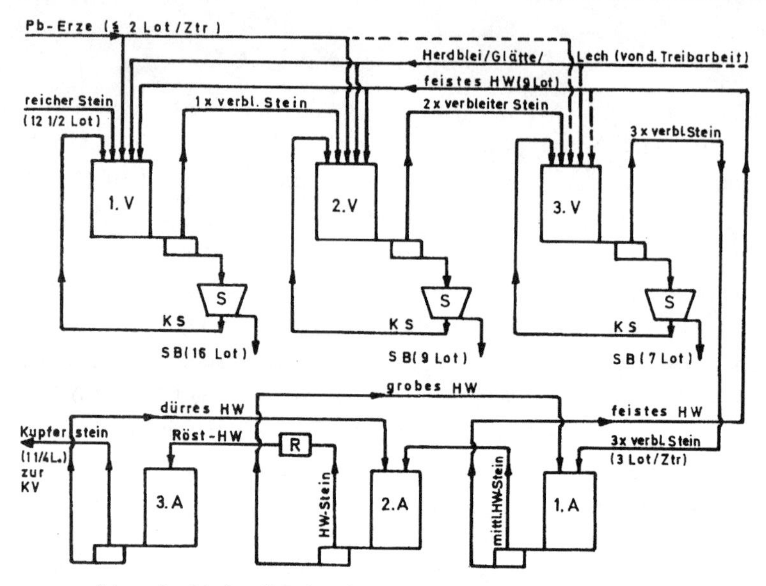


FIGURE 4.1. Schematic of the interlinked smelting operations used to separate copper and silver during the mid-sixteenth century in the Tyrol (Schwaz), demonstrating the complexity of the processes. Before this process started, the ores had been roasted and smelted into several sorts of copper stone with differing silver content. The process was followed by further smelting of the components copper, silver, and (added) lead to refine the metals (Bartels et al. 2006, 3:679; after Suhling 1976, fig. 23, p. 158).

features of the procedures are of particular interest. Suhling's reconstruction demonstrates the operations to be a process of working raw and intermediate materials in interlinked sequences of operations. What is more, it demonstrates the necessity of determining the silver content of the materials at every stage of the process. This was done by smelting small quantities of material, a procedure known as assaying.

In general, assaying duplicated the operations involved in the smelting and refining processes. It was essential to work with precise quantities—all of the ingredients had to be weighed carefully before and after each step. The assayer was an important expert. His results indicated whether or not the process worked correctly and underpinned decisions about how to treat intermediate products. Assaying was a continuous process that accompanied smelting from the beginning. The first assay revealed the overall metal content of the crude ore. Sulfidic ore was roasted to extract part of the sulfur. A first smelting process (Suhling's example was raw copper smelting) produced slag, *Kobolt*, a product rich in copper and silver that could be concentrated in a single further step, and copper matte (*Stein*) that needed to be roasted again.[51] The copper matte underwent three further smelting processes, with each step producing slag and two intermediate products: copper matte that had to be roasted again and raw copper (*Hartwerk*) of varying silver content. The silver content, as determined by assays, was decisive for subsequent treatment of the products resulting from each step. Neither the smelting nor the assaying processes were simple operations, but were sequences of operations carried out by experts.

In the early sixteenth century, assaying methods began to be committed to writing, an important step in developing this expert branch of knowledge in the field of metal production. In addition, methods for quantifying the metal content of raw materials and intermediate products were further developed. Another important aspect was the separation of metals by smelting or other means and the preparation of ingredients for these purposes. Early smelting texts from the late fifteenth and early sixteenth centuries indicate that assaying must have been part of the operation, but there are no written accounts of the technique before 1500, probably as a consequence of the traditions of secrecy operating in medieval metal production. This indicates that a significant change occurred with the first publications in this area, beginning around 1520.[52]

## LAZARUS ERCKER: A UNIVERSITY-EDUCATED ASSAYER AND CONTROLLER OF THE MINT

One of the most influential assaying texts was *Beschreibung Allerfürnemisten Mineralischen Ertzt* by Lazarus Ercker (1528/30?–1594), first published in 1574 and dedicated to Emperor Maximilian II.[53] It is a voluminous work in five chapters, covering procedures with silver, gold, copper, lead ores, and saltpeter and summarizing his expertise in this area. In his introduction Ercker emphasized the importance of the art of assaying and its links with alchemy, giving a brief description of the areas of knowledge a good assayer had to master: connoisseurship of stones and metals, know-how of smelting processes, and expertise concerning the construction of assaying instruments and minting as well as higher forms of learned knowledge such as metrology and arithmetic. Ercker is thus a paradigmatic example of the type of hybrid expert

51. For general information on smelting, see Bachmann 1993.

52. Suhling 1976, 86–88.

53. On Ercker's life, see Kubátová, Prescher, and Weisbach 1994; Technische Universität Bergakademie, Freiberg 1994.

studied in this volume. His expertise, which relied as much on apprenticeship and hands-on knowledge as on university-based education, texts, and mathematics, resembles the mixed expertise of apothecary-chemists, distillers, dyeing inspectors, ceramists, or gunners studied in other contributions to this volume.

Ercker had studied at Wittenberg University, though it is not known at which faculty. In 1552 he became a controller (*Wardein*) at the Annaberg mint in Saxony, and in 1555 he was employed as a controller of the mint (*Probationsmeister*) by Elector August in Dresden, to whom he dedicated his manuscript *Kleines Probierbuch* (A Little Book on Assaying) in 1556.[54] On the basis of this, he was appointed chief controller of the mint (*Generalprobationsmeister*) for Saxony, but as he was not successful in that position, he soon returned to Annaberg. Here he resumed his post as controller of the mint, before accepting the invitation of Duke Heinrich the Younger of Brunswick-Wolfenbuttel to become controller of the mint at Goslar, where he worked until 1566.

Ercker's interest was captured in particular by the problem of exploiting ores with a low precious metal content. Aware of the increasing difficulty of producing silver, thanks to the exhaustion of rich silver ores, he realized that a deposit with low silver content (150–200 grams per ton), as at Rammelsberg near Goslar, could be made profitable through advanced processing methods.[55] He experimented on poor ores from the Upper Harz ore veins resembling those in the Saxon ore mountains and developed a special process called *Goslarisches Schmelzen* (Goslar smelting). In 1566 he returned to Saxony, where he executed a trial on poor ores like those from the Harz region. It ended in failure, however, and Elector August withdrew his support for Ercker, in part because of intrigues by high-ranking Saxon mining officials. In 1568 Ercker became an assayer at Kuttenberg in Bohemia, the starting point of a career in which he became, successively, chief mines inspector (*Oberstbergmeister*) in 1577 and master of the mint of Bohemia in 1583. In 1586 Ercker received letters of nobility from Emperor Rudolf II, adopting the title of Ercker von Schreckenfels. There is no doubt that he achieved this ennoblement thanks to his expertise. His book went through thirteen editions between 1574 and 1745 and was thus a fundamental work in the field of assaying for more than a century.[56]

54. This manuscript was published as a facsimile in Beierlein and Winkelmann 1968, 31–214.

55. Bartels 1994b.

56. In a general discussion of advanced knowledge in the field of mining and metal production between 1450 and 1750, one highly influential publication in particular must be mentioned. In 1556, Georgius Agricola published his famous *De Re Metallica libri XII* (Twelve Books on

## CLAUS VON GOTHA: A HYDRAULIC ENGINEER

During the late fifteenth and sixteenth centuries, mining expanded across Europe.[57] The Harz example, in particular, demonstrates the extent to which this growth relied on technical expertise. After several failed attempts to drain the flooded Rammelsberg works, Claus von Gotha (biographical data hitherto unknown) began work on new hydraulic engines in 1454. Not without difficulty, he managed not only to build the necessary hydraulic machinery but also to recruit three investors from Erfurt, Eisleben, and Halle. By 1456 they had spent 3,000 talers on the construction of two pumping engines (probably *Heinzenkünste,* see figure 4.2). These pumps must have worked well, because the council of Goslar signed a contract with Claus von Gotha and his companions in 1456. The expert and the investors retained 50 percent of the mine and proposed to invest an additional sum of 600 talers. The mine worked continuously from 1456 onward, though Claus's hydraulic engines were not able to drain the deepest parts of the old workings. Up to 1565, when fundamental problems of drainage had to be solved, it was always necessary to call in experts from outside Goslar.[58]

From the late sixteenth century onward, mining in the Harz region had developed to such an extent that hydraulic engineering experts were continuously employed by the ducal mining administration, the "masters of arts" (*Kunstmeister*). Up to the nineteenth century, hydraulic machinery driven by waterwheels was referred to as "arts" (*Künste*), and the machine operators were the masters of arts, assisted by the "art servants" (*Kunstknechte*). In an informal education of some four to six years, "art apprentices" (*Kunstjungen*) learned how to handle, repair, and construct the machines. In the sixteenth century, machinery served to pump water out of smaller mining areas, but seventeenth-century growth led to the mines tapping the Upper Harz ore vein systems, in particular, becoming large technical systems, as illustrated by figure 4.3.[59] I have argued above that the use of hydraulic machinery grew rapidly following the introduction of gunpowder blasting, and a few numbers demonstrate this clearly. For the Zellerfelder Hauptzug deposit, there

Mining and Metal Production). In their famous English translation of 1912, Herbert Clark Hoover and Lou Henry Hoover described Agricola as a "pioneer in building the foundation of science by deduction from observed phenomena" (Hoover and Hoover 1950, xv).

57. For a general discussion, see Suhling 1983, 90–172.

58. Bornhardt 1931, 76–88.

59. Bartels 1989 (Turm Rosenhof Mine, Clausthal); Bartels 1992b, 126–52 (St. Anna Mine, Clausthal; this example particularly demonstrates the move from rich silver ore to galena with a low silver content).

FIGURE 4.2. *Heinzenkunst*—a pumping engine, used during the fifteenth and first half of the sixteenth centuries. A long wooden pipe was installed in a shaft reaching down to the deepest level, where water was collected in a reservoir. An endless chain (or rope) transporting leather balls was moved up in the pipe, and each ball would push a certain quantity of water to the surface. Such systems worked down to depths of some eighty meters maximum (Agricola 1556, see Hoover and Hoover 1950, 191; original and photo: German Mining Museum, Bochum).

FIGURE 4.3. The system of vertical shafts, horizontal main galleries, and hydraulic pumping and winding engines. The fencelike structures represent rod-systems transmitting the movement of the waterwheels to the shafts (Upper Harz, Bockswiese district; colored drawing by mine surveyor J. Buchholtz, 1681; original: Bergarchiv Clausthal / Upper Harz; photo: German Mining Museum, Bochum).

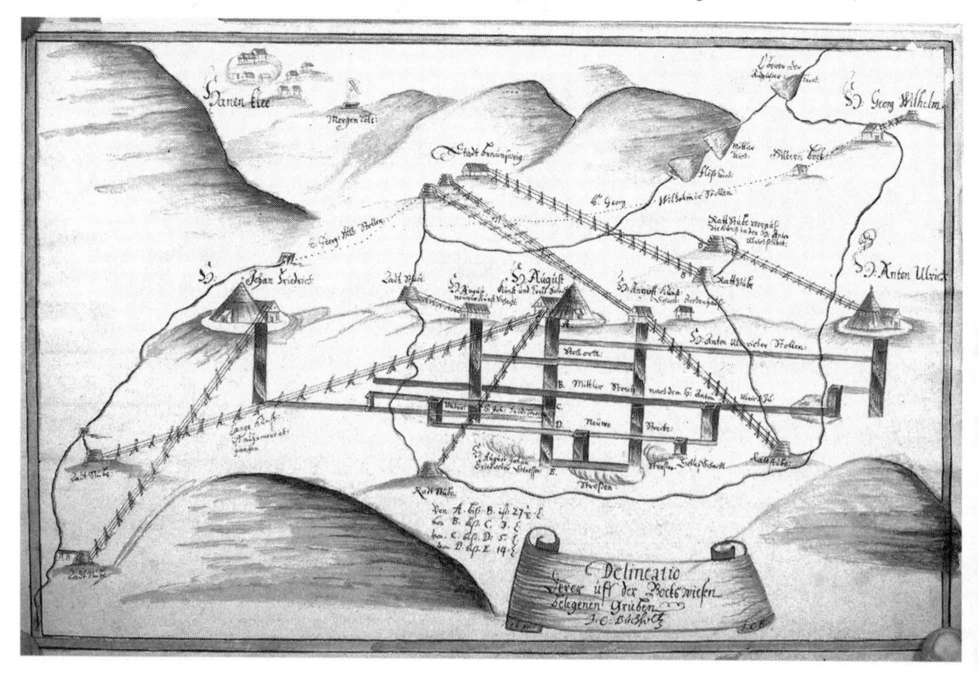

were fourteen hydraulic engines in use in 1635; by 1661 this number had risen to twenty-five, nearly an 80 percent increase. At the Burgstätter Gangzug deposit, there were only five hydraulic engines being used in 1635. By 1661 the number was thirteen, a 160 percent rise; and by 1695 the number of engines was twenty-eight, a 460 percent increase since 1635.[60]

## DANIEL FLACH AND CASPAR ILLING: SKILLED SURVEYORS

Machinery was becoming a decisive element in mining, which was no longer possible without a system of hydraulic engines and engineers who knew how to put such engines to work. The installations of hydraulic engines in the Harz mines represented high-tech, 1450s–1750s style. This is evident in an annotated, representative vertical section of the mines between Wildemann and Zellerfeld, drawn by Daniel Flach (1620–1694) in 1661, as well as a similar work by Caspar Illing (1599–1659) the same year, documenting mining activity in the Burgstätter Gangzug area at Clausthal during the seventeenth century, as well as Henning Calvör's work on machinery and technology in the Upper Harz mining district, published in 1763.[61]

The installation of large systems of hydraulic machinery on the surface and underground was impossible without highly elaborate techniques of land and mine surveying, practices that developed during the sixteenth century in particular.[62] During the seventeenth century, outstanding surveyors like Daniel Flach and Caspar Illing worked in the Harz district. The Clausthal mines archive holds several hundred historical plans and maps demonstrating the skill and ingenuity of these surveyors as well as their innovations of techniques. But it was not only tacit bodily knowledge that was required for their work. Geometry and knowledge based upon publications also played a significant role. Geometrical knowledge, the ability to use complex instruments, and good draftsmanship were indispensable in the surveying of mines. In addition to assayers and hydraulic engineers, surveyors thus formed a third expert group particularly associated with mining.[63]

60. From Bartels 1992b, 190–92.

61. Fessner, Friedrich, and Bartels 2002, with Illing's work and Flach's vertical section (1661) on CD-ROM; see also Calvör 1763. For general aspects of water engineering, see Reynolds 1983.

62. A handbook of mine surveying was published in 1574 by Erasmus Reinhold. M. Koch 1963, no. 392.

63. For general aspects of the development of mine surveying in the Harz district, see Kroker 1972.

## CARL ZUMBE: AN INGENIOUS INSPECTOR

Gunpowder blasting, introduced in 1633, is a good example of another kind of expertise involved in mining; it is associated with the reconstruction of the artillery accident occurring when a projectile lodged in the gun barrel caused it to burst. Workers drilled boreholes in the rock, around eighty centimeters deep, packed the holes with gunpowder, and closed them with a wooden peg hammered into the borehole so as to seal it tightly. The pegs, produced by the thousand, were notched to take in the sulfur thread that ignited the powder. Hammering them into the borehole was dangerous, because the gunpowder could ignite spontaneously.[64] In September 1687, a Saxon miner, Carl Zumbe (biographical dates unknown), visited Clausthal and proposed a new method for closing the boreholes with clay instead of wooden pegs. After successful demonstrations, his method was accepted. Zumbe did not rely only on oral communication of his new technique, but also sought to transmit it in written form. The foremen received a printed copy of his instructions, in addition to courses in the technique.

For his invention Zumbe received a generous reward of 760 talers (a miner earned around 50 to 60 talers a year) and a post as inspector of the smelting works. However, there was some debate as to whether his method was economical. To settle the question, a fascinating method of collecting and evaluating data was employed. The accounts of all the mines in the district were examined, so as to determine the quantities of powder, pegs, and leather cartridges used in each mine over the previous twenty-seven months. For each mine and for the district as a whole, the figures were compared to those obtained using the new method, proving beyond doubt that the new technique was indeed saving money. Additionally, the evaluation relied on oral reports by the foremen; they reported that the new method saved time and was much less dangerous.[65] In just the same systematic way, and at about the same time, the charcoal consumption resulting from dozens of different combinations of hard charcoal, made from beech, and soft charcoal, made from spruce, was carefully tested. These tests too were documented over several years. In addition, they were combined with precisely measured mixtures of ores. The main goal of the charcoal tests was to find ways of reducing coal consumption and increasing efficiency by one or two percent. The Harz mines archives are replete with documents, which demonstrate that quite modern-looking

64. In the Clausthal district alone, seven thousand pegs were produced each week.

65. For detailed references, see Bartels 1992b, 179–84.

techniques of rationalization, data collection, data processing, and business management were being used systematically in the early modern period.

## GEORG WINTERSCHMIDT: AN INVENTIVE ARTILLERY OFFICER

One important development of the eighteenth century cannot be left out, as it clearly demonstrates the ingenuity of practitioners involved in mining.[66] From around 1700 onward, partly motivated by the spectacular discovery of the ore body east of Clausthal, the search for a more efficient motor than waterwheels was on. Water power proved extremely difficult to use in the new shafts on the Clausthal plateau and famous mines such as Dorothea, Carolina, and Neue Benedicte.[67] The well-known Swedish mining engineer Christopher Polhem (1661–1751) came to Clausthal in 1709 to offer technical advice.[68] Among other things, he proposed a pump driven by pistons rather than a wheel, which he called "Sipho-Machine." The basic principle was proven not to work when debates on the topic revived in the years after 1728, but a young artillery officer, Georg Winterschmidt (ca. 1700–1770), developed a completely different idea for a machine driven by pistons, the water-pressure engine, after 1747. It followed principles similar to the steam engine, but used the nonelastic medium of water rather than steam pressure. In 1748 Winterschmidt built a prototype machine, demonstrating that his principle worked, and in 1752 a larger machine was built that could pump as much water as a traditional wheel. Between 1755 and 1757, a system combining eleven water-pressure engines was installed to drain a whole mining district near Zellerfeld. The system proved to work, but ensuring a continuous water supply was a problem. In 1763, 85 percent of the drainage of the old mines using the new machinery was complete when the systems ran out of water. From the outset, this project had been the subject of political debate between the administrations of the Einseitiger Harz and the Kommunion Harz. In 1763 the project had to be stopped when Hanover's government proved unwilling to pay the costs of an additional water supply. This example demonstrates the extent to which parts of the mining administration pursued modernization at the cutting edge of still-experimental technologies.

66. For details, see Bartels 1997a; for general aspects, see Hollister-Short 1995 and 2000.

67. For details and data, see Bartels 1992b.

68. Bartels 1992a, 1997b.

## CHRISTOPH SANDER: EXPERT IN THE STANDARDIZATION OF SMELTING WORKS

The shift in mining from the rich silver minerals to the more common ores with limited silver content and with nonmetallic components which had to be separated off yielded profound changes in ore processing.[69] At the beginning of the sixteenth century, stamp mills had been developed which crushed impure ore to powder (*Schlich*) that could be separated into the metal-bearing and non-metal-bearing minerals using gravity. Late medieval ore processing had been done by hand. This was the method used in the famous mines of the Falkenstein at Schwaz in Tyrol up to the mid-sixteenth century, when debates about whether or not to introduce the new machines divided members of the mines administration for some time. Some older, influential officials strictly rejected the "wicked novelty."[70] At the same time, it was praised in Agricola's influential *De Re Metallica* as one of the most useful technologies.[71] Indeed, ore processing using stamp mills proved to be one of the most important innovations in metal production and remained fundamental to ore processing until the early twentieth century. In the Harz district, the new ore-processing technology was introduced when mining recommenced after 1524, but it could not be used for Rammelsberg ores, whose grain size was so small that methods of separating the components before smelting were not developed until 1937.[72] This was one of the issues limiting the deposit's potential from the seventeenth century onward.

The sixteenth-century development in smelting Rammelsberg ore provides an outstanding example of the effects of expertise and technical advances on metal production between the medieval and early modern periods.[73] When mining recommenced in the fifteenth century, smelting processes did not differ fundamentally from fourteenth-century techniques. Up to the beginning of the conflict between the town of Goslar and Duke Heinrich the Younger in 1526, some fifty smelting works remained active, although they did not all produce continuously. This is little different from the general picture in around 1310. After the ducal administration took over Rammelsberg in 1552, a rapid process of concentration began, which made the ducal smelting

69. Clement and Brennecke 1975; Bartels 1992a, 54–56.

70. Bartels, Bingener, and Slotta 2006, 211n101.

71. For general developments, see Hoover and Hoover 1950, 279–83 ("Historical Note on Crushing and Concentration of Ores").

72. Bartels 1988, 52–55.

73. The processes are described in detail with reference to the sources in Bartels et al. 2007, chap. 3.6.

works more productive than the privately owned works. As discussed above, Duke Heinrich hired the smelting expert Lazarus Ercker in 1558. By 1564, five ducal smelting works with sixteen furnaces were in operation and five private smelting works with fifteen furnaces. However, the private works produced just 36 percent of the total silver output and 42 percent of the lead output. Even more telling is the fact that the ducal works produced 2.1 units of lead per unit of silver, while the private ones yielded a ratio of 2.6:1, which means that their efficiency in silver extraction was considerably lower. By 1569 the situation had changed: just three private smelting works remained, producing 26 percent of the total silver output and 22 percent of the lead output.

From 1570 onward, only the ducal works continued production, organized by an efficient administration headed by Christoph Sander (ca. 1518–1598). In the following years he reorganized the smelting works, and while the input of raw ore did not increase, the metal production per ore unit did.[74] Under Sander's administration, the smelting works developed a highly standardized smelting process. Sander introduced methods such as the calculation of differences in input and output factors. The accounts of the smelting works provide a detailed picture of quantitative progress. In 1580 one of the smelting works was changed into a *Saigerhütte,* because copper production began again at that time, and the copper contained some silver that could be separated. That the Harz experts addressed this branch of production again shows that they were now able to overcome a problem that had caused difficulties since the beginning of late medieval production in the fifteenth century. The number of smelting works was reduced from eight to seven in 1581, to six in 1593, and to five in 1599; paralleling this, the number of furnaces was reduced, while furnace capacity grew. Lead production increased up to 1600, while silver production was slightly reduced owing to the deposit. Two further smelting works were closed down in the first decade of the seventeenth century. Yet reducing production to just three works did not mean a reduction in output; on the contrary, output increased steadily from 1640 onward, having been previously affected by the Thirty Years' War. This fact clearly demonstrates the practical effects of Sander's methods of standardization and rationalization.

## Conclusion

The development of the smelting works at Goslar during the late sixteenth and early seventeenth centuries indicates the importance of both efficient orga-

74. Ibid., chap. 3.6, paragraphs 2 and 3.

nization and continuous improvement of the smelting operations by hybrid experts who possessed hands-on knowledge and skill as well as text-based and mathematical knowledge. It was a significant process of rationalization, and innovations during this time demonstrate the systematic and enduring effects of the mixed expertise of practitioners such as Sander, Zumber, or Ercker. Assaying, precision measurement, data collection, the use of mathematics, attempts at standardization, the writing of technical instructions, the writing of technical books to be published—all of this contributed to the practice of mining and metallurgy. The fact that printed literature prior to the late eighteenth century often did no more than outline the general features and methods of smelting is due to two factors: first, a more precise description of causes and effects in metal smelting was impossible without advanced chemical methods; second, there was also a tendency not to betray expert knowledge because it was a form of capital that had to be protected. Even centuries before 1450, a practice of producing, organizing, and writing down knowledge of a nonspeculative nature, based on observations of nature, experiments, and causal explanations, had already begun. Albertus Magnus called this type of knowledge *scientia,* as shown above. Mining, ore processing, and smelting could not be based upon speculation, as far as actual production is concerned. Alchemical speculations about these processes were common but should not be confused with the processes themselves.

Late medieval and early modern mining and metal production was a field in which processes of accumulating and systematizing knowledge about observed natural phenomena and techniques were concentrated, as demonstrated most clearly in printed works such as Lazarus Ercker's *Beschreibung Allerfürnemisten Mineralischen Ertz* (1574) and Georg Agricola's *De Re Metallica*(1556). The beginnings of both early modern science and technology are closely linked to mining and metal production. The complicated systems required to produce and sustain output were promoted from the fifteenth century onward, such that productive processes gave rise to an extensive literature and iconography. Without detailed accounts or reliable plans, no mine of any extent and importance could be run. Smelting processes were highly complex, involving many steps in order to produce high-quality and profitable results; it proved necessary to streamline smelting processes in a way that gave rise to early plants.[75] Such complex structures could not exist without extensive written communication. In mines, ore processing, and smelting works, not only the experts highlighted in this chapter but even every foreman had to be able to write and to do basic calculations from the

75. Burt 1984, 1991, and 1995; Bartels 2006, 201–5.

mid-sixteenth century onward. By the seventeenth century the vast majority of Harz miners were literate, and it was no longer possible to enjoy respected status in the community without being able to read and write.[76]

Metal production required experts in all its different parts and stages. Surveying was essential, and it was not only closely linked to mathematical operations but also demanded advanced instrumentation. Alchemy and early modern chemistry also hinged on mining and metallurgy.[77] The extent to which mathematics had become fundamental to machine construction by the mid-eighteenth century is demonstrated by Winterschmidt's report on his water-pressure engines, which is filled with calculations.[78] From the sixteenth century onward, a specialist literature was linked to mining and metal production, and the expert needed to be familiar with this. Investment in mining and metal production was high risk, and investors too sought to be informed in this area. The publishing of books inevitably resulted in the systematization of this information.[79] Experts' daily experience attested to the fact that only reliable knowledge could guarantee good quality products. Thus, from the sixteenth century onward, metal production was accompanied by an organized production of knowledge.

The experts who produced this kind of useful knowledge are, with a few exceptions, largely forgotten today, quite unlike some philosophers who also engaged in practical enterprises. Gottfried Wilhelm Leibniz (1646–1716), for example, the famous philosopher, mathematician, and historian of the Welfe dynasty, was repeatedly involved in attempts to improve mining technology in the Upper Harz between 1685 and 1710. Yet his plans to mobilize wind energy for means of haulage and pumping water in the shafts failed, as prior attempts had done. From hindsight we can explain why this was so; there was a fundamental demand for a steady energy flow, and wind does not blow steadily. His proposals for an improvement of winding installations in the shafts could have been realized only in strictly vertical shafts (and were indeed adopted when those shafts came into use), but they were not like that around 1700. Leibniz's attempts demonstrate a gap between some academic approaches and actual technical structures in the field of mining and metallurgy. The famous scholar was by no means better informed or equipped to solve mining problems than the mines administration experts. Without the

76. See Bartels 1992b.

77. Newman and Principe 2002; Klein 1994; Soukup 2007.

78. The report was printed in Calvör 1763, 1:159–90, and tables 16–20.

79. For details concerning the development of literature on mining and metal production, see M. Koch 1963.

advanced and elaborated knowledge of this type of experts, and without their knowledge and use of mathematical tools, it would never have been possible to construct and run the ingenious complexes of water power, nor to reach such a degree of standardization in smelting between 1550 and 1620. In terms of efficiency, resource optimization, and technical capacity, many modern enterprises would be happy to produce figures like those of the Goslar smelting works, where the technical equipment normally achieved between 98 percent and 100 percent of theoretical capacity, clearly indicating rationalization and an almost perfect management.[80] From the fifteenth century onward, the production of silver, copper, and lead revealed highly innovative technical features and at the same time contributed to the development of the experimental and engineering sciences. When looking for "the roots" of the "Age of Science," some very important elements are to be found in this branch of production.

## Primary Sources

Secondary sources can be found in the cumulative bibliography at the end of the book.

Albertus Magnus. 1569. *De mineralibus et rebus metallicis libri quninque auctore Alberto Magno summo philosopho* (ca. 1250). Cologne: Johann Birckmann und Theodor Baum.

Bartels, Christoph, Andreas Bingener, and Rainer Slotta, eds. 2006. *Das Schwazer Bergbuch.* 3 vols. Bochum: Deutsches Bergbau-Museum.

Beierlein, Paul Reinhard, and Heinrich Winkelmann, eds. 1968. *Lazarus Ercker. Das Kleine Probierbuch von 1556. Vom Rammelsberge, und dessen Bergwerk, ein kurzer Bericht von 1565. Das Münzbuch von 1563.* Bochum: Vereinigung der Freunde von Kunst und Kultur im Bergbau.

Calvör, Henning. 1763. *Historisch-chronologische Nachricht und theoretische und practische Beschreibung des Maschinenwesens und der Hülfsmittel bey dem bergbau auf dem Oberharze.* 2 vols. Braunschweig: Waisenhaus-Druckerei.

Hoover, Herbert Clark, and Lou Henry Hoover, eds. 1950. *Georgius Agricola: De Re Metallica.* Trans. from the 1st ed. of 1556. New York: Dover.

Wyckoff, Dorothy, ed. 1967. *Albertus Magnus: Book of Minerals.* Oxford: Oxford University Press.

80. Bartels et al. 2007, chap. 3.6.3.

# 5

# Ink

ADRIAN JOHNS

> And they asked Baruch, saying, Tell us now, How didst thou write all these words at his mouth? Then Baruch answered them, He pronounced all these words unto me with his mouth, and I wrote them with ink in the book.
>
> JEREMIAH 36: 16–18

Don Felipe Avadoro was the most methodical man in early eighteenth-century Spain, or so the Polish writer Jan Potocki would have us believe. Every day he lived by the same routine. Even when his beloved young wife died in childbirth and he lost himself in grief, Avadoro continued to manifest an extraordinary devotion to an intricately precise daily routine of theater-going, greetings to the neighbors, and tobacco. Ten years of this "hypochondria" produced a home life distinctly strange. And the oddest element of all was that Avadoro took to making vast amounts of ink. The habit had been incurred after a conversation in a bookshop with some lawyers, who had been bemoaning the impossibility of getting good ink anywhere; several vouched that they had tried themselves to make the stuff. The bookseller, one Moreno, had then declared that he possessed a book of recipes that contained the solution to the problem. It took Moreno long enough to find the book that when he returned nobody except Avadoro was still interested. Avadoro immediately seized the book, found the appropriate page, and "was amazed to discover that he could easily understand something which the greatest minds in Spain considered to be very difficult." All that was needful was to add a nutgall tincture to a solution of vitriol, and then mix in some gum, and a little alcohol to prevent the gum putrefying. But the text cautioned that the ink must be made in large quantities and kept hot. It must also be stirred frequently.

Avadoro bought the book and the very next day went out and got the ingredients. He bought scales, too, for weighing the proportions, and a huge flask—the largest in Madrid—to hold the ink. To his satisfaction, the process worked the first time. He brought a bottle of his ink back to Moreno's shop, where the lawyers all avowed it "excellent." From then on, gratified by the praise, he devoted himself to making ever more of this precious liquid. He obtained a large Barcelona demijohn to hold his ink, which Madrid's literati consumed as fast as he could produce it. Then, when the glassware proved too fragile, he sent to Tobosco for a clay jar (of a type used in saltpeter manufacturing) taller than a man; it could sit atop his continuously hot stove. A valve at the base of the jar allowed ink to be drawn off at will. If he clambered to the top of this assembly, he found that he could stir the mixture as much as necessary. Men of letters sent their servants to obtain his ink—and when notable books were published, Avadoro would reflect with pride on his own contributions. He was soon known far and wide as Don Felipe del Tintero Largo—Philip of the Huge Inkpot.

Meanwhile, Avadoro's eleven-year-old son—whose birth had precipitated his mother's death and his father's grief—heard of the ink making and decided to come and see it. Elaborate arrangements were made, and the son was warned not to touch anything in the room. He duly arrived one Sunday, dressed finely and in a state of nervous excitement. He could barely sit still, and especially noticed the huge jar in one corner, alongside a glass-fronted cabinet containing the other equipment. He impulsively decided to climb on top of the cabinet. Dodging his aunt, he jumped on the stove and ascended to the top. But then word came that Avadoro was coming. Panicking, the son fell headlong into the jar of ink. Only the quick wits of the aunt saved him from drowning; she grabbed the pestle and smashed the jar. Ink and fragments of earthenware sprayed across the room. Avadoro, entering at that very moment, saw a black demon destroying his prized achievement, fled, and fainted. As for the son, he endured a long convalescence and then left the shameful scene for exile in Burgos. On the way, he declared his determination to travel as an armed gentleman. This he did—and on the first day, he became so enamored of the itinerant life that he became a gypsy, taking the name Pandesowna. By the time Potocki's hero, the Walloon officer Alphonse van Worden, ran into him, Pandesowna was the leader of a bandit troupe, living the very antithesis of his father's time-tabled and circumscribed city life.

This brief episode introduces only one of the hundred-odd interleaved tales that make up Potocki's gothic/erotic/Shandeian masterpiece of the 1790s–1810s that goes by the title *The Manuscript Found in Saragossa.*[1] Compared to the others, in fact, it is rather tame—it contains no incestuously sluttish vampires, no cabalists, no ghosts or zombies, no grave-robbing neo-Paracelsians, no Wandering Jew, no Inquisition torture chambers, and no

1. Potocki 1996, 134–40. MacLean's introduction makes clear that no firm dating can be established.

grand Muslim conspiracy. Still, even lacking these features, the incident is an appealing one to introduce the topics of the present chapter. It may even be the case that Potocki had a point of his own to make with it—he was, in fact, responsible for launching a free press in Warsaw in the 1780s and quite possibly would have had to know about ink to do so. Certainly, as we shall see, he was prepared to have his ink-making patriarch bear a lot of allegorical weight.

But what attracts me about the episode here is the image of Pandesowna emerging from his near-drowning as, in effect, the Obelix of the Enlightenment. Every reader of today must surely recognize the reference to Goscinny and Uderzo's French hero. Asterix's friend famously fell into a cauldron of magic potion as a baby and incurred permanent superhuman strength, allowing him to contribute, albeit rather unawares, to a resistance movement against centralizing Roman authority. Pandesowna fell into a cauldron of the eighteenth-century equivalent to magic potion, namely, ink. Apparently it lent him a counterpart rebellious prowess, to which he was equally insouciant.[2] The first question the tale raises for us, therefore, is this: What was so magical about ink in that era?

## Invisible Ink

Ink has been an essential component of publicity, self-awareness, learning, culture, and knowledge for most of the last two millennia. In its different forms—and some of them are so different as to make the use of one term misleading—it has been one of the foundations of human culture in almost every civilization now remembered. It is more ancient than paper, although not as ancient as writing itself (which was first inscribed by a stylus rather than traced out with a pen). Every significant turning point or achievement of culture since antiquity has been at least registered, and often created, by deploying ink. And, digital innovations notwithstanding, this has not quite ceased to be the case yet.

And yet ink is all but invisible in history. This most basic of "media," to call it that, seems merely a medium—it merits no attention in its own right as the trends and events it mediates flow freely. A striking example is the recent book by a colleague of mine at Chicago, Rebecca Zorach. Zorach's *Blood, Milk, Ink, Gold: Abundance and Excess in the French Renaissance* uses the term *ink* as a synonym for *print,* and in particular for printed images. It does not

2. Hobsbawm's *Bandits* (2000) is the essential introduction to characters like Pandesowna.

examine ink itself at all.[3] I say this not to criticize the book, but merely to note it as an outstanding instance of what is, in fact, universally true across the historical profession. Most remarkable of all, if we examine the burgeoning historiography of print culture, of the book, and of writing and reading—those suddenly fashionable fields, prone to extravagant claims for their ubiquitous significance—it is hard to find ink mentioned even there. This is not a disciplinary cluster that neglects subjects traditionally seen as antiquarian or marginal—including margins themselves. Yet ink does not seem to merit attention in, for example, Elizabeth Eisenstein's *The Printing Press as an Agent of Change* (which is entirely silent on the subject, unless I nodded), or in my own *The Nature of the Book* (except for a few almost inconsequential references), or in Lucien Febvre and Henri-Jean Martin's *l'Apparition du Livre* (where it is dismissed in a few words), or in Roger Chartier, or in Donald McKenzie, or in Marshall McLuhan, or in Walter Ong. The entire canon of this field, despite its announced focus, ignores the stuff. Yet what is being passed over is a substance without which not a single character of a single line of a single page of a single book could *ever* have existed. It is as though one were to write about the industrial revolution without mentioning coal or iron.

In part, the reason for the historiographic invisibility of ink is that we are all nowadays Pandesownas. We are submerged in a world of ink at an early age, with the result that we tend not to notice it as remarkable any more, nor to notice distinctions within the category *ink* itself. Moreover, in adult life we almost never see ink in, as it were, its natural state, as a liquid, let alone as a liquid in the making. It is always safely hidden away in some kind of cartridge, refill, or ballpoint pen. Unlike paint, we see it only in the form of the traces it makes. In sum, historians of our own era would no more appeal to ink in their accounts and explanations than they would to air. Ink is merely a constituent of the cultural atmosphere, a given, and not a subject in its own right. Like air—and unlike, say, paper, printing, or coal—ink seems to have existed in all distinct cultures we tend to think about in comparative terms (such as China, Japan, Russia, and the Ottoman Empire). Furthermore, ink seems to have been to all intents and purposes the *same* thing for long periods and across large distances. It is simply a black—or at least dark—fluid that makes a permanent—or effectively permanent—trace. What does it matter if the trace is more or less black, exactly how permanent it will prove, or what its chemical composition might be? The question need detain only antiquarians, connoisseurs, technicians, and artisans—all of them social species from

3. Zorach 2006.

which the modern historical profession has assiduously distinguished itself. To attend to ink would be to mark oneself as something other, and less, than a historian.

The result is rather ironic. The essential medium of all communication and openness is itself a mysterious, closed, secret substance. But this is not only a product of our own divisions of academic labor. It also derives in part from the early modern period itself, when ink recipes were peddled in books of secrets and of natural magic or else swapped in manuscript as valuable hints, and when successful practical knacks for making it were jealously guarded.[4] As such, in the Renaissance ink was a powerful substance associated with chemical medicines, Paracelsian solvents, and the like.

This condition of invisibility nevertheless can and, I would argue, should be changed. In fact, ink has been not only an essential element in virtually all human practices of reasoning, recording, creating, and communicating, but a distinctly changeable one. It is even arguable that there is no one thing properly denoted by the word *ink*, for its properties vary enormously and it has historically enjoyed no common chemical or cultural composition. Some early inks were vegetable or animal in origin; some were mineral. Some inks were prized for the clarity and evenness of their black, while others were valued for their ability to be absolutely invisible (until conjured into opacity by some knowledgeable reader). The Renaissance knew "sympathetic" inks that used the powers of natural magic in the service of espionage and statecraft. The inks applied to textiles—a vast industry spanning the colonial world—differed entirely from those applied to paper. Among those applied to paper, printers' inks differed entirely from writers'. Woodcut techniques required different inks from copperplate engraving, and letterpress different inks again. In each case, moreover, the kind of ink employed—its constitution, hue, viscosity, and so on—carried implications for the practices to which it could be subjected. The history of inks is the history of this variety—a variety not just of substances and colors, but of places, skills, personnel, and attitudes.

## Making Ink

How and where was ink made before the industrial era? It is hard to answer that question because, if we went by textual records, we might well conclude that ink was scarcely made at all. There were no ink manufactories to speak of until the mid-seventeenth century and no sustained industry of ink making until quite late in the eighteenth. So there is no equivalent in this enterprise to

4. For such collections of secrets, see Eamon 1994.

the profusion of records allowing us to reconstruct the daily life of workers in the Elzevir, Plantin, or Aldus printing houses. In practice, of course, ink making did occur and must have done so on a large and diverse scale. But it took place in settings primarily devoted to other practices and crafts. In particular, after the invention of printing in the mid-fifteenth century, printers' ink was largely produced by printers themselves. But until the eighteenth century, when ink fell under the gaze of natural philosophers interested in the study of arts and trades, formal records of their processes remained extremely scanty, piecemeal, and sometimes even deliberately vague.

The records we do have take the form of recipes. They are enough to convey that making ink was not a straightforward business and that its associations were not those we might now assume. It was notorious among workers well into the nineteenth century that making ink involved long, arduous, "filthy," and stinking labor and that it was highly dangerous; it might also take months of waiting to produce a cask of the good-quality stuff. We can see why from the receipts that circulated in the Renaissance as "secrets."[5] Here is an early and rather involved example, taken from Theophilus's *On divers arts:*

> When you are going to make ink, cut some pieces of [haw]thorn wood in April or in May, before they grow blossoms or leaves. Make little bundles of them and let them lie in the shade for two, three, or four weeks, until they are dried out a little. Then you should have wooden mallets with which you have completely removed the bark. Put this immediately into a barrel full of water. Fill two, three, four, or five barrels with bark and water and so let them stand for eight days, until the water absorbs all the sap of the bark into itself. Next, pour this water into a very clean pan or cauldron, put fire under it and boil it. From time to time also put some of the bark itself into the pan so that, if any sap has remained in it, it will be boiled out. After boiling it a little, take out the bark and again put more in. After this is done, boil the remaining water down to a third, take it out of that pan and put it into a smaller one. Boil it until it grows black and is beginning to thicken, being absolutely careful not to add any water except that which is mixed with sap. When you see it begin to thicken, add a third part of pure wine, put it into two or three new pots, and continue boiling it until you see that it forms a sort of skin on top. Then take the pots off the fire and put them in the sun until the black ink purges itself from the red dregs. Next, take some small, carefully sewn parchment

5. A good sampling of these is in Bloy 1967. In fact, the recipe remained the standard genre for information on printers' inks too, at least until the nineteenth century. William Savage reprinted a series of them in his *On the Preparation of Printing Ink,* while proclaiming that "these recommendations would mislead every one who placed confidence in them" (Savage 1832, 21).

> bags with bladders inside, pour the pure ink into them, and hang them in the sun until [the ink] is completely dry. Whenever you want, take some of the dry material, temper it with wine over the fire, add a little green vitriol [*atramentum*] and write. If it happens through carelessness that the ink is not black enough, take a piece of iron a finger thick, put it into the fire, let it get red-hot, and immediately throw it in the ink.[6]

In its specifics, this was rather more exact than the similar recipes that appeared in Baptista della Porta, in Antonio Neri, and in Pietro Caneparius's *De Atrametis* (1612), but the broad process it describes was similar. The conjunction of a familiarity with rural nature's timetables, an arbitrary precision in terms of quantities and periods, and an appeal to specific equipment and knacks (consider what is not said in this recipe) all resemble natural-magical receipts. Generally, these ink "secrets" invoked the use of minerals, soot, and vegetable and animal colorings. One from as early as the eighth century mentioned the use of the dregs from winemaking and the rind of a pomegranate, and these ingredients were widely repeated. Such old methods survived until about the early nineteenth century. But Gutenberg's invention of the press demanded new, oil-based printing inks, and later the demands of mechanized printing machines would create another demand for innovation because of a shortage of ingredients, especially lamp-black. So a proliferation of techniques took place. Still, it remained the case that ink makers would put bread or onions into the mix, and as late as the nineteenth century a recipe testified that engravers' ink was made from "stones of peaches and apricots, the bones of sheep, and ivory, all well burnt."[7]

The variety of ingredients and protocols recorded in these "secrets" arose for a variety of reasons. Some may have been climatic—inks, as fluids, behaved differently in different temperatures and humidities, and printers knew well that their compositions must be adapted to their specific location of use (something that might extend to using different inks on the ground and upper floors of the same building). Moreover, "good" ink ought ideally to display a diverse range of qualities, some at odds with others: fineness of black, softness of varnish, thoroughness of mixing, clarity, mellowness, tone, strength of color, stiffness, resistance to filming or decomposition, and so on. The ink must have an "affinity" for paper so as to adhere, but not such as to tear off the paper's surface. (Printers apparently often complained of inks on this score when the real culprit was poorly made paper). "Many efforts have

6. Theophilus 1963, 42–43. For Theophilus and the culture of such recipes, see Long 2001, 72–88.

7. Plant 1974, 186; Carvalho 1904, 63.

been made to conquer these difficulties," Hansard would remark in the 1820s; "many printers have thought themselves possessed of this *aurum potabile.*" But almost always an advance on one front meant a retreat on another. Certainly, no one published recipe ever fit the bill. It all depended on how, where, and by whom the recipe was put into effect.[8]

As secrets, ink recipes were typically juxtaposed to instructions for producing pigments for painting or colors for illuminating manuscripts, with dyes for textiles, and with medicines.[9] Like making pigments, making ink was hard, unpleasant, long labor; and it was labor that imputed a certain grappling with nature's powers and propensities. For artisans and alchemists, color was a sign of substance, and ink making partook of this conviction.[10] Moreover, the oil-based ink required for printing originated, in all likelihood, in the experiments of artists with oil paints.[11] Ingredients—including some remarkably exotic ones—were shared with the large dye industry. And in the ancient world both Pliny and Dioscorides had given recipes for ink that explicitly aligned it with pharmaceuticals, Pliny remarking that ink "may be set down among the artificial drugs," and Dioscorides saying that his preferred mixture of smoke black from burned resin was also a good medicament against gangrene.[12] All these realms were commercially important in the period, and all bore associations with magic and practical chymistry. And the association with medicines, too, was old and shared: lapis lazuli, which illuminators had used to furnish a brilliant blue in medieval manuscripts, was an apothecary's exotic medicament. Michael Camille says that the illuminator was "part alchemist, part cook, and part botanist"; the ink maker for a long time had a similar, albeit not identical, constitution (that later evocation of the philosopher's stone was not entirely warrantless).[13] And there might even be a similar exoticism to the craft, ingredients for which might originate far afield in the realms of empire; an edition of the *Encyclopédie* was once seriously delayed because one of Paris's two monopoly ink makers found his ingredients interrupted by the war of American Independence.[14]

One aspect of the association of ink with medicaments and chymistry deserves to be highlighted here. This was the concern for adulteration. If there

8. Hansard 1825, 722.

9. For example, see Wecker 1660, 328–30.

10. See the chapters by Smith and Shell in this volume.

11. P. Smith 2004, 96–97, 110–14; Bloy 1967, 2–3.

12. Carvalho 1904, 34, 76; Bloy 1967, 8.

13. Camille 1998, 320, 322.

14. Darnton 1979, 182.

was a pathology attending early modern materials in general, adulteration was it. It was everywhere suspected. Adulteration was a problem—of knowledge as much as of substances—that pervaded the worlds of reagents and pigments. It was most evident, perhaps, in the world of medicine, where it was a major incentive driving the production and policing of printed pharmacopoeias and a leitmotif of the struggles between physicians and apothecaries.[15] But remarkably similar issues attended pigments and inks. If you bought a medicine from an apothecary, how did you know it was the genuine article, and of proper purity, and that it was not stale? If you bought ink, how did you know it was made to the correct recipe, had been prepared properly, and had stood for the requisite period? Answering that question might well be all-important, and especially so for the most prominent, high-profile publishing projects.

In large part the proliferation of these concerns reflected the dispersal of knowledge and skills. When ink making was done outside the printing house or home, a problem of trust necessarily arose, just as it did for patients and physicians requesting a remedy of an apothecary. Even a good recipe could be ineffective if the ink maker chose to dilute the ink down, reduce the intensity of its boiling, or decant it prematurely. If complaints of adulteration pervaded the world of ink in the eighteenth century, therefore, as they certainly did that of medicine, then this was largely because of a major change in the venue and personnel of ink making. Writing ink had long been produced within the household, this being one of the skills of "housewifery." Peddlers also sold ink for writing, along with pens to use it. But printers demanded a completely different substance. The early ones, of Gutenberg's generation, did make their own, either individually or in collaboration with other printers (there is some evidence in the Amerbach correspondence for this latter practice).[16] There is evidence that ink making became the focus of chapel customs, including feasting on bread that had been fried in the hot linseed oil. But in many European cities by the late sixteenth century the enterprise had begun to be carried out separately on a sustained basis. The new University Press at Cambridge, for example, did buy equipment and raw materials to make its own ink, but for the most part had supplies shipped in from London or further afield; in around 1700 Louis XIV's Imprimérie Royale tried to negotiate for the formula, only to be told that the Press's ink was produced to a secret recipe by a firm in Antwerp.[17] That kind of combination of in-house

15. There will be a longer discussion of this in my *Piracy* when it eventually appears.

16. Halporn 2000, 99.

17. Wiborg 1926, 97; McKenzie 1966, 1:48, 399.

production and outsourcing, and of commerce and secrecy, obtained toward the beginning of what was a long process of disaggregation. It occurred at a different pace in different cities. Paris had the first known dedicated ink maker in 1522. London may have had one (now anonymous) by 1660 or so, and certainly William Blackwell's plant was operational in Clerkenwell by 1755. Dublin got its first only in 1765.[18] In pre-Independence America, Isaiah Thomas recorded, most printers imported casks of ink from London, and Rogers and Fowle in Boston were "the only printers, I believe, who at that time [the 1740s] could make good ink." The bad printing in the revolutionary war, Thomas continued, "was occasioned by the wretched ink, . . . which printers were then under the necessity of using."[19] More or less everywhere, the separation was complete by the early nineteenth century. At that point, as Hansard averred, "few printers, of any eminence . . . attempt to be entire makers of their own ink."[20]

Joseph Moxon, whose *Mechanick Exercises* (1683–1684) was the first comprehensive account of printing-house practices, outlined the unpleasantness, demanding character, and complexity of ink making that gave rise to this trend. Ink's manufacture, he wrote, was "as well laborious to the Body, as noysom and ungrateful to the Sence." It demanded bespoke equipment, long attention, and dedicated space in what were often cramped quarters. And above all, perhaps, it was dangerous. The fire boiling the oil had to be kept hot constantly for very long periods, but if the oil itself ignited, then it would burn eagerly; several recipes actually stipulated that the varnish should be deliberately ignited. The combination of oil, fire, and paper in close proximity was not an auspicious one, especially in a city like London, where the booksellers' quarter stood alongside the smoldering ruins of St. Paul's Cathedral, burned down in 1666. In Germany, at least, printers used to make an excursion outside the town itself and turn ink-making day into a twice-yearly festival (the bread fried in the oil was considered a delicacy and even medicinal, if patients could stand its smell). In France, we know that the eighteenth-century authorities actually forced one maker to stop working in a residential precinct.[21] Master printers thus increasingly sought to abandon the work and instead buy their ink from ink makers.[22] Still, even in the nineteenth century, Hansard could testify that "one of the most tremendous fires that hap-

18. Pollard 2000, 647; Wiborg 1926, 100; Phillips 1998, 220; Bloy 1967, 66.

19. Thomas 1810, 121, see also 408–9.

20. Hansard 1825, 716.

21. Bloy 1967, 7–8, 48, 80.

22. Moxon 1683–1684, 82.

pened in this metropolis a few years since" had been caused by an unwary ink maker.[23]

As this separation occurred, so ink makers became a group—albeit a small one—in their own right. They began to be noticed as such and invited severally to the journeymen printers' annual wayzgoose feasts. But at the same time the scope for adulteration arose and increased with the separation of ink making from printing. Ink makers, who were largely anonymous still at this time, became in effect the apothecaries of print—or, perhaps better, the druggists.[24] They were relied on by the master printers but at the same time suspected of all kinds of underhanded and perilous malpractice. But good ink, Moxon said, depended on them. To be exact, the quality of ink was a reflection of "the Conscience of the *Inck-maker*." And as the division of labor and knowledge became firmer, so that reliance grew more entrenched. Decades later it was still being said by those in the know that ink's quality "must depend entirely upon the judgment of the printer, the liberality of the employer, and the honour of the ink-maker."[25] Conscience and honor: in other words, throughout the early modern period and on into the industrial, the permanence and readability of ink was based on the moral virtues of its maker. And this, again, was for much the same reason that the efficacy and safety of medicaments depended on the moral virtues of *their* makers. The first official standards for inks—ink pharmacopoeias, as it were—appeared in the United States in the mid- to late nineteenth century, by which time an ink manufacturer might advertise over a hundred pigments designed for specific conditions and machines, and no printer could hope to tell by inspection any more the properties of an ink.

There were various ways in which ink makers could produce poor quality ink, and Moxon listed some of them.[26] They might use inferior linseed oil, or adulterate the oil with rosin and other substances. They might use new linseed oil that had not been allowed to sit for months to separate. They might not boil or burn it for long enough, in a bid to save on fuel and time, and to produce more ink (because less fluid would have boiled away). This risked producing an ink that was "smeary" and reluctant to dry. They might not

23. Hansard 1825, 723n.

24. A druggist was a wholesaler of common medical substances, on whom the apothecaries relied for some of their ingredients. Just as the apothecaries were distrusted by physicians, so the druggists were accused by apothecaries of sharp practices. The druggists were also more anonymous figures, based in the peripheries of the metropolis, and with little contact with the population of patients.

25. Hansard 1825, 620.

26. There is also a modern summary in Bloy 1967, 88–91.

clear the ink properly. They might introduce the blacking prematurely into the still-hot varnish, which ruined its quality. Or they stinted on the blacking itself, producing an anemically pale result. Each of these, Moxon stipulated, had real consequences for the resulting printed pages. It impinged on printability, and readers—especially the most influential readers—noticed poor quality work and remembered it. The *mise en page* of the Enlightenment depended on ink as well as paper, type, and composition.

In the period from Moxon to Hansard in the early nineteenth century, ink became the subject of eager efforts by natural philosophers and lecturers to analyze and improve it. For the first time, publications included not just "secrets" and recipes, but chemical arguments about the composition of inks and suggestions for how they might be improved. Don Felipe Avadoro was presumably a fictionalized manifestation of this interest, which gave rise to a series of publications. One of the most remarkable came from the experimental lecturer William Lewis.[27] Lewis, a Fellow of the Royal Society, was one among several writers to take on the subject of ink as part of efforts to revive the long-desired links between experimentalism, commerce, and the arts. Proposing his work in 1748, he anticipated "no less than to lay the foundation of a philosophical and experimental history of arts," based on a survey of their "capital ingredients and materials." An analysis of ink was to play a major part in this project, the intent of which was to identify cases in which cheaper materials could be substituted for more expensive ones traditionally used, or domestic ingredients adopted in place of imported ones, perhaps thereby also creating new trades. And above all Lewis believed that identifying, reducing, and reusing "refuse" from certain trades, dyeing and ink making being his leading instances, would create substantial new value. "Experiments to improve printers' ink" were explicitly offered. On the other side, the genre of printers' grammars launched by Caleb Stower and Philip Luckombe (inheriting many of Moxon's discussions) made the problems and opportunities of ink more visible.[28] The first work devoted solely to printing ink appeared only in 1832, when the Printer to the Royal Institution, William Savage, issued his *On the Preparation of Printing Ink*.[29]

The story of the establishment of industrial ink making is interesting partly because it shows that industrialization began in *this* field, before it moved to the far more familiar fields of papermaking and, at length, printing itself. T. C. Hansard, a champion of the changes then occurring, described

27. Lewis 1748; 1763, 371ff.

28. Stower 1808; Luckombe 1771.

29. Savage 1832.

how. For centuries ink had relied on poor-quality lampblack, Hansard recalled, until about 1760, when John Baskerville came up with a purer black. Baskerville became widely famous for the quality of his printing, but that quality rested not on his typography (as everyone thought) so much as on his ink. His achievement spurred many rival attempts, with the first maker of ink to establish a large manufactory in England being Beale Blackwell, who did so in the mid-1780s. But the new techniques were not widely adopted until the 1790s. At that point Baskerville's old apprentice and foreman, a man named Robert Martin, began his own ink-making operation in Birmingham. Martin started buying up high-quality black from glass-pinchers and solderers, whose lamps produced it as waste, in order to produce good-quality ink for fine printing. For such printing, it proved unsurpassed—Hansard singled out Forster's *Anacreontis Odaria,* 1802, as the supreme exemplar, as well as Dibdin's *Decameron.* (Figure 5.1 is a sample sheet.) But the supply was too small for Martin to extend his operations. Worse still, once workmen realized that there was a demand, they began to adulterate the black itself.

These early attempts to reform the making of ink thus ran headlong into the problem of artisanal adulteration.[30] Another venturesome operator, William Bulmer, tried to get around this problem by making black himself, but he found it too slow and arduous. Bulmer's efforts inspired a "rage" for ink making among printers. Suddenly, generations of resentment at the apparent complacency of the ink makers boiled over: they acted as if they already had achieved perfection, the printers complained, and, rather than conducting research into improvements, preferred to put obstacles ahead of potential rivals. A conflict between the master printers and the ink makers was inaugurated. It opened the whole realm of ink to entrepreneurial competition, in an arena where an existing reputation might now be of no help at all. Several new ink makers launched enterprises, assured of at least a sympathetic welcome. Thomas Martin of Birmingham, nephew to the original Martin, proved the most successful of them. Martin, like Bulmer, resolved to make his own black. This he did using a custom-made set of glazed earthenware vessels arranged in a large premises and operating on a massive scale. Martin's new technique took the black from a specially purified coal tar, which was burned in this apparatus of tubes to yield a smoke that the apparatus collected in seventy to eighty canvas bags (see figure 5.2). He declared that the entire machine was a product of extensive experimentation and that only an operator experienced in printing itself would have seen its virtues. Ink was often the cause of "destroying the excellency of the work," Martin pointed

30. See also Klein's chapter in this volume.

FIGURE 5.1. Specimen of six-shilling ink, from the plant of T. Martin and Co., Birmingham. This specimen was intended to demonstrate the "exact shade" that the ink could produce "when worked with proper care" (Hansard 1825, appendix, no. 3).

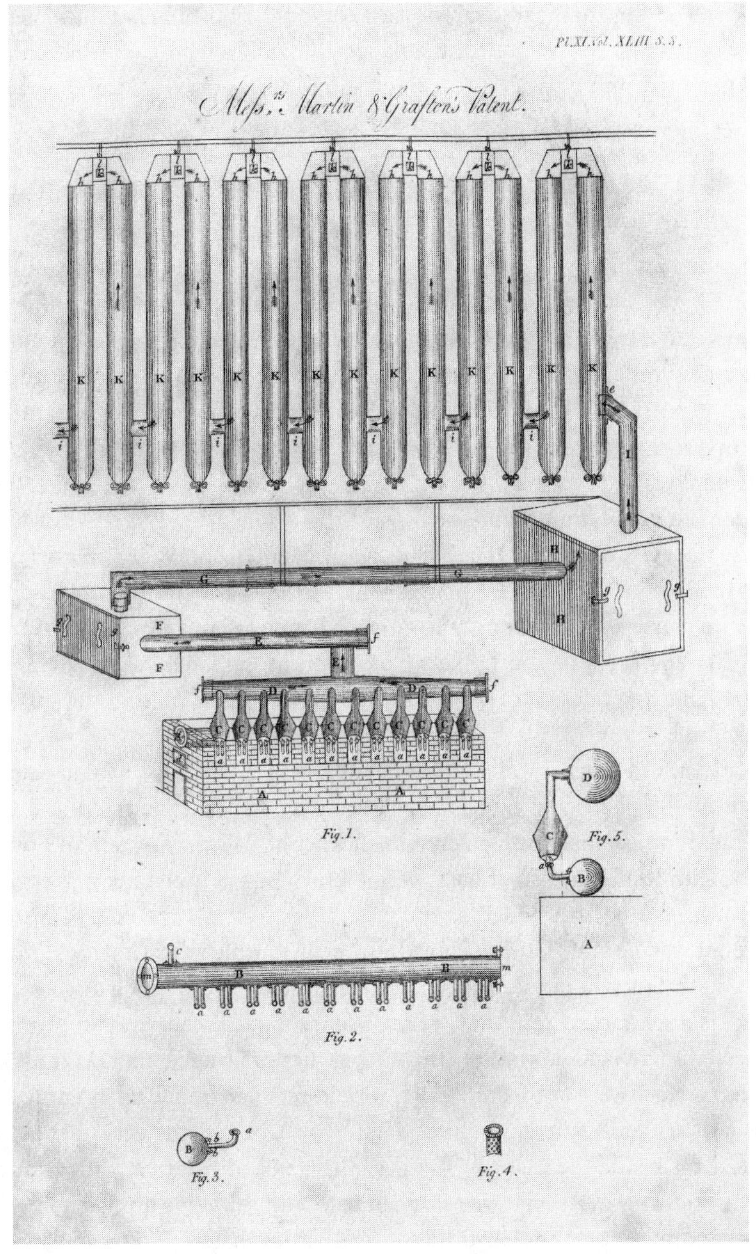


FIGURE 5.2. Martin's ink-making plant. This patent specification shows an apparatus for burning refined coal tar to produce "spirit black," from which ink is made. The coal tar is held in the tube *BB*, 19 feet long, and burns at the small outlets marked *a*. Smoke flows through the apparatus (*GG* is 30 feet long), depositing its coarser particles, until fine black collects in a series of at least 70 canvas bags 3 feet in diameter and 18 feet long, marked *K* (Martin and Grafton 1823, facing p. 260).

out in an address to the trade, because it could tear at the fiber of the paper, or not dry properly, or even turn the paper yellow. His manufacturing process was patented, and Hansard, who was a supporter, declined to say more, as it would not be proper "to lay open to the world an invention which has cost so much expense, time, and labour to perfect."[31]

Yet avoiding adulteration and uniting printers' and ink makers' skills in this one machine were not sufficient of themselves to free Martin from the combinations that were so much a feature of the printing industry in this era. (Charles Babbage would fall afoul of them shortly after Martin.)[32] His address culminated in a warning of what Martin called the "insinuations" of other newcomers to ink making, who were apparently desperate to get his secret process "by the most despicable means." This apparently meant luring his workers away from his plant and into their rival operations, so that they could launch their own ink-making enterprises. But "though they may succeed in getting our men away," Martin cautioned, "still their knowledge of the business cannot by that means become enlarged." Their problem was that they were "totally unacquainted with the nature of printing."[33] And this was the message that he sought to convey: progress in making ink must rely on master printers and ink makers *cooperating* in "research," rather than "endeavouring to render every part of their respective knowledge mysterious and undefined."

Martin did have a local rival for a while—an ex-druggist named Samuel Thornley—but his real targets here were practices that the trade had long tolerated more generally. He denounced what he called a "system of bribery and corrupt influence" that had obtained for years between ink makers and journeymen and that added as much as 7½ percent to the cost of printed pages. It was, he and Hansard agreed, an "evil" to which the whole trade must attend. And there is independent evidence for it, for it was much discussed by printers and craftsmen in these years. It does indeed seem that journeymen expected to be paid a "royalty" or "chapel money" by the ink maker and, if such payment were not forthcoming, would retaliate by alleging that the ink was poor or adulterated or even by sending it back (on one occasion adding a dead cat inside the cask).[34] Martin's and Hansard's charges thus brought claims for progress in ink itself into alliance with those for utility and political economy in printing. And these were battles that had to be fought; the

31. Hansard 1825, 721, 928–31; Martin and Grafton 1823 (filed in 1821).

32. Johns 2007, 403–28.

33. Hansard 1825, 929.

34. Bloy 1967, 77–78.

practices and customs they were assailing were widely honored and of long standing in the chapels. The same conflict thus raged over ink in these years that Babbage would encounter in the publishing industry more generally. At stake were both the role of progress in print and that of print in progress.

In short, ink as a substance in this period serves as something of a proxy for many aspects of modernization. It reflected the shifts from craft to industry, from domestic production to manufacturing, and from natural magic to collaborative research. The concerns for credit that attended pharmacy in its own development also attended ink throughout the period. For a printer (or a reader) preparing to *use* ink, these concerns might become critically important—a major project like an edition of Shakespeare might stand or fall in the market according to the quality of its printing, and ink, printers knew warily, was crucial to achieving the requisite standard.

## Using Ink

Printers testified frequently from the mid-seventeenth century onward about the paramount importance, not so much of ink itself, as of ink and skill conjoined. The achievement of a printed page depended on both being deployed harmoniously, in a specific location that might affect each. This elaborately choreographed scene is actually better documented for us than the practices of ink making, because master printers from Moxon onward repeatedly portrayed it, and modern historians have done much to restore to view the work that pressmen did.

Briefly, the part of the pressman's work that is relevant here began with the "rubbing out" of old leftover ink from the top of the vessel, where it formed a film. This must be done carefully, because residual particles of film could otherwise spoil the impression in various ways (of which printers had a taxonomy). This done, the pressman could proceed to "beating"—the application of ink itself. This seems to have been a highly refined skill.[35] Its tools were two balls. Printers' balls resembled round-headed mallets, but with broad and round surfaces that allowed for the uniform application of a layer of ink. They were made from sheepskin by the printers themselves, using skins from which the hair had only just been removed. The skin must be thoroughly soaked in chamber-lye (urine) before being removed and "curried"—that is, stretched backward and forward around a "currying iron" to remove water and lime and make the skin pliable. It was then trodden on until it stuck to the printer's foot, and spread out as thinly as possible on a convenient

35. Johnson 1824, 2:523–24.

press-stone. A used skin was put on top as a lining, using old ink to make them adhere. Wool cardings were laid atop this pairing, one by one, "knocking up" the ball itself. The ball was finally dipped in lye again, and scraped with a ball knife to clean it. It was then dried on paper until it would take up ink uniformly. The whole thing was by common consent a "filthy, troublesome, uncertain process." One contemporary described it as one of "the nastiest processes imaginable, which converted the press room into a stinking cloaca"—testimony that may have been more than metaphorical, given that printers' balls were often left to soak in urine overnight.[36]

Pressmen used balls in pairs. The ink was spread onto a stone and then taken up in the balls and dabbed onto the type in the form. There was a distinct skill to this, such that just enough ink was applied, and with the requisite uniformity. Moxon remarked that it required "a craft (acquired by use) in the Handling of the *Balls*" such that in taking up the ink they could be rotated from right to left hands so as to get a an equal distribution. "This is Handcraft," he added, "which by continued use and practice, becomes familiar to his Hands." In beating, the balls were dabbed sequentially up and down the form, being held with their handles almost vertically upward in another complex but habituated motion designed to prevent nonuniform application.[37] And all this was done as part of a routinized set of "Formal Postures and Gestures of the Body" performed at high speed in synchronization with the pulls of the other pressman on the press itself, to make impressions.

In short, the application of ink was a finely tuned, exquisitely timed practice. The appearance of pages depended on the quality of the skill with which it was conducted. But this was precisely the kind of craft phenomenon, along with its counterparts in ink making, that came under sustained pressure in the later eighteenth century from ascendant convictions about political economy, laissez faire, and the division of labor. A key component of those convictions was that progress would lie in the replacement of human skills—seen now as capricious and replete with potential for monopoly and combination—with machinery, even to the extent of automata.[38] In the case of printers' balls, there was an additional focus of criticism in that they were seen as central to a practice that was *wasteful*. Long before the Victorians made such a virtue of analyzing factory systems in terms of "work and waste," Lewis's analysis of ink concentrated precisely on the reduction and reuse of waste products in that process. In the printing house, likewise, the use of balls involved a

36. Hansard 1825, 600–601; Bloy 1967, 53–57.

37. Moxon 1683–1684, 290–91.

38. For example, Schaffer 1999.

substantial amount of ink being scraped off as dried crust and discarded. So when a new device, the roller, appeared toward the end of the century and swiftly supplanted balls in many printing offices (partly because they were less smelly), the abandonment was hailed in the light of political economy as an elimination of waste. None other than Babbage would try to elevate it into a major moment in the development of a modern communications economy.

But Babbage was not the first to see things this way. Charles Mahon, Third Earl of Stanhope, is now recalled as the inventor of an iron press (among many other devices); but in fact his printing experiments began with the desire to reduce the waste associated with the use of balls to apply ink. Stanhope carried out extensive experiments, seeking some way of using a roller to do the same job. The idea was simple (and not original), but making it work in practice was fiendishly difficult for want of a smooth material out of which to fashion the roller itself such that it had exactly the right properties of retaining and releasing ink. He reportedly tried every kind of substance imaginable, and all failed. In the end, the solution was discovered by chance coating a device then used in the Staffordshire potteries called a *dabber*. From this, printers adopted a "composition" of glue and molasses. Various recipes soon appeared for this "compo," some boasting, as Hansard dryly noted, "the recommendations which distinguish the recipes of ancient physicians; namely, a vast variety of articles with counteracting properties." Hansard himself added a virulent poisonous substance called Paris-white to his own compo. At that point, it became an artful compound of "vegetable substance," "animal substance," and "earthy substance," allowing it to display the necessary contradictory properties. This substance was then coated on a wooden roller by a special process, and permitted the replacement of the old printers' balls. It seemed an entirely progressive change, from caprice, filth, and waste to reliability, cleanliness, and efficiency, and it was widely adopted very quickly. The invention, Hansard reported, had freed the printing house from "offensive effluvia," saved large costs in wool, skins, and time, and eliminated most of the waste in ink. But it was not entirely unambiguous in practice. John Johnson, a devotee of fine printing, still pronounced judgment decisively for printers' balls and against rollers.[39] And printers found that the compo would sometimes become "sick" with some mysterious but presumably climatic "ailment" that prevented its use. At that point the predictability of the machine could only be sustained by the intervention of the craftsmen. Even the layout of the printing house could affect the susceptibility of the

39. Johnson 1824, 2:648.

tool to this sickness; Hansard himself found that being on the ground floor changed the composition he needed.[40] Still, he insisted on its positively epochal significance. Without the roller, "no machine-printing would ever have succeeded."[41]

This was very much Charles Babbage's view. In his *On the Economy of Machinery and Manufactures*—the most influential and wide-ranging account of the new age published in late Georgian Britain—Babbage made printing into the archetype of all industry. The ability of machines to produce exact duplicates, he argued, was the basis for all prosperity and progress to come. Repeatedly he returned to printing processes to make this point, nowhere more impressively than when he described an experiment undertaken at a London printing house to prove the essential importance of machinery in making economic efficiency possible.[42] "The rapid improvements which have taken place in the printing-press during the last twenty years," he began, "afford another instance of saving in the materials consumed, which has been well ascertained by measurement, and is interesting from its connection with literature." But in fact what he described was not so much an improvement in the press as a change in the use of ink. He recalled how pressmen had traditionally used balls to spread ink, but this had generated waste that had to be scraped off the balls as a crust; it had also meant that the ink layer itself was never uniform, since it varied with the skills of the men. Babbage claimed to have measured the waste with a modern press using rollers at half the level of the old process. Furthermore, this new machine was admirably suited to the introduction of steam power; it therefore had the potential to add a vast economy of time to that of material. "The most perfect economy," Babbage concluded, could only be produced "by mechanism."

## Substance and System

One industrial ink maker of the twentieth century insisted that ink is not a *substance*, properly considered, but a *system*.[43] It has a chemistry and a physics; moreover, it exists and takes effect only in symbiosis with people, places, processes, and papers. There should ideally be a kind of dynamic equilibrium between these various components of a creative industry, but the equilibrium

40. Hansard 1825, 625–29.

41. Ibid., 623.

42. Babbage 1989, 8:44–46.

43. Voet 1952, vii. (In point of fact, by declaring that inks were "physical systems," Voet was drawing attention to their physical properties, fluid dynamics, energetics, and the like; so I am stretching his point here.)

is in practice inevitably unstable and has to be shored up all the time by those most closely involved. Change the paper used in a press (or in handwriting, for that matter), and the ink and equipment might well have to be changed too if the consequences are not to be ghastly—and very costly. And this is not just a point about the relativity of perceptions. The viscosity, opacity, and color of ink varies with humidity and air temperature, and different inks have very different "affinities"—to put it in eighteenth-century terms—for different papers. So constant corrections are needed even for an unchanging quality to be the result.

In consequence, far more could and should be said about inks historically. More needs to be said about colors, for example, and about the aesthetics of the page in the Enlightenment. To cite a minor example, in 1764 a Dublin printer announced a pamphlet printed entirely in a green ink which, apparently, was "not only a Preservative, but also a Restorative to Sight."[44] From roughly that decade onward, an increasing panoply of color theories can be found aimed at printers and ink makers, and (slightly later) emerging from them. The quest was to find color combinations that would endure, that were aesthetically harmonious (on a scientific basis), and that facilitated reading. Such efforts still persist in different forms today.

More could also be said about the various diversifications that took place in inks with the creation of new arts—arts like lithography, stereotyping, linotype, typewriting, and the various forms of digital printing. In each case there is a tale to be told about the problems and opportunities furnished by the adaptation of inks in new settings. The tale severally requires elements of intellectual history, the history of technology, the history of chemistry, and even the history of aesthetics. In this realm these are inseparable constituents.

Furthermore, the traces made by ink can, of course, be read in different ways themselves. In the eighteenth century, the first attempts were made at chemical analyses of ancient inks, aimed at determining whether they were the same as those currently in use (and hoping to explain in some cases their apparent superiority).[45] Papyri recovered from Herculaneum were closely examined in this spirit, not least by Humphry Davy, who spent months at the site seeking ways to unroll the ancient documents and took the opportunity to examine their ink too.[46] Later in the nineteenth century, inks were subsumed into the forensic gaze of the law in its bid for scientific evidence, and

44. Phillips 1998, 220.

45. Blagden 1787, 451–57.

46. Davy 1821, 204–5.

a new armory of techniques was brought to bear to read them as evidence about fraud, forgery, or worse crimes. Inks became the equivalent in the domains of recording and communication of fingerprints in the domain of the body.[47] As with the making and use of ink, its construal too has a history that tracks—and perhaps underpins?—what historians have come to recognize as major currents of development in the modern era.

It is appropriate in this light to conclude by returning briefly to the bandit Pandesowna, our Enlightenment Obelix. At intervals in later life—when not being imprisoned in underground torture chambers by sadistic ducal widows, leading gangs of gypsy desperadoes in battle, cutting a dash on the king's galleys, dealing with mad Newtonian system-builders, and so on—the youth would hear tell of the continued devotion of Don Felipe del Tintero Largo to his ink making.[48] Eventually it came to a climax. Some new neighbors moved in next door to the old man—an aunt and niece named Cimiento. Avadoro found to his delight and wonder that the daughter had a glazed cupboard filled with "the brightest colours," as well as gold dust, silver dust, and lapis lazuli—in short, all the materials for an illuminator. Eventually, curiosity about this broke through the man's routine and reserve, and he discovered that she was engaged in making wax seals from these colors. A symbiosis beckoned: he could write letters in his ink, and they could be sealed with her seals! He went back to Moreno's bookshop and found the local literati equally enthused. That night, in a kind of eroticized appeal to traditional sensation theory, "he dreamed about sealing-wax."

Then the daughter gave Don Felipe three bottles of red, green, and blue ink that she had apparently made herself. At Moreno's once more, the colors impressed a finance ministry official who was accustomed to using special secret inks of these very hues in official documents. Don Felipe left in high excitement. "Once home, he fetched his recipe book and found three recipes for green ink, seven for red, and two for blue. They all became confused in his head, but the beautiful arms of Señorita Cimiento were clearly etched in his imagination. His dormant senses were aroused and made him aware of their power." To cut a long story short, he married the niece. But the Cimientos proved to be tools of an unscrupulous manipulator, who harangued Don Felipe with a diatribe about the evils that ink like his had caused in the world, broke his precious ink-making bottle—dousing Don Felipe and flooding his home in the process—and jollied him into going through with the wedding. The new Señora Avadoro immediately revealed herself to be, in fact, a

47. Compare the very general comments in Ginzburg 1986, 96–125.

48. For example, Potocki 1996, 333, 347.

"flibbertigibbet," and the manipulative relative got hold of Avadoro's savings. Faced with such disruption to his routine, and no longer able to make ink, the devastated old man fell into a fatal lethargy. He recovered only temporarily when Pandesowna himself—whom he failed to recognize—stole into the house to rebuild the precious ink flask. Don Felipe expired.

"That is the end," Pandesowna concludes, "of the story of my childhood."[49] It is a story of ink making, reason, and passion, and what is compelling about it is just how much it evokes at once: the bureaucratization of the state—registered here in its special colored inks; the bookshop as proxy for the public sphere, and Don Felipe's imagination that, in fueling that sphere with his ink, he is in some sense a substantial contributor to its culture; the rebellion of the gypsy and bandit; the storm of sensation and sensibility, in the imagery of eroticized impressions and seals; and the sempiternal debate about the moral status of ink and print themselves, voiced with no subtlety whatsoever by the unscrupulous interlocutor. All the principal cultural currents of the age converged at once. When they did, they culminated in a smashed bottle, a tragic-comic death, and an unstaunchable flood of ink that washed away all tradition, honor, and decorum in its path.

## Primary Sources

Secondary sources can be found in the cumulative bibliography at the end of the book.

Babbage, Charles. 1989. *Works*. Ed. M. Campbell-Kelly. 11 vols. New York: New York University Press.

Blagden, Charles. 1787. Some Observations on Ancient Inks. *Philosophical Transactions* 77:451–57.

Davy, Humphry. 1821. Some Observations and Experiments on the Papyri Found in the Ruins of Herculaneum. *Philosophical Transactions* 111:191–208.

Hansard, Thomas C. 1825. *Typographia*. London: for Baldwin, Cradock, & Joy.

Johnson, John. 1824. *Typographia*. 2 vols. London: Longman.

Lewis, William. 1748. *Proposals for printing, by subscription, Commercium philosophico-technicum*. London: R. Willock.

———. 1763. *Commercium philosophico-technicum*. London: for the author.

Luckombe, Philip. 1771. *The History and Art of Printing*. London: by W. Adlard and J. Browne for J. Johnson.

Martin, Thomas, and Charles Grafton. 1823. Specification of the Patent Granted to Thomas Martin and Charles Grafton, of Birmingham, . . . for Making Fine Light Black of very Superior Colour. *Repertory of Arts, Manufactures, and Agriculture*, 2nd ser., 43:257–63.

Moxon, Joseph. *1683–1684. Mechanick Exercises on the Whole Art of Printing*. Ed. D. David and H. Carter. 2nd ed. London: Oxford University Press, 1962.

49. Potocki 1996, 536–49.

Wecker, J. J. 1660. *Eighteen Books of the Secrets of Art & Nature.* Ed. R. Read. London: for S. Miller.

Savage, William. 1832. *On the Preparation of Printing Ink; Both Black and Coloured.* London: for the author.

Stower, Caleb. 1808. *The Printer's Grammar.* London: by C. Stower for B. Crosby.

Theophilus. 1963. *On Divers Arts.* Translated by J. G. Hawthorne and C. S. Smith. New York: Dover.

Thomas, Isaiah. 1810. *The History of Printing in America.* Edited by M. A. McCorison. New York: Weathervane, 1970.

6

# Blending Technical Innovation and Learned Natural Knowledge: The Making of Ethers

URSULA KLEIN

In the eighteenth century, apothecaries manufactured and traded in medicines, ranging from vegetable, animal, and mineral raw materials to ingenious preparations created from dozens of different ingredients. Among the latter were hundreds of remedies manufactured by means of chemical techniques such as distillation, dissolution, and extraction with solvents. These so-called chemical remedies, prepared in the apothecary's laboratory, proliferated during the seventeenth century to become a normal part of the stock of remedies in the eighteenth century. Distilled acids, sublimated essences, precipitated salts, extracted vegetable and animal oils, composite elixirs, ardent spirits, volatile ethers, and so on were substances produced in the apothecary's laboratory to be subsequently sold as chemical remedies.

The chemists of the eighteenth century, too, were preoccupied with making and studying material substances. Many, even most, of the raw materials and processed chemical substances these eighteenth-century chemists studied in their laboratories were mundane commodities and remedies. The same materials that apothecaries purchased from merchants and sold as simple drugs in their shops, and the same chemical remedies they produced in their pharmaceutical laboratories, were also prepared and studied in the chemists' laboratories. Materials known as remedies and invented by apothecaries were frequently reproduced in academic chemists' laboratories, and substances discovered in the academic laboratory were soon remade in the apothecary's laboratory to be sold as chemical remedies. By the late eighteenth century, even "philosophical substances" such as the different kinds of airs (or gases) soon left their site of discovery to be applied as remedies.[1] In their laborato-

1. On the medical applications of gases, see Levere 1994; Porter 1992.

ries, eighteenth-century chemists observed and tested the properties of substances, including many chemical remedies, identified them, analyzed their composition, and studied their invisible chemical reactions. They further created taxonomies of substances, ordered their reactions in tables of affinity, and explained their properties and chemical behavior in chemical theories. In so doing, they transformed quotidian materials into objects of learned natural inquiry.[2]

Material substances were not the only item that eighteenth-century chemists and apothecaries shared. Laboratories, instruments, and techniques of production and experimentation intersected as well. Around this overlapping material culture, apothecaries and chemists generated a system of practice and knowledge that is strangely at odds with our present-day disciplinary boundaries and our common divide into a history of chemistry and of pharmacy, and a history of science and of technology more broadly. In this chapter I argue that instead of dissecting this overlapping system of practice and knowledge into its chemical and pharmaceutical elements, and ascribing each of these elements to the separate domains of pharmaceutical art and chemical science, we ought to conceive of it as a loosely coherent culture involving a broad spectrum of forms of knowledge with differences only in degree; bodily skills and technical know-how constituted one extreme of this spectrum, and analytical knowledge and chemical theories the other. This shared material and epistemic culture helps to explain why eighteenth-century apothecaries were able to shift their activities from commercial production to natural inquiry, and why academic chemists were often concerned with pharmaceutical innovation. It also helps to explain why so many academic chemists began their career as apothecaries; in late eighteenth-century Germany around half of the chemists were apothecaries.[3] I will scrutinize this mixed chemical-pharmaceutical culture by following one particular class of materials: the "ethers."[4]

Ethers are interesting materials for many reasons, not least because they appear to be utterly uninteresting objects, for ethers do not lend themselves to a story about great discoveries by an outstanding genius. They were, rather, materials that were remade, varied, and studied over and over again by many

2. For a more extensive discussion of materials and their classification in eighteenth-century chemistry, see Klein and Lefèvre 2007.

3. See Hufbauer 1982, 54–55.

4. Today, only the ordinary ether, made with spirit of wine or alcohol and concentrated sulfuric acid, is designated "ether," whereas most of the other eighteenth-century "ethers," prepared with acids other than sulfuric acid, are identified as "esters."

eighteenth-century apothecaries and chemists. There was no major, decisive breakthrough in the eighteenth century that settled the many questions about the preparation and identification of ethers and the invisible reactions underlying their preparation. On the contrary, the challenges arising from the manipulation and identification of ethers induced apothecaries and chemists to continue their trials well into the nineteenth century.[5] But this is precisely why the ethers are of such great interest within the context of this volume. Ethers and practices involving ethers provide insight into the continuous, communal processes of eighteenth-century pharmaceutical innovation and chemical investigation, which proceeded in small steps over a longer period. They allow us to explore obstacles and incentives that emerged in laboratory practices as well as the manifold forms of knowledge applied and generated in these practices. As the French chemist Pierre-Joseph Macquer remarked in 1766, "The production of ether is one of the most beautiful and instructive phenomena in chemistry."[6]

Compared to milk, vegetable foods, and mineral waters, studied in this volume by Orland, Popplow, and Eddy, there was only a small market for ethers in the eighteenth century. Eighteenth-century physicians and apothecaries praised the medical virtues of ethers as anodyne and antispasmodic medicines and promoted their distribution by recognizing them as official remedies to be included in pharmacopoeias. In the early nineteenth century, the market for ordinary (or sulfuric) ether began to expand when the practice of inhaling the vapor of sulfuric ether instead of nitrous oxide (laughing gas) was popularized in pneumatic medicine. As a consequence of this new application of sulfuric ether, additional, unexpected properties of ordinary ether were discovered.[7] Like laughing gas, ether not only relieved pain but was also a drug with extraordinarily stimulating effects on the brain. Around 1830 "ether frolics" became fashionable among medical students and in well-to-do circles in Germany, England, and the United States. A decade later, sulfuric ether was further used as a modern, highly effective anesthetic.[8]

This chapter focuses on the production, commercial and experimental, of ethers in eighteenth-century pharmaceutical and academic laboratories. It studies techniques of production and forms of knowledge involved

5. For an analysis of the early nineteenth-century investigation of ethers, see Klein 2005a.

6. Macquer 1766, 1:458.

7. See W. D. A. Smith 1982; Winter 1998.

8. See Bigelow 1876; Leake 1925.

in production, as well as challenges emerging from these processes. Challenges that emerge in processes of making things are promising probes into materiality, the discovery of new properties of material things in contexts of application, like those mentioned in the previous paragraph, being another manifestation of materiality.[9] It was not the historical actors' goals and interests alone that directed their innovative preparations of ethers, but also the resistances and stimuli coming from these materials. Unwanted side effects, such as explosions, and unexpected varieties of the outcome of ether preparations provided powerful incentives to continue the trials with ethers, debate their identity and classification, and think about their invisible formation reactions. Like other material objects, ethers could not be rigidly defined once and for all. Their potential agency and applicability, rather, spurred an open-ended process of both technical innovation and learned inquiry into nature.

Taking into account national and historical differences, the following study of ethers concentrates on practices in late eighteenth-century Germany. In the first part, I will show how apothecaries tried to innovate the production of ethers and how they communicated their trials in published reports. I will first scrutinize trials of a relatively unknown apothecary, who never acquired a great reputation within pharmacy nor was recognized as a chemist. I will then demonstrate that some apothecaries carried their technical trials further toward chemical analyses and explanations of the nature of material substances and their chemical transformations. This is done by a study of activities of an apothecary who would later be acknowledged as a chemist, namely Johann Friedrich August Göttling (1753–1809). This first part also includes brief histories of the preparation of ordinary (or sulfuric) ether and of varieties of ethers in the second half of the eighteenth century. Based on the empirical case studies in the first part, the second part will discuss in a more general way the interconnectedness of chemical pharmacy and chemistry in the eighteenth century. I start with a discussion of the pharmaceutical profession in eighteenth-century Germany, including the question of how much chemistry was involved in the pharmaceutical art. This is followed by an overview of the laboratory, one of the most important links between pharmacy and chemistry, and a discussion of the permeability of the boundaries of eighteenth-century chemistry.

9. In one way or another, all of the essays collected in this volume address the issue of materiality. For an excellent discussion of this issue, see also Pickering 1995, 2005.

## Ethers: Commodities and Objects of Learned Inquiry

### A SHORT HISTORY OF ORDINARY ETHER

By the eighteenth century, ethers already had a long history.[10] Recipes for the preparation of a liquid that might have contained some ether had been published as early as the sixteenth century by the Swiss physician and alchemist Bombastus von Hohenheim, known as Paracelsus, and by the Nuremberg physician Valerius Cordus.[11] According to the recipe for *oleum vitrioli dulce verum* published by Cordus in his pharmacopoeia, spirit of wine (a distillate consisting of alcohol and water) was mixed with oil of vitriol (concentrated sulfuric acid) and the mixture subsequently distilled. In this way a liquid distillate with a fragrant odor was obtained, which was presumably a mixture of alcohol and ether along with some additional by-products. From the beginning, the distillate was used as an anodyne medicine. In the early eighteenth century, a mixture of this distillate with spirit of wine, named *liquoris anodyni mineralis,* was prepared and sold by the Halle apothecary Martmeyer. Around 1720, Friedrich Hoffmann (1660–1742), professor of medicine and chemistry at the University of Halle, began to recommend, prepare, and sell this substance as his own secret remedy. Hoffmann's *liquor anodynus Hoffmanni* or *Hoffmanns Tropen* (Hoffmann's drops) became a famous remedy, which can still be purchased in Germany today. During the 1740s, Hoffmann's secret liquor became a common remedy thanks to its inclusion in German pharmacopoeias.[12] The pharmacopoeias' recipes for this *liquor anodynus mineralis* mentioned two ingredients—oil of vitriol and spirit of wine—but presented no detailed quantitative and technical instructions for its preparation, apart from distillation.

The properties of the pure, water-free ether were first described in an article by the German chemist Siegmund August Frobenius (?–1741), published in the *Philosophical Transactions* of 1729–1730. Little is known about this German chemist, apart from the fact that he lived in London, became a Fellow of the Royal Society in 1729, and presumably worked in the laboratory of Robert Boyle's German assistant, Ambrose Godfrey Hanckwitz (1660–1740).[13] In his

10. For the early history of ordinary ether (also known as sulfuric ether or vitriolic ether), made with vitriolic acid, see Kopp 1966, 4:299–305; Priesner 1986.

11. See Kopp 1966, 4:299–302. Cordus's pharmacopoeia, published in Nuremberg in 1546, was the first German pharmacopoeia. See Schneider 1972, 96.

12. Schneider 1968–1975, 3:82, 6:32.

13. On Frobenius, see Kopp 1966, 4:302; and Partington 1961–1970, 2:546; on Hanckwitz, see Partington 1961–1970, 2:543.

paper entitled "An Account of a Spiritus Vini Aethereus, Together with Several Experiments Tried Therewith," Frobenius first introduced the name *ether* (or *aether*) and described several of its properties. Frobenius's ether was very volatile, easily flammable, possessed a fragrant odor, dissolved many different kinds of oily substances that were difficult to dissolve otherwise, and had a "wonderful harmony" with gold.[14] But Frobenius did not include the recipe for his new substance in his publication, for he wanted to protect his commercial interests. Together with his colleague Hanckwitz, Frobenius sought to sell ether as a remedy and a powerful reagent. Accordingly, he praised his ether as "the most noble, efficacious and useful Instrument in all Chymistry and Pharmacy."[15]

From its very invention, pure ether was both a commodity and an object of learned natural inquiry. More than any other chemist, Frobenius presented his substance as a rare and curious philosophical object. His choice of the name "ether" is telling in this respect; the word had cosmological connotations, traditionally denoting a pure, unchangeable substance of heavenly origin. Based on his observation that the ether was liquid, extremely volatile, highly flammable, and burned without leaving a residue, Frobenius proclaimed that it consisted of the purest fire (the cause of its volatility and flammability) combined with water (the cause of its fluidity): "Hence it appears, that this aether is both Fire and a very fluid Water, but so volatile as it soon evaporates, and that it is the purest Fire."[16] At the end of his paper, Frobenius described his ether in an even more enthusiastically alchemical manner as "the very Ens, or Being most pure of Flame."[17] Despite the fact that the name "ether" was soon adopted by chemists and apothecaries alike, to coexist for a while with older names such as "naptha" and "oleum vitrioli dulce," most eighteenth-century chemists and apothecaries did not share Frobenius's understanding of ether, as the underlying cosmology and alchemical philosophy was no longer accepted after the 1730s.

Along with his published paper, Frobenius had deposited a paper at the Royal Society that contained the recipe for preparing pure ether from oil of vitriol and spirit of wine. After Frobenius's death in 1741, Cromwell Mor-

14. Frobenius 1729–1730, 285. Frobenius's remark on the harmony between ether and gold was based on the observation that ether attracted gold out of its dissolution with aqua regia (a mixture of nitric acid and hydrochloric acid).

15. Ibid., 286.

16. Ibid., 284.

17. Ibid., 288.

timer, the acting secretary of the Royal Society, made this recipe public. From that time onward, pure sulfuric ether began to be used on a broader scale as a chemical remedy, and recipes for its preparation were included in pharmacopoeias. Mortimer's publication also initiated several attempts to replicate the preparation of sulfuric ether, leading to a cascade of publications by chemists and apothecaries in ensuing decades and well into the nineteenth century.[18] The outcome of these attempts to produce pure ether depended on many technical factors, among which the concentration, purity, and proportion of the two ingredients, spirit of wine and oil of vitriol, as well as temperature regulation during distillation, were the most important. No single trial or publication resolved the many technical obstacles involved in preparing ether; none of them settled commercial questions concerning the efficiency of the production process, nor did any of these publications receive unambiguous acclaim for its conceptual explanation of the invisible formation reaction underlying the preparation of ether.

## PRODUCTION OF VARIETIES OF ETHERS IN THE SECOND HALF OF THE EIGHTEENTH CENTURY

Among the myriad of technical variations on the preparation process, apothecaries and chemists also tried to make ether with acids other than oil of vitriol. The substitution of other kinds of acids for oil of vitriol had unforeseen consequences. The ethers produced using oil of vitriol, nitric acid, muriatic acid, or acetic acid had many properties in common, but also differed in some aspects. This raised the question of whether the ethers made with different kinds of acids were mere varieties of the same chemical substance or truly different kinds of substances. From a pharmaceutical point of view, this was an important question, because different chemical substances could have different medical effects. The question was no less interesting from a chemical-analytical point of view, as it also concerned the understanding of the chemical reactions taking place in ether formations. Did the varieties of ethers have the same composition or did they consist of different components? Was it always the same chemical reaction, or did the different acids interact differently with spirit of wine? Well into the nineteenth century, these

18. Before 1741, the French chemists Étienne-François Geoffroy and Jean Grosse had received samples of pure ether from Frobenius and Hanckwitz and subsequently experimented on this substance. See Kopp 1966, 4:303.

questions were discussed controversially, and, as we will see below, apothecaries, too, participated in this discussion.[19]

During the second half of the eighteenth century, the number of apothecaries and chemists interested in the preparation and study of the different varieties of ethers grew steadily, as did the number of publications dealing with this group of substances. As early as 1757, the French pharmacist Antoine Baumé published a dissertation on ethers over three hundred pages in length.[20] Descriptions of different varieties of ethers and their preparation also entered chemical and pharmaceutical manuals and textbooks. The new professional journals for chemists and pharmacists appearing from 1778 onward, such as Lorenz Crell's *Chemische Annalen,* Johann Friedrich Göttling's *Almanach oder Taschenbuch für Scheidekünstler und Apotheker* (Almanac or Pocketbook for Chemists and Apothecaries) and the *Annales de Chimie* became popular forums for exchanging information and ideas about ethers. Both chemists and apothecaries, many French or German practitioners among them, published the results of their trials in these journals, and often the apothecaries had the lead.

Among the apothecaries who published on ethers in the second half of the eighteenth century, there were many who never became recognized as chemists. They studied the formation of ethers, varied their preparation methods, and tested the use of new acids for making ethers. By the end of the eighteenth century, the apothecaries were the true experts in the techniques of manufacturing ethers. Hence, when the French chemist Antoine-François Fourcroy (1755–1809) and the chemist-pharmacist Nicolas-Louis Vauquelin (1763–1829) published their influential paper on the theory of etherification in 1797, they observed that the preparation of ethers "is a complicated *pharmaceutical* operation, the results of which are as well known as its theory remains obscure."[21] The designation of the operation as "pharmaceutical" rather than "chemical" demonstrates that in the 1790s much of the factual knowledge and technical know-how of the operations resided with the apothecaries. A decade later, the French chemist-pharmacist Pierre F. G. Boullay (1777–1869) observed that the "use of sulfuric ether is today quite extensive, and its consumption is considerable; it has become a true product of art produced on a grand scale."[22] How much times had changed by then can

19. On the continuation of the discussion in the nineteenth century, see Klein 2005a.

20. See Spary's chapter, this volume. In France, unlike Germany, apothecaries were designated "pharmacists" in the second half of the eighteenth century.

21. Fourcroy and Vauquelin 1797, 203; my emphasis.

22. Boullay 1807, 242. All translations are my own, except where stated.

be seen when we compare Boullay's observation with the earlier complaint of Pierre Joseph Macquer that the high price of sulfuric ether prevented its broader application in the arts and crafts. "Ether is not yet employed in the Arts," Macquer wrote in his famous *Dictionnaire de chimie* of 1766, "although it appears capable of being usefully employed in many cases, and particularly in the dissolution of certain concrete oily matters [contained] in varnishes; however its high price is a considerable obstacle to its introduction into the Arts."[23]

## AN APOTHECARY'S TRIALS TO INNOVATE ETHER PRODUCTION

Many of the technical innovations in the preparation of ethers and other chemical substances effected by eighteenth-century apothecaries are unknown to us, for the vast majority of ordinary apothecaries did not publish the results of their trials. The publication of Crell's *Chemische Annalen* and Göttling's *Almanach oder Taschenbuch für Scheidekünstler und Apotheker* in the 1780s offered new incentives in this respect. An impressive number of apothecaries reported their observations in these periodicals, among them many who are forgotten today. Göttling's *Almanach*, in particular, published numerous short reports and letters by ordinary apothecaries who were never recognized as chemists. To this latter group belonged C. F. Voigt, an apothecary from Erfurt. His case demonstrates the transition from oral to written reports, as Voigt first communicated his experience to the editor of the journal, Göttling, in person before submitting letters for publication.[24] His oral communications and letters illustrate the technical challenges emerging during the preparation of ethers, as well as apothecaries' technical innovations in this practice.

Little is known about Voigt, apart from the fact that he was a subscriber to Crell's *Annalen der Chemie* between 1787 and 1791.[25] He was the stepfather of a much better-known Erfurt apothecary-chemist, Christian Friedrich Bucholz (1770–1818), who praised Voigt in 1803 as a "skilled apothecary." Bucholz also claimed that his stepfather had invented the ether of vinegar

23. Macquer 1766, 1:461. On Macquer's work on ethers, see also Spary's chapter in this volume.

24. Göttling's position as administrator of W. H. S. Bucholtz's *Hofapotheke* from 1775 to 1785 meant that he was based in Weimar, not far from Erfurt, so it was easy for him to meet Voigt and exchange pharmaceutical news.

25. See Hufbauer 1982, 290. Voigt is not mentioned in the *DAB* (*Deutsche Apotheker-Biographie*; see Hein and Schwarz 1975–1997).

in 1783 and had published on this substance twenty years before the French chemists did so.[26]

In his *Almanach* of 1781, Göttling reported two observations by Voigt, one concerning the preparation of the ether of vinegar, which will be discussed below, and the other concerning the preparation of the ether of nitric acid.[27] The preparation of the ether of nitric acid was known to be very dangerous, and Voigt had obviously found a way to avoid the explosions frequently occurring in this process. He achieved this by using saltpeter and vitriolic acid instead of free nitric acid,[28] and by means of a specific technique for mixing the ingredients; he also carried out the subsequent distillation of this mixture at a moderate temperature. A glass retort with a tubulus (an additional neck melted onto the retort) and a sand cupel were the two more specific instruments needed for the new technique. Göttling reported on Voigt's new technique as follows.

He placed twelve ounces of dried, purified saltpeter in the retort, connected it to a distillation receiver, and then poured a mixture of five ounces of English oil of vitriol and eight ounces of highly purified spirit of wine, "mixed together beforehand with all possible care," through the glass tubulus. The tubulus was then closed with a glass stopper and the retort put into a sand cupel without heating. After one hour, the retort had heated up spontaneously, and "one observed in the laboratory a very pleasant odor of naphtha," stemming from two ounces of ether (or "naphtha") collected in the receiver. As no further ether was obtained in this way, Voigt changed the receiver and then carefully heated the sand cupel, which yielded another four and a half ounces of a liquor from which two and a half ounces of ether could be separated after adding water. The following day, Voigt repeated the entire operation once more, pouring eighteen *Loth* of the mixture of oil of vitriol and spirit of wine into the retort, which contained the residual unreacted saltpeter.[29] By this means he obtained another ounce of ether. On the third day, he prepared another chemical remedy from the remaining mixture of ingre-

26. On Buchholz, see Hufbauer 1982, 217. On the French chemists' work on the ether of vinegar, see below. For Buchholz's claims for the priority of Voigt's preparation of ether of vinegar over Vauquelin and other French "chemists," see his letter of 24 December 1802 (Bucholz 1803). According to Göttling, however, the inventor of the ether of vinegar was the apothecary and medical doctor Johann Christoph Westendorff; see Göttling 1779, 40; 1780, 7. On Westendorff, see Hein and Schwarz 1975–1997, 2:740–41.

27. On the preparation of the ether of vinegar, see Göttling 1781b, 4–10; for the report on ether of nitric acid, see Göttling 1781a.

28. The mixture of saltpeter and vitriolic acid yields nitric acid.

29. A *Loth* was 14–18g

dients, namely, "dulcified nitric acid"—a chemical remedy similar to nitric ether that was frequently described in eighteenth-century pharmacopoeias—by adding eight ounces of spirit of wine and subsequently distilling the mixture. When he changed the receiver and continued the distillation, he recovered the untransformed proportion of nitric acid (which had been formed from saltpeter and oil of vitriol). Voigt added that even more nitric acid could be recovered by prolonging distillation, but he had stopped the operation to avoid damaging the retort.

This report on Voigt's trials and observations illuminates several important aspects of eighteenth-century apothecaries' innovative chemical practice. As a rule, chemical techniques were not described in any great detail in eighteenth-century pharmacopoeias, whose recipes were restricted to basic information about the quantity and nature of ingredients and the type of chemical operation. For example, with respect to the manufacture of the ether of nitric acid, they mentioned distillation but gave no further details concerning the instruments specific to the procedure or the techniques of mixing, heating, changing receivers, separating mixtures in the receiver, and so on. Nor were they concerned with the economy of manufacture, such as recovering the untransformed proportions of ingredients. Economic manufacture was, of course, of great commercial interest to apothecaries. It was also a significant stimulus to perform chemical operations in a quantitative way. Therefore, Voigt carefully measured both the ingredients and the reaction products and developed the operation further to produce a by-product and to recover untransformed portions of the ingredients. Furthermore, he circumvented the explosions that frequently occurred during the manufacture of nitric ether by modifying existing preparation techniques.

In 1783 Göttling reported on Voigt's observation of another kind of ether obtained from spirit of wine and vinegar. Voigt's original goal had been to make vinegar by fermenting the residue of brandy distillation.[30] However, when he distilled the vinegar thus prepared in order to make distilled vinegar—a chemical remedy introduced in the seventeenth century—he was greatly surprised to obtain a liquor that smelled like the ether of vinegar, for which he had already found a quick preparation method in 1781.[31] Further distillation yielded "a true ether of vinegar" along with a "good dulcified spirit of plants (or *Liquor anodinus vegetabilis*)."[32] Its appearance raised the

30. Göttling 1783b, 1.

31. Göttling 1781b.

32. It was for this kind of ether that Friedrich Bucholz was to claim priority for Voigt in 1803. See above.

question of where its ingredients originated, which were usually an acid—in this case the acid of vinegar—and alcohol. Göttling assumed that during the fermentation process a small quantity of alcohol must have been produced along with the vinegar, which had yielded the unexpected ether of vinegar when the mixture was distilled. Although it is not possible to determine from Göttling's report whether this explanation stemmed from Voigt or Göttling, the example clearly shows that Voigt made careful observations during the manufacture of chemical remedies and was ready to recognize unforeseen new results.

In the same issue of the *Almanach*, Göttling reported yet another observation by Voigt on the manufacture of ethers, this time on an ether obtained from alcohol and the acid of ants (later called formic acid).[33] Just as for his accidental observation of the ether of vinegar, Voigt had set out to prepare a quite different remedy, namely, an infusion made from ants and spirit of wine.[34] But when he distilled a sample of the infusion, he recognized the ether by its odor. He then repeated the preparation with fresh ants, obtaining in this way a "wonderful ether."[35] In this case, the ether at stake had been known before, but, as Göttling emphasized, Voigt found a much "quicker way" to prepare it.[36]

After these reports by Göttling, Voigt obviously had no further hesitation about reporting his trials himself, for the 1784 issue of the *Almanach* included a letter in which he described his new technique of manufacturing the ether (or "naphtha") of vinegar in great detail, along with a couple of other observations. "My last preparation of naphtha of vinegar yielded excellent products," Voigt began his report.[37] Voigt's technique for the preparation of ether of vinegar resembled his technique for the production of nitric ether, inasmuch as he did not mix the spirit of wine and the isolated acid directly, but added a mixture of oil of vitriol and spirit of wine to *Sal Tartari*.[38] As

33. Göttling 1783a.

34. Apothecaries chemically extracted various kinds of chemical remedies from ants, including an oil, a spirit (*Spiritus Formicarum*), and an acid (*Acidum formicarum*); see Schneider 1968–1975, 7:34. "Infusion" was the designation for materials extracted by mixing spirit of wine or purer forms of alcohol with a plant or animal substance.

35. Göttling 1783a, 78.

36. Ibid., 77.

37. Voigt 1784, 184.

38. *Sal Tartari* contains tartaric acid as a component, but lacks acetic acid. However, acetic acid may have been procured by mixing tartaric acid with oil of vitriol and spirit of wine. Voigt did not give any hint as to why he used *Sal Tartari* as an ingredient; nor did Göttling comment on this.

the production of this ether was not dangerous, he used an ordinary retort, distilling the mixture at a moderate temperature on a sand bath for half a day. Furthermore, as in the case of nitric ether, Voigt again attended to economic issues, looking for by-products, attempting the recovery of untransformed proportions of ingredients, and carefully weighing all of the ingredients and reaction products. Besides the first product, ether of vinegar, the operation yielded "dulcified spirit of vinegar" as well as untransformed acid of vinegar (obtained from the *Sal Tartari*) at the end of the distillation. Comparing the weights of ingredients and products, Voigt wrote: "From 17 *Loth* of laminated salt of tartar, 10 *Loth* of oil of vitriol, and 37 *Loth* of spirit of wine, I obtained 19.5 *Loth* of naphtha and 22 *Loth* of dulcified spirit of vinegar."[39] He also mentioned that he had recovered three ounces of acid of vinegar. Although Voigt did not supply complete information about the untransformed proportions of the ingredients in this final comparison, the example demonstrates that there was no major difference between this kind of calculation, spurred by economic considerations, and the kind of balancing of ingredients and reaction products for which Antoine-Laurent Lavoisier (1743–1794) has always been celebrated. Lavoisier's schemes of weighing certainly required high precision, but it was not as outstanding and unusual as has often been portrayed by many historians of chemistry. Quite unknown apothecaries came close to this kind of quantitative measurement and comparison for a very different reason: namely, commercial profit.

## SHIFTS TO EXPLANATIONS OF ETHER FORMATIONS: THE CASE OF JOHANN FRIEDRICH AUGUST GÖTTLING

In the second half of the eighteenth century, many apothecaries and chemists were attempting to improve techniques for preparing ether. It would be impossible to trace in a single chapter all, or even most, of their efforts and suggestions for technical improvement. Identifying ethers prepared with different kinds of acids, analyzing their chemical composition, and explaining their formation reactions were issues that remained on the agenda of apothecaries and chemists for decades. This section examines apothecaries' ways of explaining the invisible reactions underlying the formation of ethers. Such explanations clearly transcended technical and commercial goals and interests, but they were by no means the exclusive intellectual property of academic chemists. Although apothecaries' publications on ethers were predominantly concerned with preparation techniques and ways of unambiguously

39. Voigt 1784, 186.

identifying ethers, a number of apothecaries did not restrict their reports to the technical issues relevant to their business. As the following example will show, these apothecaries went a step further, seeking a more analytical chemical understanding of their practice. In so doing, I argue, they did not have to enter an entirely new realm of higher learning that was utterly distinct from their artisanal world, but rather smoothly shifted their activities to additional inquiries.[40] Apothecaries' shifts toward an analytical understanding of chemical preparations were spurred not least by the fact that such forms of knowledge contributed to the identification of chemical products. As the unambiguous identification of chemical remedies was of genuine commercial interest to apothecaries, contributing to the standardization of medicines and the avoidance of adulteration, explanations of chemical reactions could also be useful for pharmaceutical practice.

Johann Friedrich August Göttling, the editor of the *Almanach,* who has been mentioned several times above, is a good example of an apothecary who extended his pharmaceutical-chemical trials to more systematic and analytical chemical inquiries. Göttling spent his pharmaceutical apprenticeship (from 1767 to 1775) in the shop of the renowned apothecary and chemist Johann Christian Wiegleb (1732–1800), and from 1775 to 1785 he administered the *Hof-Apotheke* (court apothecary's shop) in Weimar.[41] During this time, he became more widely known to the learned public through a book on chemical pharmacy, *Einleitung in die pharmaceutische Chemie* (Göttling 1778), an introductory text on pharmaceutical chemistry. Two years later, still a practicing apothecary, he began to edit his *Almanach,* whose main goal was "to make pharmacists (*Pharmaceutiker*) acquainted with proper chemical knowledge."[42] In 1785 Göttling relinquished his position at the *Hof-Apotheke* in return for a stipend from Duke Carl August of Saxe-Weimar, allowing him to study at the University of Göttingen for two years; by this time he was generally recognized as a chemist. From September 1787 until February 1788, he traveled to England at the Duke's expense. Like the stipend, the journey was part of an arrangement between Johann Wolfgang von Goethe (1749–1832) and the Duke, who planned to appoint Göttling to a new chair of chemistry and technology to be established at the University of Jena. In 1789 these plans were realized, and Göttling received a salaried extraordinary professorship at

40. On such shifts, see also Klein 2005a. For the opposite view, see Hufbauer 1982, 55–56.

41. Wiegleb belonged to the group of leading German chemists in the last third of the eighteenth century. Like Göttling, he was an apprenticed apothecary, and he remained a practicing apothecary throughout his career. On Wiegleb, see also Hufbauer 1982, 190–191; Krafft 2002.

42. Göttling 1792, "Vorbericht."

Jena University, shortly after being awarded a doctorate at the same university in acknowledgement of his earlier publications.[43]

Like many other apothecaries of his time, Göttling began his studies of ethers by exploring technical problems of preparing them and ways of identifying them. In his 1778 pharmaceutical manual, he mentioned that pharmaceutical journeymen had told him that they did not understand why the distillation of spirit of wine sometimes exclusively yielded Hoffmann's anodyne liquor, but sometimes also produced true ether.[44] The reason for this, Göttling explained, was of a purely technical nature; it was necessary to know the right proportions of spirit of wine and vitriolic acid and to change the receiver after the first distillate, consisting of nearly pure spirit of wine, had occurred. "Frequent repetition" of the operation and "painstaking attention," he stated, were the best teachers, leading eventually to success in production.[45] In this text, Göttling further described the different varieties of ethers that had been prepared to date using different kinds of acids, and he identified them by means of their observable properties.

A year later, Göttling published a lengthy paper on ethers in Crell's brand-new chemical journal that was concerned with the different varieties of ethers. At the time, Göttling was still a comparatively unknown apothecary, which prompted Crell to add some information about the author in a footnote: "The author of this remarkable essay is known to the learned world through his *Einleitung in die pharmaceutische Chemie.*"[46] The format of the essay—especially its historical introduction and its subsequent systematic division into eighteen "experiments" (*Versuche*)—manifests the apothecary's systematic approach to the preparation of different ethers, which relied to a considerable extent on his reading of the most recent publications on the issue. Göttling left no doubt that he wanted to "confirm" the "general proposition" (*allgemeiner Satz*) about the preparation of ethers that Crell had stated the previous year, namely, that ethers could be prepared not with the strong mineral acids alone, as common belief had it, but also with the weaker acids derived from the plant and animal kingdoms.[47]

To this end, he examined the question of whether vegetable acids other than vinegar, such as the acids obtained through the distillation of wood, tartar, bread, sugar, and so on, also yielded ethers. For his long series of eighteen

43. The biographical information comes from Hufbauer 1982, 207–208; and Möller 1962.

44. Göttling 1778, 67–68.

45. Ibid., 69.

46. Göttling 1779, 61.

47. Ibid., 43.

experiments, he used "the cheapest [vegetable acid], namely, wood acid."[48] Such economic considerations played a role not only with respect to expenditure on the experiments, but also regarding the possible commercial uses of the experimental products. "Herr Westendorff praises his naptha [i.e., ether] of vinegar," Göttling remarked at the end of his paper, "as anodyne and antispasmodic, especially against whooping cough in children; might not this naphtha of wood [newly prepared by Göttling] have the same effect?"[49] As usual, commercial and intellectual goals were not worlds apart.

The apparent success of the experiments in yielding pure "ether of wood" seemed to confirm the "general proposition" that ethers could be prepared with vegetable acids other than vinegar. Hence, Göttling stated at the end of his essay that this positive result encouraged him to perform similar experiments with another vegetable acid, tartaric acid, in the near future.[50] But the apothecary did not restrict his inquiries to confirming a more or less technical proposition. He also sought a chemical explanation for the invisible formation reactions of the different varieties of ethers, or "naphthas." "The right explanation for the formation of naphthas," he observed at one point of his experimental description, "is still a secret for us."[51]

It is clear that Göttling's attempt to explain the formation reactions of ethers was not derived from his own local experience alone. He was also a keen reader of chemical texts, as becomes clear from his discussion of the different opinions that well-known chemists had published on the issue. In the second half of the eighteenth century, many academic chemists assumed that spirit of wine was composed of a flammable principle and water. This conjecture was based partly on the observation that spirit of wine was very flammable, burned without leaving a residue, and could be mixed with water in any proportion.[52] It further relied on the chemical theory of compounds and "principles," which stated that the observable properties of material substances were caused by their hidden components, or "principles."[53] The renowned French chemist Pierre Joseph Macquer (1718–1784) further concluded that ordinary ether "is nothing else than spirit of wine that is deprived,

48. Ibid., 44.

49. Ibid., 61.

50. Ibid., 61.

51. Ibid., 57.

52. Some chemists, such as Macquer, held that the flammable principle was the pure phlogiston, whereas others conceived of it as a more compounded oil that contained phlogiston (see Macquer 1766, 1:413); see also Spary's chapter in this volume.

53. For this theory, see Klein and Lefèvre 2007.

by means of the vitriolic acid, of a part of its water principle; in this way it acquires the nature of an oil."[54] He thus explained the reaction between spirit of wine and oil of vitriol (or vitriolic acid), and the formation of ordinary ether, as a chemical decomposition (or "analysis") that was achieved by the acid's attraction of the component "water" contained in the spirit of wine. Göttling observed that many chemists unjustly generalized this explanation of the formation of ordinary ether, claiming that all kinds of acids would act in the same way as vitriolic acid. "The opinions of most chemists (*Scheidekünstler*)," he stated, "agree about the assumption that a naphtha is nothing other than the oily part of spirit of wine."[55] In other words, most chemists believed that the apparently different varieties of ethers were in reality the same kind of substance.

Göttling confronted the former understanding with an alternative theory proposed by his pharmaceutical master, Johann Christian Wiegleb. He presented a long quotation from a recent publication by Wiegleb, according to which the ethers consisted of the oily principle contained in spirit of wine *plus* a part of the acid used to prepare the ether.[56] Hence, they were products of both a decomposition, or analysis (namely, of the spirit of wine by means of the acid), and recomposition, or synthesis, of its oily principle with a certain proportion of the specific acid. Wiegleb therefore also conceived of the ethers as "artificial oils."[57] His theory further implied the assumption that the observed varieties of ethers actually were truly different kinds, or chemical species, of substances. For this latter view he invoked, in particular, the different smell of the ethers prepared with different kinds of acids. Göttling did not hesitate to embrace this theory of his pharmaceutical master. What is more, he offered additional quantitative evidence for this theory. Putting it in his own words, he "refuted" the theory accepted by the majority of academic chemists by way of an experiment: "Hence, I dare say that those who claim that a naphtha is merely the oily part of spirit of wine are completely refuted by the foregoing experiment. I have used 10 drachma (*Quentgen*) of concentrated wood acid and the same quantity of alcohol [and] produced 13.5 drachma of naphtha. Where would the naphtha be

54. Macquer 1766, 1:461; see also 409 and 459. This understanding was derived from the observation that ordinary ether was highly flammable, burned without a residue, and could be mixed with water—but to a lesser degree than spirit of wine.

55. Göttling 1779, 57.

56. Ibid., 57–59. It should be noted that in his dictionary Macquer (1766) did not rule out this alternative explanation.

57. Göttling 1779, 43.

derived from (*herleiten*) if it were nothing else than the oily part of the spirit of wine?"[58] Göttling argued that, because the weight of the ether, or "naphtha," of wood acid was greater than the weight of the ingredient spirit of wine, the ether could not be merely a product of the decomposition of spirit of wine.

How shall we interpret Götting's move to analytical considerations? Was it a huge step from the artisanal realm of recipes toward the academic world, as some historians would argue?[59] Was it a step that was characteristic of his unusual individual chemical talent, recognized by Goethe and others only some six years later? Or was it a smooth shift that required additional intellectual efforts, such as the reading of chemical publications, but did not presuppose a break away from the interests of an apothecary? I argue for the last alternative, not least since the type of theory at stake here had direct significance for pharmaceutical practice. It was useful knowledge, because it provided additional arguments for the identification of material substances. The question of whether the ethers prepared with different kinds of acids were the same chemical species, which might be slightly altered by impurities, or different kinds of chemical compositions, was a crucial one in the context of the medical application of these substances. The unambiguous identification of substances was one of the most recurrent concerns of apothecaries, linked with their struggle to standardize remedies and avoid adulteration. In this context, it was a sacred principle of chemical-pharmaceutical manufacture that its outcome depended first of all on the different kinds of ingredients, and that the ingredients of a prescribed remedy were never to be altered. Eighteenth-century apothecaries continually attempted to standardize the mode of producing chemical remedies in order to ensure that a named remedy was always the same kind of substance. Much of their effort to generate new techniques for preparing chemical remedies was driven by this interest and by the fact that the recipes contained in pharmacopoeias were not precise enough to guarantee the identity of a given remedy. Yet these recipes contained one basic, reliable instruction: the kinds of ingredients that were to be used to prepare a given chemical remedy.[60] The apothecary's professional ethos thus further endorsed the view that different ingredients meant different products.

58. Ibid., 58.

59. See Hufbauer 1982, 55, 56.

60. In most German states, apothecaries were forbidden by medical law from substituting a single ingredient prescribed by the recipe contained in the pharmacopoeia with another.

## Discussion: Pharmacy and Chemistry in Eighteenth-Century Germany

Our scrutiny of trials involving the ethers of two apothecaries has provided a glimpse of the ingenuity and the chemical knowledge of these experts. These material and epistemic activities may be divided into four main groups:

1. Trials for improving the techniques of preparing chemical remedies; these trials were spurred by attempts to standardize medicines.
2. Search for methods of unambiguous identification of chemical remedies, including avoidance of adulteration.
3. Experimental analysis of the chemical composition of remedies, including quantitative considerations.
4. Exploration and explanation of the invisible reactions underlying the preparation of a chemical remedy.

The analyses in the previous section have also shown that chemical knowledge, ranging from technical know-how to certain kinds of theories, was useful in pharmaceutical practice and that relatively unknown apothecaries were innovative in developing chemical knowledge. We must not assume, of course, that all apothecaries carried out all four types of activities, in particular not the third and fourth. We also ought to bear in mind that the majority of eighteenth-century apothecaries never published their observations and that only a small percentage of them were ever acknowledged as chemists in Germany. However, as I will describe in more detail in this section, the eighteenth-century chemists were not a homogeneous group either. What is more, in late eighteenth-century Germany, half of this group were in fact apprenticed apothecaries, and many of them practiced as apothecaries for a long time in their career. I argue not that eighteenth-century chemists and apothecaries were identical personae, but rather that they shared an enormous range of forms of knowledge, activities, and material items, most importantly laboratories. Well into the nineteenth century, there was certainly no gap between chemical science and pharmacy, as some historians of science have argued.[61] Instead, eighteenth-century chemical pharmacy and chemistry ought to be conceived as a coherent culture, a differentiated continuum of forms of producing, experimenting, and knowing. In what follows, I wish to flesh out this argument by comparing some aspects of eighteenth-century German pharmacy and chemistry in a more systematic way.

61. See Hufbauer 1982, 55, 56; J. Simon 2005, 7, 167.

## THE PHARMACEUTICAL PROFESSION

In most eighteenth-century German states, apothecaries were not organized in guilds, as they were in France and Italy. Instead, governments regulated their commercial rights and duties through medical and pharmaceutical laws defining the relationships between apothecaries, physicians, and other medical practitioners. Physicians and the official *collegia medica* compiled pharmacopoeias (the collections of officially sanctioned medical compositions) and official tariffs for remedies, or *Arzneitaxen,* examined prospective apothecaries, and regularly visited apothecaries' shops to inspect the quality of their remedies. But as a rule, the certified apothecaries, who had passed through an apprenticeship, were the only ones with the privilege to run an apothecary's shop and to prepare and sell remedies. The medical laws thus protected apothecaries from competition with other vendors of remedies, such as physicians, herbalists, distillers, preparers of chemical remedies (*Laboranten*), and grocers (*Materialisten*). To some extent, they also regulated the apprenticeship period and the rights and duties of masters, journeymen, and apprentices.

But eighteenth-century apothecaries' daily practice of purchasing natural simple drugs, dispensing Galenic composite medicines, and producing chemical remedies could not, of course, be externally regulated throughout by laws, governments, and medical doctors. It must also be studied in the context of the contemporary system of trade and manufacture, and from the perspective of the actual practice, goals, and interests of apothecaries, who were not only personnel of the medical system, but also merchants and manufacturing artisans. The latter becomes particularly evident from the fact that eighteenth-century German apothecaries manufactured chemical remedies in their own laboratories and also sold and produced a plethora of commodities other than remedies, such as coffee, tea, tobacco, and spices; confectionery and syrups; pigments and tints; soaps, hair powder, and pomades; wines, brandy, and liqueurs.[62] Trade and manufacture required knowledge and skills that far exceeded medical expertise. As debates about pharmaceutical apprenticeship in the second half of the eighteenth century show, most German apothecaries agreed that a knowledge of chemistry and botany was particularly useful to their art, even though there were considerable local differences in botanical and chemical training.[63]

62. Coffee, tea, and tobacco were used as remedies in the seventeenth century but became transformed into luxury articles in the eighteenth century; the same is true for certain spices and for spirit of wine.

63. See Klein 2007b.

In eighteenth-century Germany, pharmaceutical apprenticeship predominantly meant training in the apothecary's shop. The main way for an apprentice to learn chemical techniques was through watching the practice of masters and journeymen and practical exercise. Pharmaceutical apprentices also occasionally read manuals and other books on botany, chemical pharmacy, and chemistry more generally, as well as *Arzneitaxen* and pharmacopoeias, which were written in Latin. Their apprenticeship thus displayed characteristics of both artisanal apprenticeship and higher education. The apprenticeship normally lasted five to six years, followed by six to eight years of service as a journeyman.[64] Apprentices were accepted at the age of fourteen or fifteen after attending school, a primary-school (*Volksschule*) education being enough if supplemented by some instruction in Latin. Apprentices were trained in the dispensary and the laboratory by a master apothecary (*Principal*), either the owner of the shop or a head administrator (*Provisor*), and by the older journeymen. During their years of apprenticeship, pharmaceutical apprentices acquired knowledge of hundreds of materials and their provenance—raw materials of vegetable, animal, and mineral origin, Galenic composites, chemical reagents, and chemically prepared materials—as well as know-how in storing natural drugs and preparing Galenic and chemical remedies, and skills in weighing, measuring, and calculating.

In the eighteenth century, most medicines were still either natural drugs or Galenic composites. Activities such as purchasing, storing, and selling herbs, and weighing and mixing natural drugs or Galenic medicines in the apothecary's dispensary, belonged to the daily routine of apothecaries. But as a consequence of the introduction and proliferation of chemical remedies during the sixteenth and seventeenth centuries, pharmacy was also in a state of persistent change and innovation in the eighteenth century. There was hardly a single recipe for the manufacture of chemical remedies that was not questioned, varied, improved, or replaced by a new one. And there was hardly a single chemical remedy that was not on the testing bench as a possible adulteration or a material that had been unambiguously identified.[65] Apothecaries had performed simple chemical operations such as distillation and decoction long before "chemical remedies" were promoted by sixteenth- and seventeenth-century chemists and physicians. Hence, pharmacy, as a craft,

64. For general outlines of the pharmaceutical apprenticeship in eighteenth-century Germany, see Kremers and Urdang 1951, 120–122; Adlung and Urdang 1935, 133–134; Beyerlein 1991.

65. For the practical problem of adulteration and the form of knowledge linked with this problem, see also the contributions of Johns and Nieto-Galan in this volume.

was by no means unprepared for the surge of innovations in the manufacture of remedies encouraged by the chemical reforms of Paracelsus and his followers in particular. Nevertheless, the manufacture of chemical remedies entailed the introduction of many new materials, instruments, and techniques into an existing artisanal tradition, and the way in which German apothecaries met this challenge, integrating chemistry into their art and their apprenticeship system, is not yet well understood. What has been known in this respect from medical edicts, pharmacopoeias, and drugs tariffs, as well as from inventories of apothecaries' shops, surviving pharmaceutical-chemical instruments, and publications by apothecaries, is that chemical remedies were indeed generally accepted in Germany by around 1700.[66] By this time, the expression "chemical remedies" was as ubiquitous as the division of remedies into *simplicia* and *composita*, and the subdivision of the latter into Galenic composites and "chemical preparations" or "chemical remedies." Equally ubiquitous was the apothecary's "laboratory," that is, a distinct room for the manufacture of chemical remedies, equipped with furnaces, distillatory apparatus, and numerous other chemical instruments, vessels, and materials.

## LABORATORIES

Pharmaceutical laboratories proliferated in the seventeenth century, and by the early eighteenth century it was quite common in Germany for apothecaries to establish laboratories in their apothecary's shops. The official medical and pharmaceutical edicts and pharmacopoeias of late seventeenth- and eighteenth-century Germany took pharmaceutical laboratories for granted (see figure 6.1). The Brandenburg medical edict of 1698, for example, ordered that "chemical remedies" (*chimische Medicamenta*) must not be purchased from "vagrants" and "preparers" (*Laboranten*) of chemical remedies but prepared and sold by apothecaries in their own "laboratories."[67] The frequent use of the term *laboratory* in contemporary descriptions of apothecaries' shops, autobiographies, and correspondence, as well as in publications describing their chemical-pharmaceutical operations, further demonstrates the omnipresence of pharmaceutical laboratories. Today, pharmaceutical museums

66. For a list of the chemical remedies mentioned in German pharmacopoeias from the sixteenth century onward, see Schneider 1968–1975, vol. 3. Wolfgang Schneider and Erika Hickel have provided ample evidence, based on their detailed analysis of pharmacopoeias and experimental reconstruction of recipes, that the manufacture of chemical remedies was a firmly established part of the pharmaceutical art in early eighteenth-century Germany; see, in particular, Hickel 1973; Schneider 1972, 1968–1975. See also Krüger 1968; Beisswanger 1996.

67. See Stürzbecher 1966, 49.

FIGURE 6.1. The small laboratory of the Berlin *Hofapotheke* (eighteenth century), equipped with furnaces, a chimney, alembics, retorts, and other instruments (Hörmann 1898, 224).

exhibit a large range of chemical instruments and vessels used in eighteenth-century pharmaceutical laboratories. The German Pharmaceutical Museum (*Deutsches Apotheken-Museum*) in Heidelberg, for example, exhibits a collection of instruments that were in normal use in apothecaries' laboratories from the seventeenth to the nineteenth centuries (see figure 6.2). Apart from instruments and vessels, our knowledge of eighteenth-century pharmaceutical laboratories relies on verbal descriptions, inventories, drawings, and the conclusions we can draw from the architecture of existing apothecaries' shops established back in the early modern period.

An important precondition for the frequent shifts of eighteenth-century apothecaries from immediate pharmaceutical production to chemical inquiry into the nature of substances and their reactions was the strong

FIGURE 6.2. Chemical-pharmaceutical instruments from the seventeenth to the nineteenth centuries. Courtesy of the Deutsches Apotheken-Museum, Heidelberg.

FIGURE 6.3. The laboratory of the University of Altdorf (Puschner, ca. 1720).

correspondence between the pharmaceutical and the academic chemical laboratories of the time.[68] All of the instruments and vessels used in eighteenth-century pharmaceutical laboratories are also familiar from academic chemical laboratories, as can be seen in illustrations such as the famous plate in Diderot's *Encyclopédie* or the drawing of the laboratory of the University of Altdorf (see figure 6.3).[69] Eighteenth-century apothecaries had more than a few isolated instruments in common with academic chemists. The types of furnaces, retorts, receivers, alembics, jars, bottles, crucibles, and balances that academic chemists used in their laboratories were the same as those used by apothecaries to make chemical remedies.

What is more, the techniques of chemical-pharmaceutical production were the same as those involved in chemical experimentation. Dissolutions, distillations, evaporations, precipitations, calcinations, combustion, smelting, and so on were both experimental techniques and methods of material production. In his essay on early modern metallurgy, contained in this volume,

68. For further details on this issue, see Klein 2007a, 2008a, 2008b.

69. See Diderot and d'Alembert 1965, vol. 24.

Christoph Bartels has pointed out that metallurgical assaying involved the same techniques as the production of metals from ores, the difference being merely a difference in scale. With a few exceptions, even scale was the same in the case of eighteenth-century pharmaceutical production and chemical experimentation.[70] Like apothecaries, academic chemists used large furnaces and large, immobile distillation apparatuses as well as the same types of smaller vessels made of earthenware and of glass, which were important for systematic observation (see figures 6.1 and 6.2).[71] It is clear that such strong similarities between the pharmaceutical and academic-chemical laboratories facilitated shifts from pharmaceutical manufacture to chemical analysis and vice versa.

## THE PERMEABLE BOUNDARIES OF EIGHTEENTH-CENTURY CHEMISTRY

Compared to eighteenth-century chemistry, pharmacy was a well-defined art and craft in the eighteenth century, regulated by rules, either governmental or established by cooperation among apothecaries, and performed at a distinct site, the small-scale apothecary's shop. Likewise, apothecaries were a distinct professional group, recognizable by the pharmaceutical apprenticeship, rules of access to their profession, professional ethos, and so on. No such thing existed in the chemistry of the time. Historians of chemistry have long known that there was no professional education for eighteenth-century chemists, nor an unambiguous type of academic-chemical institution comparable with today's university chemistry departments and laboratories devoted to chemical research only. They also agree that it is extremely difficult to provide a historically adequate outline of the system of knowledge and practices of eighteenth-century chemistry. Nevertheless, some historians of chemistry have proposed

70. One exception is the distillation of essential oils and spirits of wine, which were often done on a larger scale in eighteenth-century pharmaceutical laboratories than in the academic chemical laboratories. Another exception is large-scale pharmaceutical manufacture as performed in the London Apothecary's Hall; on the latter, see A. Simmons 2006.

71. This began to change slowly in the second half of the eighteenth century, as chemists began to produce and identify new kinds of substances, such as the different kinds of air, in their laboratories and began to make increasing use of new types of precision instruments. The second edition of Macquer's chemical dictionary, published in 1778, manifests these changes. Whereas Macquer had remained silent in the first edition on the subject of air pumps, thermometers, pneumatic apparatuses for collecting gases or kinds of air, and many other philosophical instruments, in the second edition all of these instruments are mentioned (see Macquer 1778, 2:1–9). For the introduction of precision instruments into eighteenth-century chemistry, see also the essays assembled in Holmes and Levere 2000.

to define eighteenth-century chemistry as a teaching discipline.[72] This definition effectively questions earlier depictions of pre-Lavoisierian chemistry as a mere handicraft, but it also tends to overemphasize the scholarly, textual tradition of chemistry, and of the earlier alchemy, at the expense of its locally contextualized practices and hands-on experience. The sites of chemical practice were extremely diverse in the eighteenth century, and they must not be fully equated with academic experimentation. As in the alchemical period, in the eighteenth century, chemical practice also included artisanal practices and sites, such as the laboratories serving pharmacy, metallurgical assaying, gunnery, perfumery, or the distillation of spirit of wine and liqueurs.[73] Chemists' "experimental histories" continually referred to such artisanal practices, collecting facts from them and repeating artisanal production techniques in their own laboratories.[74] In addition to these artisanal sites, chemical experiments were also performed in public places such as botanical gardens and salons as well as in laboratories at universities and academies and in the private laboratories of learned chemists.

The term *chemistry* was an actors' category, and we may conclude from its use during the eighteenth century that the historical actors themselves recognized some coherence of the field. However, we should also bear in mind that such terms, like many others, are abstractions that do not convey the full variety of actual activities, resources, forms of knowledge, social interactions, institutions, and sites. Adopting a principle of arrangement often used in eighteenth-century chemical textbooks, many historians of chemistry to distinguish between chemical practice and theory when they refer to eighteenth-century chemistry as a whole. This is helpful, but this is still a quite abstract definition. As a clearly discernible map, the whole system of chemistry existed only in textbooks and lecture halls. Beyond the realm of paper and teaching, the contours and internal order of eighteenth-century chemistry were fluid. Far from being a unified discipline that would have subordinated different practices and forms of knowledge under one dominant goal and within a single epistemic frame, early modern and eighteenth-century chemistry resists clear classification as either a scholarly or an artisanal enterprise. It was neither unambiguously an academic and teaching discipline, nor merely a mechanical art, but all of these at once.

72. Hannaway 1975 is the classic formulation of this view.

73. On eighteenth-century laboratories in connection with gunnery and the distillation of liqueurs, see the chapters by Mauskopf and Spary in this volume.

74. On "experimental history" as a distinct style of experimentation in eighteenth-century chemistry, see Klein 2005a, 2005c; and Klein and Lefèvre 2007, 21–31.

Operations considered to be "chemical" by the historical actors were performed in arts and crafts such as pharmacy, mining, and metallurgy; the fabrication of oil of vitriol, nitric acid, and other chemicals; dyeing, bleaching, and calico printing; brewery and the making of wines and liqueurs; the making of porcelain and ceramics; the manufacture of different kinds of soap and glass; gunnery and the making of gunpowder and fireworks; the extraction of sugar from beets; agriculture and horticulture; and so on. In the second half of the eighteenth century, authors of chemical texts classified these chemical operations under labels such as "chemical pharmacy," "chemical metallurgy," or "the chemistry of dyestuffy." All of these different artisanal practices fed into chemistry as a whole. Other significant parts of eighteenth-century chemistry were experimental chemical analyses of the invisible composition of material substances, experimental investigations of chemical reactions, formulations and uses of chemical concepts and nomenclature, classifications of material substances and of their reactivity in chemical tables, chemical modeling, as well as the elaboration of specific "chemical theories," that is, theories referring to a particular, clearly demarcated domain of observation and experiments involving material substances. The much more comprehensive matter theories such as atomism and the philosophy of chemical principles, which had predominated much of the theoretical discourse in seventeenth-century alchemy, played a less significant role in the chemistry of the second half of the eighteenth century, though their significance varied from country to country and was sometimes the subject of debates aiming at a demarcation of proper chemical theories from metaphysics.

I have pointed out above that pharmacy was in a state of rapid transition in the eighteenth century, and I have distinguished four types of chemical activities by eighteenth-century apothecaries. As a consequence, it makes no sense to circumscribe an ideal persona of the apothecary in eighteenth-century Germany, merely because pharmacy was a profession at the time. Even more diverse were the people designated "chemists."[75] In the second half of the eighteenth century, almost half of the one to two hundred Germans acknowledged as "chemists" were apothecaries who had served an apprenticeship, and about the same proportion possessed a medical doctorate; among the latter group, a considerable number supplemented university education with training in mining and metallurgy to become mining officials.[76] The situation was similar in France, as can be inferred from the chapters by

75. My discussion concerning the category "chemists" applies to Continental Europe but not to Great Britain, where apothecaries were quite often designated "chemists."

76. See Hufbauer 1982; Klein 2007a, 2007b.

Orland, Spary, and Nieto-Galan included in this volume. Many chemists held several different posts simultaneously within the artisanal and scholarly worlds. Neither the teaching professor nor the skilled apothecary-chemist can be taken as the "typical" eighteenth-century chemist. There was a whole range of forms of chemical education and training, with the artisan-chemist who had served an apprenticeship located at one pole of this spectrum and the university-based philosopher-chemist at the other.

Throughout Europe, eighteenth-century chemists varied widely in their educational level, occupation, and social status. Yet despite this diversity, they also shared a common practice and culture in which they participated with varying degrees of learning and sophistication.[77] To become acknowledged as a chemist in late eighteenth-century Germany required the following minimal conditions. A chemist had to be a skilled experimenter who knew how to perform a variety of experimental tests on the chemical properties of substances and how to carry out chemical analyses on a wide range of materials. This experimental know-how involved not only bodily skills, but also certain conceptual assumptions about the invisible dimension of substances. A chemist also had to be a connoisseur of hundreds of different material substances, far beyond the local artisanal use of materials, and he had to be skilled in their identification and classification.[78] Furthermore, a chemist also needed to possess knowledge of chemical theories, although in the Germany of the second half of the eighteenth century the famous chemists were empirically minded and not much interested in the active development of theories. As we have seen in the case of Göttling, publication promoted acknowledgement as a chemist, though it was not an absolute requirement; the apothecary Andreas Sigismund Marggraf, for example, became a member of the Berlin Society of Sciences on the basis of his experimental work, without prior publication.[79] Teaching was also a highly esteemed activity, but by no means all eighteenth-century chemists were teachers. Membership in scientific societies was important as well, and many, but not all, German chemists were members of academies and other scientific societies. An academic education was even less important, as can be seen by the fact that approximately half of German eighteenth-century chemists were artisans (mostly apothecaries) who had undergone an apprenticeship and not a university education. In late eighteenth-century Germany, a chemist's reputation rested mainly on par-

77. In her work on late medieval and early Renaissance medicine, Nancy Siraisi observed a similarly diversified medical community (see Siraisi 1990).

78. I refer here to Michael Polanyi's concept of connoisseurship. See Polanyi 1994.

79. On Marggraf, see Hufbauer 1982, 180–181; Klein 2007a, 2007b, and 2008a.

ticular expertise with materials and with experimental methods of testing and analyzing substances. Compared to France and Britain, the German chemical community of the time offered significantly smaller rewards for theoretical activities and was extremely reluctant to adopt any chemical philosophy and matter theory aiming at a comprehensive explanation of nature.

In keeping with a famous definition of *science* in the sociology of scientific knowledge—"scientific research is what scientists collectively do"—one might thus argue that eighteenth-century "chemistry" was what eighteenth-century "chemists" collectively did.[80] This is to some extent a working definition, as it is less difficult to determine who was recognized as a chemist in the eighteenth century than to map the various sites and activities that historical actors labeled as constituting the domain of "chemistry" (or "chymistry," "*Scheidekunst*," and so on). Weighing against this definition is, however, the fact that it was not only acknowledged "chemists" who contributed to chemical practice and knowledge in the eighteenth century. On the contrary, in a time before the professionalization of chemistry, many different social groups contributed to chemistry, and one of the goals of this chapter is to demonstrate this fact. As has been shown above, the pharmaceutical art in particular was an immense reservoir of new chemical facts, techniques, and material objects. This is the reason why the editors of the new professional chemical periodicals founded from 1778 onward, such as Crell's *Chemische Annalen* and Göttling's *Almanach oder Taschenbuch für Scheidekünstler und Apotheker* encouraged apothecaries to contribute to their journal and to report every observation they made during the manufacture of remedies.[81] These invitations by Crell and Göttling explicitly acknowledged that any apothecary could make a valuable contribution to chemistry. The manufacture of chemical remedies involved a great variety of different chemical techniques, instruments, and materials, and often required a complex series of chemical operations. As Crell and Göttling pointed out, any apothecary could contribute to their journal and to chemistry at large by improving chemical techniques, inventing new materials, and observing curious phenomena.[82]

I have discussed eighteenth-century pharmacy and chemistry at some length mainly to provide some historical context for the concrete examples of chemists' and apothecaries' studies of ethers in the first section of this chap-

80. Barnes, Bloor, and Henry 1996.

81. Göttling in particular encouraged apothecaries to "pay attention in the future to incidents that often occur accidentally in apothecary's shops" and to report their observations in his journal. He insisted that he did not expect all contributions to his journal to be "extended treatises" and that he was willing to publish "every little remark." See Göttling 1785.

82. See Klein 2007b.

ter. This discussion also ought to clarify my argument. Focusing on the intersection of pharmacy and chemistry with the arts and sciences more generally, I am not suggesting that eighteenth-century chemistry and pharmacy were fully identical cultures or that there was no difference at all between apothecaries and chemists. As my outline above has demonstrated, eighteenth-century chemistry comprised many areas of practice and knowledge other than chemical pharmacy, and pharmacy was an art and branch of commerce extending beyond *chemical* pharmacy. Instead, my argument is that there was a domain of doing and knowing within which chemistry and pharmaceutical art overlapped, forming a mixed system of chemical-pharmaceutical practice and knowledge with differences only in degree. Outside that boundary zone, chemistry and pharmacy also differed from one another.

## Conclusion

In their chemical investigations, the late eighteenth-century apothecaries did not have to bridge a huge gap between their own artisanal practice and the world of university-based chemists. Well into the nineteenth century, the production and experimental studies of chemical remedies meant a continual and quite immediate back and forth of techniques and learned knowledge from the pharmaceutical to the academic laboratory and vice versa. This circulation of knowledge between the laboratory located in the apothecary's shops and the academic chemical laboratory was embedded in a shared material culture and practice in which the same kinds of materials were produced and reproduced, either for economic or intellectual goals. It was further spurred by intellectual and social incentives such as the reading of books and the existence of journals for both apothecaries and academic chemists.[83]

Historians and philosophers of science have often focused their inquiries into the interdependence of science and technology on the role played by scientific theory for the latter. In so doing, they have obliterated forms of expert knowledge developed in academic institutions and practices that were not theoretical, such as experimental techniques, methods of analysis, connoisseurship of substances, and methods of their identification, classification, and naming. Aside from a few exceptions, they have also rarely studied the question of how much the evolving early modern sciences owed to the arts and crafts, or technology, prior to the more visible institutions of modern technoscience. My analysis has instead demonstrated that late eighteenth-century German apothecaries shared with academic chemists techniques

83. For further social incentives of this kind, see Klein 2007a, 2007b.

of chemical preparation and connoisseurship of substances, as well as basic chemical concepts and some types of specific theories that provided useful knowledge. In their publications about chemical remedies, late eighteenth-century German apothecaries were concerned mainly with improving techniques of preparing and ways of unambiguously identifying material substances, along with standardizing medicines and avoiding adulteration. Some individual apothecaries carried their inquires further to include chemical analyses and explanations of chemical transformations.

In the eighteenth century, and even after the "chemical revolution" of the 1770s and 1780s, a significant part of pharmacy concerned with the manufacture of chemical remedies, or "chemical pharmacy," and a large domain of chemistry, namely, the experimental practices and theories dealing with materials that doubled as chemical remedies, constituted a loosely coherent chemical-pharmaceutical culture, which foreshadowed the technoscientific networks of the twentieth century. I have argued that instead of taking this mixed technoscientific culture apart in order to highlight the analytical and scientific dimension of eighteenth-century chemistry and the artisanal and commercial dimension of pharmacy, we ought to approach it as a differentiated continuum of scientific and artisanal ways of knowing, producing, and experimenting.

**Primary Sources**

Secondary sources can be found in the cumulative bibliography at the end of the book.

Bigelow, Henry J. 1876. A History of the Discovery of Modern Anaesthesia. *American Journal of the Medical Sciences* 71:164–84.

Boullay, Pierre François Guillaume. 1807. Mémoire sur le mode de composition des éthers muriatique et acétique. *Annales de Chimie* 63:90–101.

Bucholz, Christian Friedrich. 1803. [No title.] *Allgemeines Journal der Chemie* 10:219–22.

Diderot, Denis J., and Jean LeRond d'Alembert. 1965. *Encyclopédie, ou Dictionnaire raisonné des sciences, des arts, et des métiers (1751–1780).* 35 vols. Stuttgart: Frommann.

Fourcroy, Antoine-François de, and Louis N. Vauquelin. 1797. De l'action de l'Acide sulfurique sur l'Alcool, et de la formation de l'Ether. *Annales de Chimie* 23:203–15.

Fourcroy, Antoine-François de. 1801–1802. *Systême des connaissances chimiques: et de leurs applications aux phénomènes de la nature et de l'art.* 11 vols. Paris: Baudouin.

Frobenius, Siegmund August. 1729–1730. An Account of a Spiritus Vini Aetherius, Together with Several Experiments Tried Therewith. *Philosophical Transactions* 36: 283–89.

Göttling, Johann Friedrich A. 1778. *Einleitung in die pharmaceutische Chymie für Lernende.* Altenburg: Richter.

———. 1779. Chymische Versuche mit der Holzsäure, in Absicht vermittelst derselben eine Naphtha zu verfertigen. *Chemisches Journal für die Freunde der Naturlehre, Arzneygelahrtheit, Haushaltungskunst und Manufacturen* 2:39–61.

———. 1780. Eßig-Naphte. *Almanach oder Taschenbuch für Scheidekünstler und Apotheker* 1780:7.

———. 1781a. Besondere Bemerkungen bey Verfertigung der Salpeter Naphte. *Almanach oder Taschenbuch für Scheidekünstler und Apotheker* 1781:39–42.

———. 1781b. Sonderbare Bemerkungen über die Eßig-Naphte. *Almanach oder Taschenbuch für Scheidekünstler und Apotheker* 1781:4–10.

———. 1783a. Ameisenaether. *Almanach oder Taschenbuch für Scheidekünstler und Apotheker* 1783:77–78.

———. 1783b. Beobachtungen über den Eßigaether. *Almanach oder Taschenbuch für Scheidekünstler und Apotheker* 1783:1–3.

———. 1783c. Vorbericht. *Almanach oder Taschenbuch für Scheidekünstler und Apotheker* 1783.

———. 1785. Vorbericht. *Almanach oder Taschenbuch für Scheidekünstler und Apotheker* 1785.

———. 1792. Vorbericht. *Almanach oder Taschenbuch für Scheidekünstler und Apotheker* 1792.

Lavoisier, Antoine-Laurent. 1789. *Traité élémentaire de chimie: Présenté das nun ordre nouveau et d'après les découvertes modernes.* 2 vols. Paris: Cuchet.

Macquer, Pierre Joseph. 1766. *Dictionnaire de Chymie, contenant la Théorie et la Pratique de cette Science, son application à la Physique, à l'Histoire Naturelle, à la Médicine et à l'Economie animale.* 2 vols. Paris: Lacombe.

———. 1778. *Dictionnaire de Chymie, contenant la Théorie et la Pratique de cette Science, son application à la Physique, à l'Histoire Naturelle, à la Médicine et aux Arts dépendans de la chymie.* 2nd ed. 2 vols. Paris: L'Imprimerie de Monsieur.

Puschner, Johannes Georg. [ca. 1720]. *Amoenitates Altdorfinae oder Eigentliche nach dem Leben gezeichnete Prospecten der Löblichen Universität Altdorf.* Nuremberg: Michahelles.

Schnurr, Balthasar. 1676. *Vollständiges und schon aller Orten bekanntes Kunst-, Haus- und Wunder-Buch Aufs neue wiederumb verbessert und vermehrt.* Frankfurt a.M.: Zubrodt u. Haaß.

Schrader, Johann Christian Carl. 1791. Vom Hrn Schrader in Berlin. *Annalen der Chemie* 1791:351–52.

Voigt, C. F. 1784. Auszug eines Schreibens von Herrn Apotheker Voigt zu Erfurt. *Almanach oder Taschenbuch für Scheidekünstler und Apotheker* 1784:184–90.

# PART TWO

# Materials in the Market Sphere

## Introduction to Part 2

The chapters by Barbara Orland, Matthew D. Eddy, and E. C. Spary, in this second part illuminate new facets of activities involving materials, in addition to challenges emerging in experimental and commercial production processes. They put the market sphere on center stage with a cast of materials such as milk, mineral waters, and liqueurs—the former a widespread food, medicine, and artisanal raw material, the latter one of the most fashionable drinks of the luxury market. All three chapters focus on the eighteenth century, when the consumer economy was thriving and the marketplace was a site where farmers, artisanal vendors, merchants, rich connoisseurs of luxury products, and less rich buyers of more mundane goods regularly met. Milk, mineral waters, and liqueurs stimulated a host of inquiries, debates, and commercial activities among these different groups of experts. The experts on milk, mineral waters, and liqueurs had different professions and social ranks, and they possessed a broad range of forms of knowledge. Among them were pharmacists knowledgeable in chemistry as well as in the making of butter and cheese and in wet nursing; polite connoisseurs of liqueurs and apprenticed distillers; academic chemists analyzing alcoholic distillates, milk, or mineral waters; and practicing physicians performing chemical analyses and contributing to the commodification of mineral waters. These experts were interested both in the usefulness of their objects and in improving natural and medical knowledge. Hence, experiments in the laboratory, chemical analyses, as well as theories play an important role in all three chapters.

Barbara Orland's chapter addresses the mundane material of milk, which participated in a broad variety of eighteenth-century practices. The chemists who studied milk most intensively, such as Antoine-Augustin Parmentier, Nicolas Déyeux, and Samuel Ferris, were also pharmacists or physicians who continued to address questions concerning its practical uses. These chemists collected experimental histories about milk from many different areas, including the arts and crafts, and they also participated in an Enlightenment discourse on milk and breast-feeding. In their laboratories they further subjected milk to experiments, trying to identify and classify it more precisely and to study its components. Yet milk resisted accurate chemical analysis. There were not only different varieties of milk, depending on its natural origin; even milk of the same, say, human origin sometimes varied according to factors such as time of day or duration of lactation. Chemists thus considered analytical knowledge about the components of milk to be uncertain. Nevertheless, Orland argues, their experiments stimulated new questions about organic materials and new insights into the technique of organic-chemical analysis. They thus contributed to both animal chemistry and the then thriving field of analytical chemistry. As a result of their different approaches, interests, and sources of knowledge, the experts studied by Orland constituted milk as an ambiguous object that was identified and explained from chemical, medical, physiological, artisanal, agricultural, and consumers' points of view.

A similar picture emerges from Matthew D. Eddy's chapter on mineral waters and E. C. Spary's study of liqueurs. Mineral waters from spas were fashionable medicines in the eighteenth century. They attracted many patients and visitors, therein spurring tourism and the economic success of towns and villages. Using Peterhead Spa in Scotland as a case study, Eddy excavates the therapeutic theories that motivated provincial experts to commodify mineral water at the dawn of the nineteenth century. Throughout his chapter he points out that chemical language and experiments played a notable role in late-eighteenth-century pamphlets and articles (both popular and academic) that addressed the curative power of mineral water. One of the key characters in his account is Rev. Dr. William Laing. Like so many late Enlightenment polymaths, Laing had many interests. In addition to being an ordained Episcopal priest, he was also a practicing physician and self-trained chemical experimenter. After chemically analyzing the contents of the spa's water, Laing wrote a pamphlet that detailed how his results connected to the neohumoralist model of disease that guided late-Enlightenment pharmacology and therapeutics. In particular, he argued that the water could be used to cure nervous disorders, that is, a disease that affected many upper- and middle-class tourists at the time. Paying close attention to Peterhead's social, economic, and medical context, Eddy uses Laing's work to show how local experts functioned as important mediators in the commodification of a provincial medical commodity.

In E. C. Spary's chapter, eighteenth-century Parisian food shops, the marketplace, laboratories, apprenticed distillers and vendors of liqueurs, buyers and connoisseurs of such luxury products, physicians, and pharmacists as well as chemists of the French Academy of Sciences are brought together as actors in an intriguing drama about knowledge, power, and ethics concerning food. "Polite science" is her term that glues all of these different elements together. Why "polite science"? What is meant here is certainly not our modern notion of "natural science," entrenched in a highly specialized university system and often practiced at remote research laboratories. "Polite science," rather, means a complex system of knowledge that included systematic observation and subtle experiments without being severed from mundane concerns about fashion, standards of politeness, social rank, and commercial profit. As in the case of milk and mineral waters, chemistry was a significant basis for liqueur makers' and distillers' claims to innovate and to possess enlightenment as well as for the marketing of their products. Spary points out that these practitioners worked in laboratories, often in a room behind the public area of the café. Like apothecaries, chemist-dyers, and comptrollers in the production of gunpowder (see Klein, Nieto-Galan, and Mauskopf, this volume), they contributed to the development of laboratory techniques and experimental knowledge, especially in chemistry. Spary further argues that liqueurs, like other prepared food, had no existence independent of the diverse groups of actors dealing with them and that all of these groups contributed to their identification, meaning, and value. She concludes that "no epistemological superiority can be assumed for scientific objects over everyday ones in the eighteenth-century city."

# 7

# Enlightened Milk: Reshaping a Bodily Substance into a Chemical Object

BARBARA ORLAND

Shortly before the outbreak of the French Revolution, in 1785 and 1787, two medical institutions—the Royal Medical Society of Edinburgh and the Royal Society of Medicine in Paris—commissioned investigations into the physiological and chemical qualities of the milk produced by humans, cows, goats, donkeys, sheep, and camels. In Edinburgh, the prizewinner was Samuel Ferris (?–1831), a physician with strong interests in chemistry, and in Paris the competition was won by the pharmacists and chemists Antoine-Augustin Parmentier (1737–1813) and Nicolas Déyeux (1753–1837).[1] Both studies were published in English, French, and German, but the French chemists' memoir became especially well-known. For decades, it was treated by specialists and the wider public as a source of reference for all relevant information on milk and milk products from the viewpoint of chemistry, medicine, and rural economy. For instance, the entry "Milk" in Krünitz's economic and technological encyclopedia, some 250 pages covering chemical analyses and everyday knowledge of agricultural and household milk processing, mainly drew upon Parmentier and Déyeux's book.[2]

However, neither Ferris nor Parmentier and Déyeux were the only or even the first scientists to perform milk analyses.[3] Throughout Europe, milk, as one

1. See Ferris 1785; and Parmentier and Déyeux 1798. I had only the German translations at my disposal (Ferris 1787; Parmentier and Déyeux 1800). All translations, unless otherwise indicated, are by the author. I wish to thank Christine Luisi and E. C. Spary for their help in revising my English. Furthermore, the comments by one of the anonymous referees of the University of Chicago Press and Ursula Klein helped to improve my argument.

2. Krünitz 1803, 339–571.

3. For the broader communal context of these chemical analyses, see also the contributions by Nieto-Galan, Spary, Eddy, and Klein in this volume.

of the most important fluids of the animal kingdom, gained more and more interest among the growing community of chemists.[4] Pierre Joseph Macquer (1718–1784), professor of chemistry at the Jardin du Roi in Paris, summarized various milk studies in his influential *Chemical Dictionary,* first published in French in 1766 and translated into German in 1781.[5] Johnson's *History of the Progress and Present State of the Animal Chemistry* of 1803 devoted about 80 pages to presenting the milk chemistry of the last two decades.[6] Chemical investigations of milk represented a new type of scientific fascination, since up until then dissertations and academic treatises on milk had been concerned mainly with medical questions and were directed at the medical community (physicians, midwives, apothecaries, and so on). Chemical experiments—if they were undertaken at all—were subordinated to medical discourses.[7]

Against this backdrop, the studies by Parmentier, Déyeux, and Ferris were remarkable for several reasons. First, they claimed to draw together from different fields of knowledge (philosophical, medical, and agricultural) all valuable information about the whitish fluid. Second, these authors discussed recent political concerns about milk. Whether they referred to Enlightenment advocacy of maternal breast-feeding or physiocratic concerns with agricultural improvement, in fact, the texts did not draw sharp boundaries between what was then called the "nature" of the fluid and everyday experience or practical use of it. Finally, they offered a book-length description of chemical experiments on milk. Chemical investigations served to map the "principal nature and properties" of milk, but should also contribute to "the enrichment of animal chemistry" as a specific branch of chemistry.[8]

In the following, I first examine late eighteenth-century chemists' attempts to transform milk into a chemically defined object. These attempts included experimental histories, that is, collections of facts and beliefs from many different traditions, as well as experiments in the laboratory.[9] I will also show that, based on their experimental knowledge, chemists tried to improve artisanal practices, and I will discuss the many challenges they faced in this

4. Young 1769; Bergius 1772; Voltelen 1779; Stiprian-Luiscius and Bondt 1787–1788; Hermbstädt 1808.

5. See Macquer 1781–1783, 3:539–73.

6. Johnson 1803, 105–80; also Gren 1794, 2:403–10.

7. An illustrative summary of these discourses is given by Albrecht von Haller in his contribution to the *Encyclopédie* (Haller 1773). I thank Hubert Steinke for making this source available to me.

8. Parmentier and Déyeux 1800, 3.

9. On "experimental history," see Klein and Lefèvre 2007.

endeavor. The second part follows milk in quite a different context, studying its uses as food along with maternal nurture and the disorders it sometimes caused. In this context, I will address, in particular, the Enlightenment debate about wet-nursing. Part three studies different ways in which milk was commodified in the second half of the eighteenth century and looks at the enlightened societies that sought to improve the agricultural output of milk. This is followed in part four by an examination of some theoretical discourses on milk, especially in connection with the understanding of digestion. The final part summarizes and discusses issues of particular relevance to this volume, such as the mixed context in which milk was constituted as object of inquiry.

## Milk in the Laboratory

In the introduction to their book, Parmentier and Déyeux asked why chemists had not hitherto shown any interest in such a "charitable" fluid as milk. Nobody would dispute its utility as a food (and not only for the very young), as a medicament, and as a chemical used in different arts, including textile bleaching, meat preservation, and the distillation of liqueurs, brandy, or vinegar, but chemistry had nothing valuable to say about it. Of course, this was scarcely more than a rhetorical question, because the authors knew the extensive corpus of writings on milk all too well and occasionally referred to them. But, like Samuel Ferris, they found just one reference inspiring, a medical dissertation by the Scottish physician, obstetrician, and professor of midwifery Thomas Young (1726?–1783).[10] There is a simple reason why they judged other treatises less valuable, one which simultaneously explains the core concept of their investigations. Young had tried to systematically compare the milk of various animal species with human milk. Most of the other investigations apparently considered different sorts of milk without drawing such comparisons. When they explored the natural history of the animal economy or pursued medico-pharmaceutical interests, they used cow's milk, as the easiest to obtain; when, by contrast, they sought to understand the

10. See Young 1769. Young, elected professor of midwifery on 18 February 1756, was the founder of the first school of midwifery in Scotland, and "instead of confining his attention to the education of females in this necessary branch of medical practice . . . opened a class for the students at the University, and thus was the means of preventing it from being engrossed by a very ignorant and credulous set of practitioners" (Bower 1830, 6). A short biography is given by Hoolihan 1985.

pathology of the substance, they studied human milk, said to be the most sensitive and finicky kind.[11]

Systematic experimentation upon different forms of milk, then, posed the biggest challenge to the authors. And not only this. Most experimenters, according to Parmentier and Déyeux, even neglected the task of identifying the different conditions to which the fluid produced by just one animal was subjected.[12] Depending on the time of day, the duration of lactation, the age of the animal, the season, the constitution and temperament of the animal, and a host of other factors, the character of the milk could alter rapidly. Different life circumstances could materially affect the animal material. This was common knowledge and, not surprisingly, chemists expected the constituents of milk to vary in their relative quantities, not only between different animals, but even in the same animal. For the purposes of comparative analysis, the milk sample, therefore, had to be selected carefully. In other words, the first step by which a vital fluid could be fit into the environment of a chemical laboratory entailed classification, selection, reduction, and standardization, what Bruno Latour and Steve Woolgar once called "the creation of order out of disorder."[13]

But accuracy also involved attempts to improve methods of analysis. Chemists had hitherto utilized a spectrum of methods for examining materials in order to identify their organoleptic (color, form, smell, taste, etc.), physical (firmness, solubility, softness, liquidity, etc.), and chemical (behavior in acids and caustic solutions, fermentation and decomposition properties, etc.) characteristics. Operations such as distillation and dissolution stood at the heart of each analysis.[14] Nonetheless, some significant changes had taken place in the course of the eighteenth century. At the beginning of the century, only "strong" distillation methods had been available, whereas at the end of the century there were distillation procedures at low temperatures and

11. In Sweden, the well-known physician and botanist Peter Jonas Bergius (1730–1790), a pupil of Carl Linnaeus, had performed some experiments with reagents on the milk of a single wet nurse (Bergius 1772). Only a limited sample was also available to the Dutch physician Floris Jacobus Voltelen (1754–1795), professor at the University of Leyden after 1784 (Voltelen 1779). Abraham von Stiprian-Luiscius and Nicolas Bondt compared human and nonhuman milk at precisely the same time as Parmentier and Déyeux undertook their investigations; however, their publication was not mentioned by the latter (Stiprian-Luiscius and Bondt 1787–1788).

12. Parmentier and Déyeux had fifteen cows at their disposal in order to perform a comparative analysis on the milk of one species.

13. Latour and Woolgar 1979, chap. 6.

14. See Holmes 1971; Klein 2003a.

solvent extractions that avoided denaturing organic substances.[15] Parmentier and Déyeux also declared that the widespread use of fire in the study of composed bodies led only to the decomposition of organic substances and that this resulted in mistakes and contradictory perceptions. What kind of instruments did our laureates use in their investigations? In fact, except for employing an areometer, they stuck to methods commonly used by all chemists at work in the 1780s. Analysis by distillation (at low temperature) and solvent extraction remained the mode of practice for the study of animal matter. The greater accuracy expected from the areometer, which served to measure the density of the liquid, did not yield the expected results. After much trial and error to determine the density of milk, the authors gave up, losing interest in this particular instrument.[16] It was not so much the laboratory instrumentation as the substance under consideration that was worked on in an innovative way. This, however, meant nothing more nor less than bringing common knowledge into an academic format. There existed a widely shared body of empirical knowledge about the fluid and its various conditions, and chemists needed to challenge this popular knowledge in order to construct scientific facts. They did so by following a two-pronged strategy. First, scattered artisanal and everyday knowledge was to be gathered together in a general survey. Second, as many experiments as possible were to be carried out, rendering a complete picture of milk as a chemical object.

As to the first strategy, chemists took up, for example, the old assumption that the milk of different animals varied materially according to the species of origin.[17] As we shall see in more detail below, this issue was part of a much older debate in medical practice and physiological theory over the most suitable milk for nourishing babies and the sick. For the moment, I merely want to emphasize that the quality of food had long been understood as a function of easy digestion and conversion into blood. The sanguification of food was central to basic theories about human or animal physiology. Would the nutritious matter evaporate quickly? Would a food ultimately produce "good" blood? Such concerns also underlay evaluations of the nutritive and medicinal value of food: if a milk was dry and hot because it contained plenty of cheesy and sticky matter, like goat's milk, then it was harmful for people full

15. See Klein 2005b.

16. Parmentier and Déyeux 1800, 10. As Bernadette Bensaude-Vincent points out, Lavoisier had similar problems with the hydrometer and did not consider this tool to be a reliable instrument (Bensaude-Vincent 2000).

17. On evaluating the nutritious value of different animal foods in general, see Albala 2002, 66ff.

of spirit. Asses' milk, by contrast, was moist and watery, thus a very meager milk, which should not be given to an adynamic, convalescent patient. Human milk, by the same logic, was the best milk of all, because it was—a merely similar substance to the human flesh—qualitatively tempered, sweet, and easy to digest.[18] Countless bits of advice like these had been handed down since antiquity and could more or less be confirmed from a chemical point of view. The milk of women, asses, and mares exhibited the largest amount of aqueous matter (whey) and milk sugar, an observation associated with the fact that they coagulated more slowly than the milk of cows, goats, and sheep. Consequentially, the last three contained more cheesy matter than the three former.[19]

Yet much of this well-known information about the composition and quality of milk could not be confirmed or refuted through laboratory experiments. Perhaps for this reason, much effort was expended on describing the experiment being undertaken and the events taking place in it, along with a discussion of its results. In general, the books under consideration followed a similar pattern: Ferris started with a chapter on the "natural use of milk" (its relevance for feeding infants), what one can observe about its composition by the senses, and the artisanal use of milk (butter and cheese making). Then, under the heading "chemical analysis in general," he described distillation and other experiments carried out on various milks. Much of the account reads like a description of milk processing undertaken under laboratory conditions. The next chapter is dedicated to the medicinal virtues of milk, and the book ends with a comparison of blood and milk. Parmentier and Déyeux divided their book into three main chapters: "Observations on Milk from a Chemical Point of View," "Milk Observed in Relation to Pharmacology," and "Observations on Milk in Regard to Agriculture." They similarly offered extensive descriptions of their chemical analyses (including the knowledge obtained from the senses and from everyday manipulations of the material), before moving on to discuss medical questions (including the problem of feeding infants), ending with the relevance of milk for agriculture.

This concise and descriptive summary of contemporary knowledge about milk then was enriched with extensive descriptions of experimental procedures. According to its various properties and constitution, the milk of different animal species was distilled in variant states. Milk was analyzed when it was still warm and fresh, but also after it had changed in state, from liquid to semisolid (cream) or solid (cheese). The effect of temperature on milk coagu-

18. See Martin 1684, 33–37; Albala 2002.

19. See Ferris 1787, 36–38.

lation was measured, as well as the mode of action of well-known artisanal substances, such as rennet, or artificial chemical reagents, like lime water or caustic alkalis. Among the primary questions to be proven by experiment were: Does the fluid contain a specific matter responsible for fermentation? What kind of changes may be observed in the physical properties of milk during coagulation? How does the residue of different samples react with organic or inorganic acids? Which acids dissolve the cheesy matter? Is it possible to produce an alcoholic drink from cow's milk, as the Tartars are said to do with mare's milk?

## ATTEMPTS TO PRODUCE NEW USEFUL KNOWLEDGE

One example of chemists' extensive investigations demonstrates that late eighteenth-century milk analyses not only reflected the state of the art of chemistry, but also tried to innovate the traditional art of milk processing. What happened during the process by which butter was isolated from the fluid and transformed into a solid state? If one puts cream in an ampulla and shakes it for a period of time, Ferris asserted, one can observe the emission of air.[20] Is that a manifest sign of spontaneous fermentation? Is butter a product of fermentation? If this were the case, Thomas Young had already reflected, then butter formation could not take place in a vacuum.[21] Both Ferris and Parmentier and Déyeux repeated this experiment. In doing so, the latter authors supposed the influence of oxygen, hydrogen, and nitrogen, only recently recognized by Lavoisier as elemental components of organic substances, to play a role in the process of butter formation. They made two assumptions: either, they wrote, the combination of atmospheric air with the volatile matter in cream led to evaporation and consolidation of the milk's creamy parts, in which case it must be possible to produce butter by aerating rather than shaking the cream; or else the formation of butter was simply the consequence of the motion of matter, as assumed in natural philosophy. In this case, the oily particles of butter must already exist within the fluid. Attempts to prove these hypotheses by, for example, shaking the cream in a vacuum, did not support the claim that common air played a role in the formation of butter. The experimenters concluded that it was impossible to produce "butter without shaking the cream." Within "the cream, . . . the spirit which isolates the butter particles must have the ability to generate a

20. Ferris 1787, 51.
21. Young 1769, 15.

compound which without doubt is not very strong, since a mere blow or vibration can destroy it."[22]

Parmentier and Déyeux's experiments did not yield clear results that could be applied in the practice of butter making. Though the authors sought to produce new knowledge regarding the management of the unstable and volatile state of the fluid, they only could confirm what was already known from agrarian milk processing, just like their predecessors: "Animal milk is a matte white liquid, a color which derives from the mixture of three substances, butter, cheese and whey. These three substances are intimately connected in fresh milk. The whey is the only liquid part of milk; butter and cheese, which are mixed with the whey, both have a certain degree of thickness, and are not dissolvable in aqueous entity," Pierre-Joseph Macquer had written in 1766.[23] Twenty years later, there was nothing to add to this account. Because it was in the very nature of milk to separate of its own volition, the claim that the fluid is principally composed of an oily matter (butter), of curd (cheese), and of serum (whey) remained undisputed. Chemists, like farmers, physicians, cameralists, believed that the firm components of milk (whey, butter, and cheese) were already contained and loosely connected in the liquid part.[24] Each of these three compounds should be a substance of its own, only loosely cohering in the fluid, and it was by no means clear how this functioned.[25]

Since at least the seventeenth century, doctors, apothecaries, and farmers had further known that milk contained an essential salt of opaque white color, called milk sugar and described as a crystal with "an insipid sweetness and somewhat earthy flavour."[26] This property induced chemists to suspect that, like other sugars, milk sugar would produce acids, something experimentally confirmed in 1780 by Carl Wilhelm Scheele (1742–1786).[27] Questions as to the intrinsic transformations of milk had engaged the attention

22. Parmentier and Déyeux 1800, 32–34. Samuel Ferris, too, came to the conclusion that air did not play any role in the process of butter formation. See Ferris 1787, 56.

23. Macquer 1781–1783, 539.

24. A similar definition of the components of milk can be found in the cameralistic work of Darjes 1756, 182. Krünitz, too, wrote that milk contains three parts that were "deeply combined within the fresh milk" (1803, 341). Like many other details in contemporary knowledge, this substantial concept can be traced back to ancient authors. Galen had already taught that milk contained two "humidities," whey and a fatty juice, named butter. A compact and earthy body was added as a third part (Galenos 1952, 24–27).

25. Physicians therefore discussed the intrinsic powers of the bodily fluids, see Rosner 1756.

26. Johnson 1803, 113. In 1694, the Venetian physician Ludovico Testo dedicated a whole book to the analysis and pharmaceutical use of milk sugar that was said to be the standard reference (see Lichtenstein 1772; Fleischmann 1910, 1–19).

27. See Benninga 1990, 7.

of scientists long before the 1780s. At the forefront of their inquiries was the problem of what caused acetous fermentation and coagulation. That fresh milk could suddenly sour in response to changes in the atmosphere, such as during stormy weather, was a mundane experience. Cheese making had also proven that many vegetable and animal substances—the gastric tissues of various animals and the plants known as cleavers—were able to separate milk.[28] In Europe, the matter generally employed was either the second stomach of the calf or its contents, known as rennet.[29] But what caused these spontaneous changes, and how did the curd, this white, jellylike and somehow elastic substance, solidify? As far back as 1669, the Paris Academy of Sciences had awarded a prize on the topic of milk coagulation.[30] Robert Boyle, like many other natural philosophers, argued that some sort of "coagulate spirit" (a sort of *spiritus rector*) must be the motor of the whole process and found it "less like that of the Cicatricula of an Egg, or the more Seminal part of the Spawn of the frogg, when they fashion the liquid Yolk or the soft Gelly into Bones."[31] But if this was a fundamental principle of all natural things, what was the general manifestation of this fermenting spirit, its material basis?

The debate about the cause of coagulation of milk was stimulated by Carl Wilhelm Scheele's discovery of lactic acid in sour milk in 1780; chemists then considered milk sugar a component responsible for the slow acidification of milk. Thanks to these investigations, methods were developed to produce fairly large quantities of lactic acid by fermentation from ordinary sugar as well as from milk sugar. But the problem of what caused fermentation remained unresolved.[32]

## Dangerous Milk and Maternal Nurture

The milk of the natural world was neither pure nor uncorrupt. Rather, it was both vulnerable and unstable. In the popular imagination, milk existed as a

28. Boerhaave had undertaken several operations in his laboratory with the aim of producing sour and solidified milk from fresh milk, based on the assumption that the phenomenon must be the result of an acidic or acetous fermentation (Boerhaave 1732, 2:299–301). On Boerhaave's fermentation theory in general, see Fruton 1999, 123–24.

29. Even artichoke florets could do the job of coagulating the milk (Young 1769).

30. See Bachelard 1987, 112.

31. Hunter and Davis 2000, 372. Boyle discussed the issue in a paper concerning the generation of minerals, where he argued that ferments and fermentation not only explained a variety of physiological processes within living beings, but tended to be a fundamental principle in the generation of all natural things. For the history of this theory, see Clericuzio 2003.

32. See Benninga 1990, 1–30.

food with many virtues; but in practice it could be dangerous, and noxious too. After milk left the body, it underwent a natural life cycle in which it aged, quickly soured, and hardened. It might contain a life spirit while flowing through the body, but once removed it lost its inherent heat, becoming cold and rapidly corrupting. Thus, it seemed to be alive, a substance with a specific identity, and people were extremely cautious about the quality of milk. Fresh milk, in particular, was not in practice an ideal food, for without processing, it "died" too fast.

Hence, there were many reasons why mothers should take care regarding their infants' food, doctors regarding their dietetic advice, and farmers regarding their milk-giving animals—and all of these problems had to be debated by chemists. The dangers arose not only after milking, but existed even within the female body. First of all, flaws of temperament and virtue could be transmitted through milk. Even though natural philosophers writing after Descartes (1596–1650) argued for a disconnection of body and mind, during the eighteenth century it was nevertheless common to assume a strong—and not just metaphorical—connection between them. Physical health and mental health were interwoven and could not be separated. Just as passions and the imagination could cause deformities in children *in utero,* they could affect breast-feeding.[33] Temperament, character, and strong emotions or passions could corrupt the bodily matter, and even animals, though lacking a rational soul, transmitted their characters through their flesh and milk.[34] A cow whose calf is taken away sometimes will stop giving milk, said a farmers' rule.

Organs could become irritated too. Thus, the breast and udder were sensitive as well. Parmentier and Déyeux referred to the report of the physician Théophile Bordeu (1722–1776) that the milk of a nursing mother had coagulated within her breast after she had seen her baby fall to the ground. Fear, they argued, constipates the breast, just like any nervous change.[35] While studying postpartum illnesses, Bordeu described the process as soured milk that had curdled after the blockage, so that the woman possessed a kind of "cheese under the skin." The buttery part of the milk had become rancid and the greasy nature of the casein made the malignancy virtually insoluble. This fermented milk could produce several ill effects: "if it reaches the head or the nervous centre; if it gets to the chest; if it floods the womb

33. Indeed, these issues were highly contested. See Blondel 1727; Rhades 1786.

34. The French ethnologist Françoise Loux noted that, according to rural folk medicine, many people held that goat's milk made children agile and sanguine, but also nervous. Sometimes they even became hysterical, shameless, and susceptible to sexual love (Loux 1980, 149). Ken Albala (2002, 149–50) found similar advice in Renaissance dietary books.

35. Parmentier and Déyeux 1800, 151.

(where nature loves to carry it), a thousand phenomena occur, all dependent on that one cause: *la cachexie laiteuse*."[36] In other words, excessive acidity could sour the humors, causing particles to clot. And if the lactating function was blocked or unbalanced, nature would respond with numerous illnesses: postpartum fevers, lacteal tumors, headaches, and even mental disorders like melancholy.

That milk could cause serious disorders within the woman's metabolism led to another essential consideration, namely, the external factors turning milk into a bad, even poisoning, substance. Of greatest importance was the food eaten by a nursing woman or given to a lactating animal. Only a well-nourished body could give healthy milk in sufficient quantity. In consequence, extensive dietetic advice was available for nursing women.[37] Likewise, the soil and pasture of the animals had an almost immediate impact on human health. According to an old farmers' rule that "cows give milk through their mouths!" examining the animal's fodder plants was of great importance.[38] But the quality of soil and pasture was also a question of taste and aesthetics, demonstrated by the appearance of milk products. Parmentier and Déyeux noted that people found yellow butter to be the best and believed that the "more juicy, spicy and aromatic the pasture, the more yellow the butter will be."[39] Ferris quoted Boyle's claim that many farmers were able to determine from the taste of a donkey's milk whether the animal had been properly groomed or not.[40] The taste, color, and odor of milk all depended on the fodder, season, location, and—not to be forgotten—the constitution of the different species.[41] Ferris suggested that, given a choice, the milk of a donkey or mare would be preferred, because they lacked the bitter and sharp taste of ewe's and goat's milk.[42] Parmentier and Déyeux concluded that every abrupt change in nutrition or habit would cause "a kind of revolution" within the animal and decrease the quantity and quality of milk. So, if there was anything to learn from the dairy business, it was that nursing women should watch what they eat.[43] Like all other problems, this advice was old hat, barely worth mentioning in treatises that promised to address the milk ques-

36. Quoted in Sherwood 1993, 34. Bordeu blamed the resulting illnesses on *la cachexie laiteuse* or what might loosely be translated as "clotted milk."

37. See Jörg 1812.

38. Rüger 1851, 7; Albala 2002, 123, 124.

39. Parmentier and Déyeux 1800, 52.

40. Ferris 1787, 16.

41. More details on this can be found in Orland 2004.

42. Ferris 1787, 170.

43. Parmentier and Déyeux 1800, 132.

tion from an innovative chemical viewpoint. Yet chemists could not ignore the guiding principle of all consumption patterns: milk behaved as a natural bond between mother and child, transmitting numerous properties from one body to another.

Milk, moreover, circulated within every body (not only the female), and as such represented the corporeal identity of a body. Over the centuries, humoral explanations of nutrition and dietetic advice had not changed radically. Every body consisted of solid and fluid parts, manufactured during the continual process of generation, growth, and restorative nourishment. Movement and circulation were the key issues of physiology; only explanations of the driving forces behind the circulatory system have changed over time. In most medical treatises written by eighteenth-century physicians, it was taken for granted that the machine of the body behaved like a set of fluid hydraulic devices.[44] In the words of Boerhaave: "The bodies of living animals are continually wasting and repairing in all their parts. . . . Animals, therefore, must necessarily be composed of what they take in as aliment; which by their vital powers, is converted into their own substance."[45] Just like the color of a face or the temperature of the body, the circulating fluids could be read as significant signs of the integrity and disorders of a human being, of his or her humoral balance, habits, and spiritual, moral, and physical well-being. The constitution of the whole body was the very foundation of individual well-being; a healthy body (and fluid) needed stimulus and moderation, temperance and balance.

But why did chemists literally leave the chemical laboratory and confuse their own writings with discussions of a diffuse, highly complex, and ancient knowledge on milk? Why did they find medical or agricultural knowledge attractive, and how did they integrate it into their chemical work? Earlier I argued that they had the goal of utility and completeness in mind, with regard both to chemical operations and to knowledge of society. Every kind of "expert" knowledge was to be brought together in a general survey and further explored with respect to its utility. To this I would add that they also sought to be up-to-date and, even more importantly, to be progressive.

44. After William Harvey had demonstrated in *De Motu Cordis* that the heart functioned as a pump, the study of the circulation and perpetual motion of fluids received new impetus, in medicine as in economics. Harvey's model favored existing mechanistic views of the body much more than chemical accounts (see Porter 2003, 51).

45. Boerhaave 1727, 179. Quotation from Shaw's unauthorized version of Boerhaave's chemical textbook. The original text is Boerhaave 1732, 2:294–304.

## THE WET-NURSING QUESTION

The second reason why chemists debated the mechanics of milk physiology was rather an ideological one. It was the Enlightenment that radicalized philosophical, medical, and chemical investigations into milk and claims about how to handle it in practice. Maternal nurture was part of a political agenda for infant survival.[46] For the first time, philosophers, naturalists, and physicians turned their attention to the child as an object of scientific study, producing anatomical, physiological, and psychological descriptions of the child, elaborating debates on child-rearing customs and generating a huge amount of hygienic advice and prescription.[47] Weaker than any animal, unable to move or make use of organs and senses, the child was "an image of misery and pain," according to Georges-Louis Leclerc de Buffon (1707–1788), writing in 1749 in his *Natural History of Man*. Even more famous were the claims of Jean-Jacques Rousseau, who based his influential account of child-rearing in *Émile* (1762) on Buffon's work. Nature, he informed the enlightened public, produces strong and robust children "who are well constituted and makes all the rest perish."[48] Far from being natural, children's weakness is an instance of society's corrupting effect on human nature, complained the man who himself had placed his own five children in foundling homes. To return to nature and its laws was the surest way to end corruption and regenerate society. *Émile* began this policy of regeneration by replacing the unnatural practice of wet-nursing with the figure of the maternal nurse, the guarantor of the family and of an incorruptible system of signs for the republic of parents.[49] "Let mothers deign to nurse their children," Rousseau preached, "morals will reform themselves, nature's sentiments will be awakened in every heart, the state will be repopulated."[50]

This newfound duty of mothering and breast-feeding, which had already figured in the relevant articles of the *Encyclopédie* in 1751, fostered an unprecedented campaign against wet-nursing.[51] There can be no doubt that the old custom of employing a wet nurse for infant nutrition had increased tremen-

46. See Fildes 1988; Jacobus 1995, chap. 4; Schiebinger 1993, chap. 2; Lastinger 1996; Golden 1996.

47. See Benzaquén 2004.

48. Quoted in Benzaquén 2004, 37.

49. As argued by Jacobus 1995, 209.

50. Quoted in Schiebinger 1993, 70.

51. See Lastinger 1996, 605.

dously in the cities of the seventeenth and eighteenth centuries. Especially in France, wet-nursing was both a social institution and a state-regulated industry. Around 1780, fewer than one thousand babies of the twenty-one thousand newborns in Paris were fed by their mothers; a further one thousand had a wet nurse in their parental home (from a population of approximately eight hundred to nine hundred thousand).[52] All the other newborns were given to wet nurses outside the city, in the peripheral areas and up to 200 km away in Normandy, Picardy, or Burgundy. A widespread feature of urban life, it was also popular in smaller towns and common among craftspeople, traders, bourgeois, physicians, and even laborers. The silk workers of Lyons were said to have given away all of their babies. Only the poorest refrained from this practice.

What is important here is that the well-known dangers of milk became radicalized within these campaigns. The threat of dangerous milk inspired both the popular and the medical imagination. Wet-nursing was named a "strange abuse," not only perverting nature but "corrupting body and spirit with milk defined as both alien and illegitimate."[53] The milk of wet nurses was the embodiment of a "bastard" milk. According to the advocates of nature, wet nurses were often drunk, malnourished, and without good character, and their bad moral conduct was transmitted in their milk. It was not morally acceptable to sell one's own milk. Similarly, it was no longer acceptable to engage a wet nurse. In fact, the campaign had a reciprocal effect on mother's milk. Whatever reason a woman might have had, if she did not nurse her own child, she ran the risk of being termed "unnatural," because she was rejecting her newborn. Doctors' recriminations involved direct warnings concerning the health of a woman's body: "Milk which has been held back becomes mixed with blood, courses through the body in an attempt to be secreted. But as the milk has a very great analogy with the lymph, it is soon absorbed by this humor and consequently all lymphatic excretions of the human body are suffused with milk, . . . obstructing normal receptacles, producing glandular tumors and causing swelling in feet, hands, face; pain occurs similar to rheumatism in neck and other membranes of the body . . . there may be pressure on the brain, . . . [even] coughs and difficulties in breathing."[54] Giving advice about suitable wet nurses, postpartum illnesses, or breast-feeding became a

52. Gélis, Laget, and Morel 1980, 164. Similar data are given by Sussman 1982, 22, who also describes the success of the anti–wet-nursing campaign (110–11). Around 1800, about half of all Parisian babies were nursed by their own mothers, although the state-regulated business still existed right through the nineteenth century.

53. Jacobus 1995, 214.

54. Quoted in Sherwood 1993, 32.

new business for male physicians, who increasingly took over the business of midwives. Now, men were obliged to test the milk of women (mainly performing their analyses on a drop of milk on a fingernail) and to treat the lactation process, mostly by giving dietary advice.[55]

Chemists' work on milk also needs to be read against this backdrop. Female problems of breast-feeding were highly politicized, and methods drawing on traditional beliefs were already on their way to becoming scientific. As members of the intellectual elite, the authors of the milk studies could not fail to be aware of the debate among the enlightened public about the benefits and dangers of the widespread practice of wet-nursing. A considerable literature by philosophers, political scientists, and academic physicians raised the "good" mother—her care, love, and education of children—to a national element of absolutist population policy. In the words of one exponent of the German cameralist police science (*medizinische Polizeywissenschaft*): To give birth to "good, happy, industrious and healthy people" should be the main task of a true woman, while the state must offer appropriate support and information with the help of doctors.[56] Faith in progress was buttressed by the findings of what a new science could bring to the understanding of lactation.

Parmentier and Déyeux, as well as Samuel Ferris, praised mother's milk as "without any doubt the best nourishment" for all young children.[57] Ferris devoted sixteen pages to a discussion of the role of chemical analysis in producing evidence of the beneficence of nature. Insisting on the importance of breast-feeding, he offered advice on the choice of a wet nurse and also discussed the medical problems faced by mothers intending not to nurse. On the basis of his deep concern about the high rate of infant mortality, Ferris demanded that all "foolish" women should take over nursing as "the tenderly business of nature."[58]

But how could chemical testing contribute to the question? Not surprisingly, the "maternalization" of breast-feeding and the polemic against wet-nursing motivated chemists to compare women's and animal milk. Chemical analysis was expected to provide political support for the value of mother's milk. To this end, Ferris took the same amount of milk from a woman, a

55. On the professionalization of male obstetricians, see Labouvie 1998.

56. Hebenstreit 1791, 96. During the Revolutionary years, chests and breast-feeding women were in vogue as media of national representation. Both the mother-child icon and images of the breast were appropriated as symbols of the unity of nation and nature, and breast-feeding women became a favored part of Revolutionary spectacles. See Jacobus 1995; Baxmann 1989, 77–84.

57. Parmentier and Déyeux 1800, preface.

58. Ferris 1787, 10.

donkey, a mare, a cow, a goat, and a ewe. The pints of milk were put in pottery containers and stored in a single room at 65 degrees Fahrenheit. The time taken for the cream to precipitate and the milk to turn sour was then measured for each sample. However, these and other experiments using solvents (vitriolic acid, salt acid, and so on) produced ambiguous results. The cream of the samples precipitated at different rates, with goat's, ewe's, and cow's milk reacting faster. This observation conformed with the results of heating the different sorts of milk. Again, human, ass's and mare's milk reacted much later and did not coagulate below a temperature of 100 degrees Fahrenheit. The only interesting observed effect was an apparent similarity between these milk sorts.

Parmentier and Déyeux, who undertook similar experiments, were likewise disappointed by the results. Because successive tests produced slightly different results, they could not give a chemically proven answer to the question of why babies should receive only mother's milk. Under these circumstances, both books went back to repeating what was dictated by the *Zeitgeist.* As a physician would have done, the authors gave instructions such as: pregnant women should avoid absinthe tincture and asparagus at the end of their pregnancy, because their milk could become bitter. Conversely, they noted several "milk-stimulating" instruments, herbs, foods, and remedies. They did not omit prescriptions handed down from ancient authors and concluded that "those substances alone have a milk-producing power which possess an affluence of nutritional fluids, from which the vigors of digestion can extract those parts needed by the mammary organs to collect all necessary matter to produce a milk."[59]

## Milk as Commodity

Enlightened societies were practically obsessed with milk. Trying to reach a better understanding of the nature of milk was of no little interest to agronomists, too. During the 1770s and 1780s, in both Britain and France, there were calls for sweeping agricultural reforms. Because the situation differed between the two countries, and because Ferris, as a Scottish physician, was less interested in agriculture, in what follows I shall focus my attention on the French case.

In France, expertise possessed proportionally greater importance than in Britain, where invention for private profit and entrepreneurialism played a

59. Parmentier and Déyeux 1800, 150.

major role, as Charles C. Gillispie has put it.[60] He described developments in France at the end of the old regime as a kind of rationalization, an initiative to educate and learn in agriculture (reserving the revolutionary mode for political and social change). Without going into detail, in Gillespie's work it becomes clear that public administrators expected new expertise from science, while for scientists the political interest offered a good chance to launch themselves into nutritional politics and public health reform. Of the trades connected with science, pharmacy was by far the most prosperous. According to Gillespie, "its thriving condition helps to explain why chemistry was the most active and populous sector in science just prior to the Revolution."[61] This statement can be confirmed from the biographies of Parmentier and Déyeux. Both of them were pharmacists and, like other scientists of the time, polymaths who combined scientific interests with the economic and political trends of the day.

## HYBRID EXPERTS

Just like the assayers addressed in Christoph Bartels's chapter or the German apothecary-chemists studied in Klein's contribution, Parmentier and Déyeux were hybrid experts who defy neat classifications as either learned men (or "scientists") or artisans. Nicolas Déyeux had run his own pharmacy in Paris for more than twenty years when, in 1795, at the age of 50, he was appointed professor of *chimie médicale* at the École de médecine. In 1797 he was elected to the chemistry section of the *Institut National des Sciences et Arts*. Under Napoleon I, he became a most influential pharmacist and, like Parmentier, a member of the Paris health council. As an enthusiastic botanist, he studied in the research collections of the Jardin du Roi (and presumably attended the courses offered there in chemistry), writing mainly about vegetables.[62] However, he lectured—with Fourcroy among others—and wrote extensively on all branches of chemistry. He worked in collaboration with Parmentier over two decades, carrying out studies not only on milk but also on blood. Together they coedited the *Bibliothèque Physico-économique*, a journal dedicated to announcing all inventions in agriculture, trade, and industry.[63]

60. See Gillispie 1980.

61. Gillispie 1980, 87. See also Klein's chapter in this volume.

62. See Löw 1979, 183, 188; Smeaton 1962, 70; Corlieu 1896, 346. Unlike the case for Parmentier, there is no biography of Déyeux.

63. Bibliothèque physico-économique 1782–1831.

Antoine-Augustin Parmentier had made a name for himself as a pharmacist with strong interests in nutritional chemistry; he has gone down in French history as a reformer in nutritional politics and is known as the "Father of the Potato."[64] As a military apothecary in the Seven Years' War (1756–1763), Parmentier was imprisoned in Prussia, where he was fed almost exclusively with potatoes. Back in France and working as a pharmacist at the veterans' hospital, the Invalides, he followed courses at the Jardin du Roi, for example the chemistry course of Guillaume-François Rouelle, and started to promote his food of distress to the French people, largely unfamiliar with the tuber. When the academy of Besançon, alarmed by rising food prices, offered a prize for a foodstuff that could replace corn in the event of famine, Parmentier's potato won (1773). He gained the patronage of Louis XVI, who assigned fifty acres of land to Parmentier in order to cultivate potatoes. Later, Parmentier was granted a permit to found a bakery school, which opened on 8 June 1780, with Parmentier as one of its professors. At the laboratory he established there, Parmentier conducted a wide range of experiments on wine, bread, mineral waters, and other foods. He investigated the gross chemical properties of gluten and starch and in his later years extracted sugar from grapes. Like most French agronomists, he belonged to the Society of Agriculture, but he was also a member of the French army's Pharmaceutical Service and, as an expert in public health questions, took part in the vaccination program established by Napoleon.

From this brief outline, it is clear that neither Déyeux nor Parmentier were scientists in a modern sense. Without any formal academic preparation, they became chemical experts through on-the-job-training and by attending lectures at the Jardin du Roi, where reforms in chemistry were being contemplated. From the point of view of career patterns, they also differed from the mainstream of agronomists. Although centering their research on agricultural phenomena, they had no personal profit motives for performing experiments. Many agricultural reformers had been landowners who, in keeping with the physiocrats' view that agriculture was the principal source of wealth and ought to be the main productive activity of the nation, rebuilt their own properties in order to set up model farms, where one could perform experiments, introduce new agricultural techniques, and encourage peasants to reform their methods. In contrast to these agronomists in the field, Parmentier and Déyeux (like many other Parisian agronomists) were townspeople, experimenting on agricultural questions mainly in their

64. In French cuisine nowadays, *Parmentier* refers to dishes made with potatoes. On Parmentier, see Kahane 1978; Balland 1902.

laboratory and rarely in fields. In other words, they sorted things out from an urban market perspective, considering questions of food supply and quality.

Public health reform, to them, required firsthand observation and experimentation and, at the same time, an understanding of a problem's history and recent practice. Both the technicalities and the application of knowledge were interesting to them. In the absence of personal economic interests, they dedicated their research to the public in general. As Parmentier put it: "When you want to be of service to your fellow-beings, it is not enough simply to tell them once and for all what you have found, what you have done, and what they need to do. You must never tire of repeating it in every way possible and in every form—except that of authority. It is only by means of popularizing science that it can be made possible."[65] Such was the experience Parmentier gained from the case of the potato. To reach a wide audience by popularizing scientific results remained a central goal and may explain not only the discursive style of Parmentier's writing, but also his success as a public expert. In any event, Parmentier and Déyeux were among the early instigators of the French public health movement, which gained momentum within the sociopolitical context of the Bourbon Restoration and July Monarchy.[66]

## AGRICULTURAL INNOVATION

Parmentier's and Déyeux's interest in milk analysis was thus motivated by the use of milk as a raw material of agricultural practice and—in the form of butter and cheese—as an expanding commodity. For centuries, milk production had been linked to regional living conditions. It was part of local farming systems, depending on the perception of the landscape and its potentialities. There existed no market for animal fodder; available fodder was, rather, dependent on the way the land was being used. For a region to produce a surplus of dairy products, it was first necessary to solve the omnipresent problem of finding adequate fodder.[67] Around 1760, the new physiocratic school of economic thought deemed land to be the key to economic progress. Its proponents demanded that agriculture be modernized to this end. Im-

65. Parmentier 1789, 9–10. Translation from Gillispie 1980, 374.

66. On the French public health movement, see La Berge 1992.

67. Besides the coastal countries of Holland, Denmark, Ireland, Sweden, and so on, dairy farming was particularly important and well developed in the northern Alpine countries, beginning in Switzerland (see Orland 2004). In France, the Alps, the Massif Central, the Vosges, the Pyrenees, and Corsica are traditional cheese-making regions. On the industrialization of milk in France, see Vatin 1990.

proving farmers were to clear their abandoned lands, drain marshland, plant fodder crops like clover and sainfoin, and introduce summer stall feeding to facilitate the collection of cattle dung and liquid manure. The promotion of animal farming and especially of milk production in new regions were unintended side effects of these new cultivation methods, in that the cultivation of fodder and the intensive use of meadows and fields increased the number of cows and the milk yield. During the late eighteenth century, as a result of the physiocratic program of land reform, this "old" system of dairy farming slowly broke down. Throughout Europe, big estates and landowners tried to implement this situated knowledge of dairy farming, to shift from arable farming to permanent meadows and pastures and to increase milk production. The trade in milk cows was intensified, and cattle breeding became an issue.[68]

Parmentier and Déyeux discussed these economic trends, noting that all French provinces engaged in animal farming relied heavily on the dairy trade, which was evidently considered more lucrative than arable farming. The chemists therefore attempted to establish quality standards and to find out why some regions developed better craftsmanship in cheese and butter production than others. They did so, for example, by assessing milking utensils, including the equipment suited to a farm dairy, and animal fodder. Both books listed well-known herbs said to have a bad influence on cows' milk. Samuel Ferris mentioned acrid clover, wild garlic, mustard, and lovage and furthermore explained that madder colors milk red. In nearly the same way as household manuals and receipt books had been doing for centuries, he discussed the influence of the seasons, locations, and production methods to the quality of meadows. Descriptions of plants and their uses were followed by experiments proving their different effects on the animal's body (astringent, resolving, digestive, and so on).

Chemistry in the service of agronomy was, so to speak, a form of applied botany and a compendium of agricultural knowledge. As self-appointed agronomists, chemists searched for tools of technical innovation. They monitored the market, discussed the state of the art of dairy knowledge, and identified the "best ways of doing things." As one of Parmentier's biographers noted, the dairy and cheese industry was favorably impressed by these studies.[69]

68. See Orland 2003.
69. Balland 1902, 60.

## MILK AS *MATERIA MEDICA*

Finally, chemical expertise was affected by the practice of using milk as a medicament. In the eighteenth century, the isolation of substances from plants and animal organs and tissues to obtain medicinally active components was one of the principal motivations of most pharmaceutical inquiries.[70] I have already mentioned that many remedies for bringing on and stopping lactation existed. But milk meant even more as an object of pharmaceutical knowledge and practice. Like other human and animal substances, it was intensively used as a "materia medica". Milk was regarded as a powerful chemical agent. It counteracted "even the strongest and corrosive" poisons and therefore should be seen as the "best antitoxin of nature," wrote Parmentier and Déyeux, already known and used by Hippocrates.[71] Due to its mucilaginous and oily texture, it blunted the harmful spikes of noxious substances and softened acerbity. For this reason, miners drank it in order to prevent diseases of the lungs and other occupational ailments.[72]

Its main virtue, as described by the authors, was its demulcent and emollient character, which, together with a smooth sweetness, could balance all the humors. As a demulcent, milk was said to be exceedingly valuable in combating irritation of the pulmonary and digestive organs. Its emollient virtue could be used to cure any kind of inflammatory disease, because inflammation was regarded as the result of an unnatural change in the consistency of blood. If there was, for any reason, a surplus of blood in any part of the body, a degree of stagnation would result and the blood would pool, increasing in viscosity or thickening. In order to reestablish proper circulation, the blood had to be diluted. Milk could thus cure numerous current diseases and could also calm the nervous, strengthen the convalescent, and prevent epidemics. Moreover, milk was a perfect vehicle for administering pharmaceutical substances and was an important part of herbal medicine for that reason. A mixture of milk with dried red roses, cinnamon, pomegranate, and bark might be the best remedy against dysentery, recommended Ferris. To cure venereal diseases, one should administer "a corrosive sublimate of mercury" together with a sip of milk. Feeding animals with curative herbs might also prove useful; in addition, the milk of some animals, especially goats, was a

70. See Holmes 1989b, 68–83; Klein and Lefèvre 2007, 195–253.

71. Parmentier and Déyeux 1800, 181.

72. Ferris 1787, 136.

natural purgative. Lastly, the color, odor, and taste of milk could be modified by various plants to make it more effective as a medicine.[73]

Remedies might also be prepared with butter, buttermilk, soured milk, milk sugar, or whey. Ferris reported a market-woman who praised her butter as "arcanum." For all the world to see, she ingested poisonous substances, then vomited them after eating a portion of butter. Besides such quackeries, physicians told of numerous cases in which butter, used fresh or in poultices, had helped as an antidote or in the treatment of injuries. And Ferris presented a list of the various milk extracts apothecaries had produced as remedies since the early eighteenth century. Friedrich Hoffmann (1660–1742), a professor of medicine at Halle, had distilled different kinds of milk using a complex procedure and made a milk extract he sold under the name of "Franchipane." Diluted with a little water and slowly sipped, it was meant to combat diverse stomach and intestinal complaints. His published prescription was used by many apothecaries in Europe to prepare *Hoffmannsche Molken-Arzney*.[74] As a milk cure and milk diet, this remedy had been very popular, Parmentier and Déyeux reported, though because of the cost of its production, it had gone out of fashion by the end of the century.[75]

In medical usage, the most popular derivative of milk was whey, a by-product of cheese production. Although farmers produced it in large quantities and sufficient quality, whey was not utilized for human nutrition. On the farm, it might be fed to pigs, while in medicine whey was given as a melting and sweetening substance for various intestinal remedies and as a solvent, emollient, and refreshment. In other words, if freshly milked whole milk was a restorative substance, whey was the more regenerative part of milk. Jean-Jacques Rousseau loved to bathe in whey and drank it in plenty to cure some of his many indispositions.[76]

Whole-milk diets did not always work, and therefore whey was the favored substance for cures. Many medical writers on the subject of the milk diet called attention to the fact that their patients were constipated on the milk cure but tolerated whey. Chemists could not say why this happened, but from the mechanics of circulation, it was easily to explain.[77] Some people were not able to digest the hard and rough cheesy matter, which clogged the

73. Because adults did not generally drink fresh milk before the late nineteenth century, many of them disliked using it as a medicament. On milk consumption practices in premodern Europe, see Lysaght 1994.

74. See Hentschel and Hoffmann 1725.

75. Parmentier and Déyeux 1800, 19–20.

76. Damrosch 2005, 117.

77. See Ferris 1787, 129–33.

body's passages and hence caused constipation. In the words of the French physician Anne-Charles Lorry, cofounder and later president of the Société Royale de Medecine in Paris, the density of nutrient matter was an obstacle to digestion, but the nondigestible was always more nourishing, because nourishing matter had to be compact.[78] For the formation of flesh, it was necessary to receive sufficient hard dry matter, but for good circulation and proper refinement of the blood and other fluids, solvents were needed.

From a medical point of view, the milk of women was supposed to be the best whole milk, because it offered both nutritional value and softness and sweetness. It was followed by ass's and mare's milk. Whey, rather, was the most fluid and solvent part of *every* milk. As such, of course, it was of no outstanding nutritional value; its function was to promote digestion. It made the nourishing parts of milk fluid, so that they could be digested.[79] In search of whey's "true elements," Friedrich Hoffmann had described it as a "delicate mucilaginous matter." Whey remained the *aqua vitae* of milk and was used as a refreshment or, according to its emollient virtue, as an anti-inflammatory substance. Only after milk sugar (called lactose today) was discovered by accident during cheese production was the nutritional value of whey reassessed. Milk sugar, a fairly soluble white substance with a sweet-sour taste, also described as a salt and soon found to be the only fire-resistant part of milk, began a career as the "quintessence" of milk.[80]

In some respects, however, chemists did not believe in the medical effects of milk, and they questioned milk medication. Their assessment of the colostrum, postpartum milk, for instance, challenged the prevailing view among physicians. This position, which can be traced back at least to the times of Galen and Soranus, took colostrum to be a poison, milk which was not yet properly prepared, and proclaimed that newborns should not drink their first milk until the second or third day after birth.[81] The opposing view, espoused by Parmentier and Déyeux as well as Ferris, was that nature had evolved this premilk in order to purge the meconium. Why would nature make the colostrum fattier if not in order to extirpate the nonvi-

78. Lorry 1785, 35. Friedrich Hoffmann wrote in his popular dietetic guidebook: "Only the very best food should be eaten, which is the sort our nature can dissolve inside the stomach and which can move rapidly through our body" (1715, 121).

79. See Macquer 1781–1783, 567.

80. According to medieval and early modern alchemy, the true ethereal nature of a substance and the distinctive principles of vitality could be extracted by means of distillation. These substrates of the alive some called "Médecine par excellence," others "Elixir" or "the fifth element." On the development of these concepts, see Fruton 1999, 118–21.

81. Garnsey 1999, 106–7.

able tissues that newborns carried in their bowels? Chemical operations that showed that colostrum was about three times as fatty as a mature milk underlined this opinion. Parmentier and Déyeux noted that the fatty portion of milk was reduced during the first days of lactation, giving way to the growth-producing, substance-forming cheese material.[82] But beyond this, it was difficult to find any chemical operation that evaluated the medical effects of milk and its several components. Authors could not really criticize medical knowledge about the curative forces of milk. Thus, there was hardly more to do than cross-reference medical advice and instructions. In 1803 Krünitz summed up their attempts by saying that the belief in the impact of milk extracts had volatilized because no one could measure or prove its effectiveness.[83]

## Animalization: A Chemical Theory of Digestion

Questioning the nutritive or medical value of milk forced chemists to bring milk into line with theories on the digestion and assimilation of matter. In addition, it was necessary to take a position on the still debated question of where, when, and how an animal's body produced milk. Of course, the female body would generally give milk after giving birth. But two contradictory theories of milk formation within the body existed. Since the time of Aristotle, it had been said that milk was the surplus blood no longer needed after delivery, transported from the uterus to the breast. From this point of view, milk was nothing other than white blood. Growing anatomical knowledge, and in particular the rise of Cartesian physiology, led many physicians to reject the idea that milk was menstrual blood which had whitened en route from the uterus to the mammary glands. Based on Cartesian ideas of matter and motion, in the mechanistic view of the body, milk was a modification of chyle, the fluid produced during the process of digestion and which, in its final stage, was transformed into blood. Obviously, there still existed an affinity between blood and milk, but now milk was a derivative of chyle and no longer of menstrual blood.

But what was going on during digestion? And how could this dynamic process be described in chemical terms? Since the appointment of Franciscus Sylvius (Franz de le Boë, 1614–1672) as professor of medicine at Leiden in 1658, it had become common among chemically oriented physiologists to explain the perpetual motion of fluids during digestion through the interaction

82. See Parmentier and Déyeux 1800, 176.

83. Krünitz 1803, 450.

of acids and alkalis.[84] Sylvius, who considered this interaction one of the fundamental principles of nature, had classified humors according to their acidity: saliva was neutral, pancreatic juice slightly acid, bile alkaline, and so on. Thus, the chemical act of digestion was revealed as a balancing of acid and alkaline materials incorporated through nutrition. In a healthy body these would be balanced and thus neutralize one another. Foods could therefore also be classified according to their acid or alkaline properties. Knowing that the distillation of all plant substances yielded some kinds of acid, foods from the vegetable kingdom were generally judged to be acid-producing, foods from the animal kingdom, by contrast, as alkalescent. The acid-alkali hypothesis was widely accepted and changed only after Robert Boyle argued that some substances are neither acid nor alkalescent. In order to show the presence of acids or alkalis, he developed a color indicator test that turned out to be very persuasive. In later decades it would be shown, to the surprise of some, that not all vegetable substances were acescent, so that the acid-alkali theory had to be revised.

It was Herman Boerhaave who pointed out that, in view of the ability of animals to exist entirely on acescent vegetable food, the predominantly acid chyle derived from it must gradually be changed into animal tissues with their associated alkalescent qualities.[85] In other words, nutrition was always a gradual transformation from sour via neutral to alkaline. He named this process the animalization of plant foods and regarded the force of circulation and the heat of the body to be the major agents fostering these changes. According to this view, the principal animal substances did not consist of animal matter by nature, but were converted into it from plant substances during the process of digestion and assimilation. The animal transformed the crude matter of the vegetable kingdom into the fluids and solids of its own animal nature. A cow produces its humors from water and hay (or the softer grass), wrote Boerhaave. Its milk must closely resemble human milk, because humans sustained by nothing more than cow's milk could transform this to produce their own fluids.[86]

Boerhaave continued with the claim that, although there existed two kinds of animals, those which fed entirely upon vegetables and those which devoured the flesh of other animals, in the end the bodies of all animals (including humans) consisted either mediately or immediately of vegetables. Therefore, it was evident that in order to obtain chemical knowledge of ani-

84. See Debus 2001; Foster 1970, 200–223.

85. See Carpenter 1994, 9–10; Knoeff 2002, chap. 4; Holmes 1975, 137.

86. Boerhaave 1983, 135.

mals, one should start the analysis with the solid and fluid parts of vegetables, which—like all other living beings—generate these parts from their food (earth and water) during digestion. But the next step should be the analysis of that animal part which has "received the least alteration in their bodies, or but just begins to lose its vegetable, and put on an animal nature. . . . This part must be such a fluid, as proceeding originally from a vegetable, has felt the vital forces of the body, mixed with the blood, passed thro' the arteries and the veins, and been soon separated again. And this can be no other than chyle from vegetables, turned to milk, and separated in the breasts."[87]

Milk appeared to differ from chyle only in having been more concocted. Any food would have to enter the gut, be absorbed by the chyle vessels, and—driven by the force of circulation—travel up the long route to the *thoracici ductus,* the major part of the lymphatic ducts, where chyle was formed. Vessels, capable of encompassing, directing, changing, separating, collecting, and secreting liquids, were the body's instruments able to produce certain movements in order to chew and mix the solid foods and convert them with the help of gastric juice and bile first into chyme (alimentary mash) and then into chyle.[88] Further, the kinetics of blood circulation provided the condensation of the aliments as well as its transportation to the tissues, where the mixture was fixed in the fibers. All solid parts of the body developed through many transitional stages, and as the transformation process progressed, the initially distinct materials converged until they reached that final state of the living organism, which was nothing other than "bare light earth."

But before the chyle entered the blood of the subclavian vein to be distributed with the circulation of blood, some part of the mixture could be directed to the breast if necessary (or, if it found no opportunity of passing off in this most natural form, it might turn to fat, or be shed in urine, which was most commonly the case in men). Consequentially, milk seemed to be very similar to chyle; like this substance, it was an oily juice of vegetable origin, circulating in plants as well as in animals. However, if milk was a derivative of chyle, then, according to the acid-alkali hypothesis, it should have been an

87. Boerhaave 1727, 180.

88. Boerhaave 1983, 102. The dogma held to be true in antiquity, and still considered valid in the seventeenth century, according to which the liver was the place where chyle was transformed into blood, was challenged both by William Harvey's discovery of the circulation of blood and by Gaspare Aselli and Thomas Bartholin's discovery of the vessels of the lymphatic system. After this, the heart was identified as the place where hematopoiesis took place, and the organs of the abdomen were believed to be responsible for preparing blood formation, predominantly through the mechanical preparation of chyle (see Mani 1961).

acescent substance, but it was not. As we know from Peter Shaw's (unauthorized) translation of Boerhaave's lectures, as early as the 1720s Boerhaave had performed chemical operations on milk in order to address this problem. A pint of new milk, still warm, did not react with either alkalis (e.g., oil of tartar) or strong acids, "whence we may fairly conclude, that new milk contains neither acid nor alkali; for if either of these were latent in it, the process ought to make the discovery, as being generally allow'd the proper criterion."[89] Thus, in fresh milk, the proportion of acid-alkali particles was balanced or neutralized.

Boerhaave claimed that the character of the food an animal had taken in could nevertheless be demonstrated by experiment. A quantity of new milk should be taken from animals about eight hours after feeding, with some fed only upon vegetables and some on flesh alone. The samples should be left to digest in a clean, open glass vessel at body temperature. Different reactions could then be observed: at first, both milks remained sweet, but after some time the milk from the herbivores would turn sour and a thick, creamy part would rise to the top, while the milk from the carnivores turned rancid, tasted saline, ran into ichor, and tended to putrefy.[90] To Boerhaave, such reactions demonstrated whether milk, as obtained during chylification, was more vegetable or more animal in nature. Only after some time would it take on the nature of the animal, because the longer it circulates in the animal body, the more it loses of its own vegetable nature. In its formation, milk was, therefore, a link between the plant and animal kingdoms. It was literally a metabolic substance, produced by the nurturing input of nourishment and becoming a fluid that nurtured. This remained, more or less, the standard theory of milk formation up to the mid-eighteenth century, and it presented the conceptual background and current status of knowledge at which Ferris as well as Parmentier and Déyeux undertook their research.

From the 1750s onward, the mechanical-chemical theory of digestion was challenged in two directions. First, the observation that, from a chemical perspective, nature's kingdoms could not easily be divided into two (or three) distinct worlds inspired comparative analyses of vegetable and animal substances. When Denis Diderot (1713–1784) asked what plants and animals were in his *Eléments de physiologie* of 1778, these were by no means rhetorical questions; it was the quarrel of the day. The very idea of nature's kingdoms was under attack, on the one hand by the theory of the animalization of plants, on

89. Boerhaave 1727, 182.
90. Boerhaave 1727, 183.

the other hand by recent discoveries of what Diderot called "plant-animals" (e.g., mushrooms) and "animal-plants" (e.g., polyps).[91] A close relationship had also recently been found at the level of chemical compounds. In 1742 the Italian physician and chemist Giacomo Bartolomeo Beccari (1682–1766) had broken down flour into gluten and starch. Because gluten gave off volatile alkali, Beccari named it "animal substance." Diderot asserted that this had been debated among the chemists of the Jardin du Roi and had inspired Guillaume-François Rouelle and Pierre-Joseph Macquer among others to repeat Beccari's operations. Parmentier, who had studied chemistry at the Jardin du Roi, took Beccari's work as a point of departure for his research on the potato.

Second, chemists of the late eighteenth century became accustomed to negotiating the question of animalization by judging components instead of whole substances. Macquer had already written that the cheesy matter in milk was similar to the fleshy gelatin that made up animals, while the oily parts were of vegetable origin.[92] Parmentier and Déyeux confirmed this view. The general trend of the time for explaining the properties common to a given class of substances by invoking an underlying principle contained in all of them was thus adopted in the study of milk.[93] It became commonplace to draw analogies between different milk components and other generic substances in the vegetable or animal kingdoms, such as oils and gelatin.

Furthermore, both Ferris and Parmentier and Déyeux confirmed the assumption that carnivore milk exhibits different chemical reactions than herbivore milk. On the one side of the spectrum could be found "pure" animal matter; on the other side, "pure" plant materials. Between them, however, existed a spectrum of gradations. What about those animals that ate vegetable as well as animal matter, or those that ruminated? How, in particular, should one examine human milk? On an operational level, the theory of animalization turned out to be highly inspiring. Chemistry, in fact, was combined with feeding experiments. Because human milk, in contrast to animal milk, did not coagulate or only very slowly turned from a liquid into a solid mass, a woman was asked to eat nothing other than vegetables for a period of eight days. The woman accepted, and after a week Ferris found what he was looking for: the milk started to coagulate in the expected time.[94]

91. See Quintili 2004.

92. Macquer 1781–1783, 69. For the nutrition of working people, Macquer wrote, cheese could substitute for meat (71).

93. Holmes 1989b, 38.

94. Ferris 1787, 127.

## Concluding Discussion: Shifting Epistemic Regimes

How should we put the chemistry of milk of the 1780s in its place? Was it a scientific investigation or an attempt by certain practitioners to grasp knowledge of the everyday world?[95] Should one describe only the experimentation as genuinely scientific and the rest as technical, in the literal sense? Or would it be more reliable to say that the authors transformed knowledge that was widely shared into knowledge that was exclusively scientific? And what about the object under consideration? Is it a story of a scientific object coming into being?[96] Or should we call milk a material that circulated through the lives of many people (including some apothecaries and physicians with scientific concerns), a material that could be experienced and needed to be handled with care, a material that could be processed both on the farm and in the laboratory? Furthermore, milk was presented as an object that incorporated older knowledge about the generation and regeneration of living beings. Was it, then, primarily a biological entity? Or did it belong fundamentally within the framework of medical knowledge, owing to the fact that it caused some health problems and cured others?

In fact, the studies under consideration in this paper provide some support for all of these questions. Milk chemistry of the late eighteenth century was a conglomerate of different ways of acting and thinking, and it was rooted in many sources of practice and knowledge. Observations from next door are followed by case studies from medical practice; precise descriptions of chemical operations are framed by quotations from older scientific literature; folk beliefs, craft knowledge, and references to natural philosophy alternate with one other. At any rate, chemists showed a clear tendency toward the systematic comprehension of a bodily fluid that was to be treated in an innovative way. To them, laboratory investigations meant developing an inventive set of experiments that was intended to advance chemical science and at the same time to produce useful knowledge.

Where earlier chemical textbooks and treatises had discussed the state of animal chemistry, they had always complemented their description of the multiplicity of natural things with the experimental study of selected examples of animal substances. In the chapter "Processes upon Animals" in his *New Method of Chemistry,* Boerhaave treated milk, urine, sal ammoniac

95. I owe these questions to Andrew Pickering, who gave fruitful comments on an earlier version of this chapter.

96. On the "coming into being" of scientific objects, see Daston 2000a.

prepared from urine, egg white, blood, and more.[97] To him, milk was an important animal substance, but he would never have claimed to have known everything of value about the fluid as explicitly as his followers did half a century later. Boerhaave and his contemporaries were persuaded that the study of the inner parts of the animal body was primarily the duty of medicine: anatomy described the parts and structure of the body; the "mechanic" examined the solids; "whilst Hydrostatics helps us to the laws of Fluids in general; and the actions of them, as they move through given canals, are explained to us by that beautiful Science Hydraulics; and lastly, let the Chemist add to all these, whatever his Art, when fairly and carefully apply'd, has been able to discover: And then, if I am not mistaken, we shall have a complete account of the physiological part of Physic."[98]

Medicine and chemistry were deeply intertwined in Boerhaave's approach. Since the sixteenth and seventeenth centuries, when the Paracelsian and iatrochemical traditions were well-known for their sophisticated theories of chemical physiology, there were many debates about the philosophical, anatomical, mechanical, and chemical contributions to the clarification of what was then called the "animal economy," concerning bodily processes such as digestion, assimilation, generation, and putrefaction.[99] Within this older framework, milk, blood, and other fluids had been examined in order to understand their relations and metamorphoses within the body. Thus, milk studies served the deeper understanding of the physiological conditions and processes of milk formation and lactation, and milk was judged mainly from a physiological point of view. As a substance that maintained multiple connections to the rest of the body, it vividly revealed the processes of bodily change. In other words, experiments were not performed in order to challenge existing knowledge about the physiological system of order, but were, rather, assimilated to it.

During the final decades of the eighteenth century, the epistemic status of animal materials changed, and the milk studies addressed in this chapter contributed to these epistemic shifts.[100] In 1800 the French chemist Antoine-François Fourcroy (1755–1809) observed, "Armed with exact instruments and ingenious methods for determining the true differences that exist between plant and animal materials, [animal chemistry] shows what happens

97. See Boerhaave 1727, 179.

98. Boerhaave 1727, 53.

99. On Paracelsianism, see Debus 2001. On the term "animal economy" as used in the eighteenth century, see Spary 1996.

100. On the general trend, see Klein and Lefèvre 2007.

to the state of the former when they are transformed into the latter in animal organs."[101] Such a statement could not have been made before the end of the eighteenth century. By then, organic materials were viewed as stable chemical objects, with clear boundaries, which were of interest for their own sake. Instead of referring to empirical phenomena coded in the language of common sense, animal chemistry promised to derive the fundamental principles that were essential for all substances from chemical analysis. Philosophers and physiologists were no longer in sole charge of the smallest units of the body below the level of visibility. Chemists too could contribute to knowledge about the constitutive structure of the invisible. With this revaluation of the importance of chemistry, the intellectual framework of medicine diminished in importance.

In the past, many historians have argued that chemistry dealing with organic materials had to refer to medical knowledge because of its lack of valid instrumentation. Frederic L. Holmes argued that chemists were able to abandon physiological explanation only once the elementary composition of organic substances had been discovered by means of the procedures for collecting and identifying the gaseous products of chemical reactions, beginning in the 1780s with the work of Berthollet, Fourcroy, and Lavoisier.[102] Up to this time, it is argued, organic chemistry remained disoriented, complicated, and puzzling, had no criteria for purity, was unable to identify the gaseous state of matter or elements, and could handle only solids and liquids with its apparatus.[103] Noel Coley has even suggested that one cannot speak of an elaborated animal chemistry before Lavoisier, whom he sees as its founding father.[104]

Such arguments result from a focus on improvements in experimental techniques in the laboratory, with the main emphasis upon advances in analytical knowledge. Furthermore, authors have judged the state of the art of eighteenth-century animal chemistry from a perspective of the organic chemistry developed in the early decades of the nineteenth century. Organic chemists of that time, among them influential scientists like Jöns Jakob Berzelius or Justus von Liebig, described the relationship between medicine, physiology, and chemistry in terms of the "application" or "uses" of chemical knowledge, confirming the view that until recently physicians did not benefit

101. Fourcroy, quoted in Klein 2003a, 45.

102. Holmes 1963, 55.

103. Rosenfeld 1999, 29.

104. Coley 1973, 24. By contrast, Klein has argued that pre-Lavoisierian plant and animal chemistry created its own order (see Klein 2003a).

from the advancements in chemistry but instead relied on imperfect methods of research.[105]

There is no doubt that chemical exploration of organic materials underwent important changes between 1785 und 1840. The oxygen theory and quantitative elemental analysis radically altered understandings of bodily matter. It might thus be argued that, as the milk chemistry of the 1780s was not at the core of the "Chemical Revolution," authors were not able to conduct their experiments in an innovative way. Such an assumption, however, misses the point of the milk chemists' concerns. First, it would be wrong to see the milk chemistry of the 1780s as lacking in new empirical content. Parmentier and Déyeux were no strangers to the French scientific establishment, and one can assume that they were well informed of the latest developments in Lavoisier's chemistry.[106] Samuel Ferris, by contrast, was an apprentice to a surgeon-apothecary in Hertfordshire and went on to study medicine in London. As a student, he was an active member of the Royal Medical Society in Edinburgh, serving as its president from 1783 to 1784 before becoming a licentiate of the Royal College of Physicians of London in 1785. Elected a Fellow of the Royal Society in 1797, Ferris was strongly in favor of the medical art becoming more scientific.[107]

Of vital importance is the fact that the scope of studies concentrating on substances like milk encompassed questions extending far beyond the scope of the inquiries that historians have long regarded as cutting-edge research in chemistry. In the case of milk, the experts were certainly scientific but had broader goals than chemistry in mind. They wanted to make the arts and medicine scientific and to intervene in and improve public health and economy. This becomes already clear from the authors' professional backgrounds in pharmacy and medicine. One might name them "hybrid savant-technologists"

105. See Liebig's polemic against medicine: Liebig 1844.

106. Worth mentioning is the authors' remark on the occasion of the translation of their book into German. In the preface, they first apologize for not having converted the weights and measures into the postrevolutionary metric system. Second, they explain that because of the revolutionary turmoils, the book could not be published earlier. Other than this, they offer not a word of comment in support of the evidence of their research (Parmentier and Déyeux 1800, v).

107. The Royal Medical Society of Edinburgh was a student association. Prior to winning a prize medal for his dissertation on milk, Ferris had obtained a doctorate in medicine at Edinburgh University in 1784. The title of his thesis was "De Sanguinis per Corpus vivum circulantis putredine." His only other publication treated the relation between chemistry and physics (Ferris 1795). As a supporter of the medical theories of William Cullen, he was engaged in the Royal Medical Society's struggle against Brunonianism (see Risse 2005, 138). I thank John Dallas of the Royal College of Physicians of Edinburgh for his help in the search for biographical details.

in regard to the aims they pursued.[108] Being rather pragmatic, spurred by immediate social, economic, and commercial concerns, their knowledge was diverse but not nonscientific. Because milk was both a chemical object and a bodily material, investigating milk demanded knowledge related to dairy practices, physiology, medicine, and child care. The milk analyses to which I have referred in this chapter discussed and explored a whole set of questions and problems in the light of ongoing political developments, common practices, and future perspectives. Assumptions, meanings, beliefs, and traditions were offered simultaneously with descriptions of concrete experiments. Old and new knowledge thus converged, and classical authors were mentioned as often as contemporary literature. This all ended in a prosaic narrative about a concrete and mundane object, experimenting with a new style of presenting mundane knowledge, bringing it into an academic format. They designed what, a few decades later, would be labeled public health knowledge.

Milk of the late eighteenth century thus was an object of inquiry that remained to some extent elusive, resisting chemical analysis. Although chemists divided the fluid into parts and separated those parts into several components, the object was supposed to be a substance with its own identity and physiological meaning. Milk was a material object with a sensual, ascertainable complexity, a substance with multifarious properties, affinities, and abilities. By no means primarily an aggregate of elements, it was an entity in its own right, signifying metabolic processes as well as enlightened beliefs about mother's milk as nature's only and best infant food. Consequently, the interplay of political, cultural, and personal symbolic meanings was constantly present in the laboratory; the laboratory object mirrored the outside world.

## Primary Sources

Secondary sources can be found in the cumulative bibliography at the end of the book.

Bergius, Jonas Peter. 1772. Versuche mit Frauen-Milch angestellt. *Der Königl. Schwedischen Akademie der Wissenschaften neue Abhandlungen aus der Naturlehre, Haushaltungskunst und Mechanik* 1:40–55.

*Bibliothèque physico-économique ou Journal des découvertes et perfectionnemens de l'industrie nationale et étrangère, de l'économie rurale et domestique, de la physique, la chimie, l'histoire naturelle, la médecine domestique et vétérinaire, enfin des sciences et des arts qui se rattachent aux besoins de la vierédigée.* 1782–1831. Paris: Buisson.

108. Ursula Klein has introduced this term into the history of science in order to emphasize the interconnectedness of systems of science and technological practices in the eighteenth century (see Klein 2005c, 228–232).

Blondel, James August. 1727. *The Strength of Imagination in Pregnant Women examin'd: and the Opinion that Marks and Deformities in Children arise from Thence, demonstrated to be a vulgar Error.* London: s.n.

Boerhaave, Herman. 1727. *A New Method of Chemistry; Including the Theory and Practice of that Art: Laid down on Mechanical Principles, and accomodated to the Uses of Life, translated from the Printed Edition, Collated with the best Manuscript Copies, by P. Shaw, M.D. and E. Chambers.* London: Osborn / Longman.

———. 1732. *Elementa Chemiae, quae Anniversario Labore Docuit, in Publicis, Privatisque, Scholis.* Lugduni Batavorum (Leyden).

———. 1983. Boerhaave's *Orations.* Trans. with intro. and notes by E. Kegel-Brinkgreve and A. M. Luyendijk-Elshout. Leyden: Leiden University Press.

Bower, Alexander. 1830. *The History of the University of Edinburgh; chiefly compiled from original papers and Records, never before published.* 3 vols. Edinburgh: Waugh & Innes.

Corlieu, Auguste. 1896. *Centenaire de la Faculté de médecine de Paris (1794–1894).* Paris: Alcan et al.

Darjes, Joachim Georg. 1756. *Erste Gründe der Cameral-Wissenschaften darinnen die Haupt-Theile so wohl der Oeconomie, als auch der Policey und besondern Cameral-Wissenschaft in ihrer natürlichen Verknüpfung zum Gebrauch seiner academischen Fürlesung entworfen.* Jena: Wittwe.

Ferris, Samuel. 1785. *A Dissertation on Milk. In which an Attempt is made to ascertain its Natural Use; to investigate experimentally its General Nature and Properties; and to explain its Effects in the Cure of various Diseases.* London: T. Cadell.

———. 1787. *Ueber die Milch. Eine Harveysche gekrönte Preisschrift der königlichen Gesellschaft zu Edinburgh, aus dem Engl. übersetzt und mit einigen Anmerkungen begleitet von Christian Friedrich Michaelis.* Leipzig: Jacobäer.

———. 1795. *A General View of the Establishment of Physic as a Science in England by the Incorporation of the College of Physicians in England.* London: Royal College of Physicians.

Galenos. 1952. *Werke. Übersetzt und zeitgemäß erläutert von Erich Beintker und Wilhelm Kahlenberg.* Vol. 4, *Die Kräfte der Nahrungsmittel,* book 3: *Gute und schlechte Säfte der Nahrungsmittel, Die säfteverdünnende Diät, Die Ptisane.* Stuttgart: Hippokrates-Verlag.

Gren, Friedrich Albrecht Carl. 1794. *Systematisches Handbuch der gesammten Chemie.* Bd. 2: *Die botanische und zoologische Chemie.* 2nd ed. Halle: Verlag der Waisenhaus-Buchhandlung.

Haller, Albrecht von. 1773. Lait. In *Encyclopédie, ou Dictionnaire universel raisonné des connaissances humaines, mis en ordre par M. [Fortuné-Barthélemy] de Felice,* ed. Denis Diderot and Jean le Rond d'Alembert. Yverdon: s.n., 1770–1779, vol. 25, Sp. 522a–523b.

Hebenstreit, Ernst Benjamin Gottlieb. 1791. *Lehrsätze der medicinischen Polizeywissenschaft.* Leipzig: Dyk.

Hentschel, Gottlob, and Friedrich Hoffmann. 1725. *De saluberima seri lactis virtute.* Halae Magdeburgicae: s.n.

Hermbstädt, Sigismund Friedrich. 1808. Untersuchungen über die Milch der Kühe. *Abhandlungen der physikalischen Klasse der Königlich preussischen Akademie der Wissenschaften zu Berlin.*

Hoffmann, Friedrich. 1715. *Gründliche Anweisung wie ein Mensch vor dem frühzeitigen Tod und allerhand Arten Krankheiten durch ordentliche Lebens-Art sich verwahren könne.* 3 vols. (1715–1728). Halae Magdeburgicae: s.n., vol. 1.

Johnson, W. B. 1803. *History of the Progress and Present State of Animal Chemistry.* London: J. Johnson.

Jörg, Johann Christian Gottfried. 1812. *Diätetische Belehrungen für Schwangere, Gebärende und Wöchnerinnen, welche sich als solche wohlbefinden wollen: In zehn an gebildete Frauen gehaltene Vorlesungen.* Leipzig: Carl Cnobloch.

Krünitz, Johann Georg. 1803. *Ökonomisch-technologische Encyklopädie, oder allgemeines System der Staats-, Stadt-, Haus- und Landwirthschaft, und der Kunst-Geschichte, in alphabetischer Ordnung, zuerst fortgesetzt von Friedrich Jakob Floerke, nunmehr von Heinrich Gustav Flörke.* Neunzigster Theil. Berlin: Pauli.

Lichtenstein, G. Rudolf. 1772. *Abhandlung vom Milchzucker.* Braunschweig: s.n.

Liebig, Justus von. 1844. *Bemerkungen über das Verhältniss der Thier-Chemie zur Thier-Physiologie.* Heidelberg: Winter.

Lorry, Anne Charles. 1785. *Abhandlung über die Nahrungsmittel, als Commentar über die diätetischen Bücher des Hippokrates.* Leipzig: J. G. Müller. (French orig., 1754–1757).

Macquer, Peter Joseph. 1781–1783. *Chymisches Wörterbuch oder Allgemeine Begriffe der Chymie nach alphabetischer Ordnung. Aus dem Französischen nach der zweyten Ausgabe (1777) übersetzt und mit Anmerkungen und Zusätzen vermehrt von Johann Gottfried Leonhardi.* 8 vols. Leipzig: Weidmann.

Martin, Barthélemy. 1684. *Traité de l'Usage Du Lait.* Paris: Denis Thierry.

Parmentier, Antoine Augustin. 1789. *Traité sur la culture et les usages des pommes de terre, de la patate, et du topinambour.* Paris: Barrois.

Parmentier, Antoine Augustin, and Nicolas Déyeux. 1798. *Précis d'expériences et observations sur les différentes espèces du lait, considérées dans leurs rapports avec la chimie, la médecine et l'économie rurale.* Strasbourg and Paris: Barrois.

———. 1800. *Neueste Untersuchungen und Bemerkungen über die verschiedenen Arten der Milch in Beziehung auf die Chemie, die Arzneykunde u. die Landwirthschaft.* Jena: Voigt.

Rhades, Fridericus Henricus. 1786. *Animadversiones circa Temperamenta Humana Imprimisque ea quae Lactatione Comunnicata habentur.* Diss. inaug. med. Sistens. Halae: s.n.

Rosner, Johann Georg Emanuel. 1756. *Dissertatio qua nonnulla circa vires lactis.* Lugduni Batavorum: s.n.

Rüger, D. 1851. *Die neue chemisch-praktische Milch-, Butter- und Viehwirtschaft.* Löbau: Dummler, vol. 1.

Stiprian-Luiscius, Abraham von, and Nicolas Bondt. 1787–1788. Diss. qua respondetur ad quaestionem propositam: Ut determinetur, per examen comparatum, proprietatum physicarum et chemicarum, natura lactis muliebris, vaccini, caprilli, asini, ovilli et equini. *Histoire ét Mémoire de la Société royale de Medecine de Paris.* 525–30.

Voltelen, Floris Jacobus. 1779. *De Lacte humano eiusque cum asinino et ovillo comparatione observationes chemicae.* Lipsiae: Beuschel.

Young, Thomas. 1769. *Dissertatio medica inauguralis de lacte.* Edinburgh: s.n.

# 8

# The Sparkling Nectar of Spas; or, Mineral Water as a Medically Commodifiable Material in the Province, 1770–1805

MATTHEW D. EDDY

*Invitation to Peterhead*
Here Health her bath's enlivening tide,
And fountain's sparkling nectar pours;
Fields fluctuate in flower pride,
While cool gales fan the quiet shores.
Her friendship warms, her smiles engage,
Her converse, quiet, learning, leisure,
Feed mirth, sooth care, afford the Sage
Instruction, and the Poet pleasure.

JAMES HAY BEATTIE, 1793, excerpt
from William Laing, *An Account of Peterhead, Its Mineral Well, Air, and Neighbourhood,* 1793

A man may know the chymical analysis of all the articles in the materia medica, without being able properly to apply any one of them in the cure of diseases. One page of practical observations is worth a whole volume of chymical analysis. But where are such observations to be met with? Few physicians are in the situation to make them, and few still are qualified for such a task. It can only be accomplished by practitioners who reside at the fountains, and who, possessing minds superior to local prejudices, are capable of distinguishing diseases with accuracy, and of forming a sound judgment respecting the genuine effects of medicines.

WILLIAM BUCHAN, *Cautions Concerning Cold Bathing, and Drinking the Mineral Waters,* 1786

I wish to thank Alistair Durie for giving me copies of several eighteenth-century spa-related texts and Andreas-Holger Maehle, Ursula Klein, and Pamela Anderson for sending me drafts of unpublished essays relevant to the subject matter of this chapter. I also received helpful suggestions from participants of the "Making of Materials" workshop hosted at the Max Planck Institute for the History of Science (Berlin) in August 2006. Finally, my understanding of eighteenth-century beliefs regarding the therapeutic value of heat and cold were enhanced by informal conversations with John Christie, Georgette Taylor, Hasok Chang, David M. Knight, and Robin F. Hendry at the "Matters of Substance" workshop held at Durham University in August 2006.

In recent decades historians have devoted a notable amount of research to the economic, social, and experimental relevance of the mineral water available for consumption in the spas that appeared across Europe from the seventeenth to the nineteenth centuries. Although studies have shed light on the links between tourism and commodification, the role of patient authority, and the isolation of chemical substances, they have not offered a clear account of how the chemical composition of the wells connected with the medical theory that legitimated their commodification and use as a remedy. In this chapter I address this gap by focusing on Peterhead Well, a provincial spa located on the northeastern tip of Scotland. In particular, I focus on several authors who wrote about the well, including Rev. Dr. William Laing, the local Episcopal priest. The spa was a popular northern health resort, and in 1793 the town's Freemason's Lodge celebrated the installation of a new pump room in which residents and visitors could enjoy a drink of the town's wine-colored water. After paying an admission fee, customers could sip at their leisure as they played card games and browsed the newspapers. Although the salubrious effects of the water had been known for centuries throughout the north of Scotland, Peterhead's new trading links to Holland, England, and the Baltic ensured the arrival of guests who knew nothing of the well's virtues. For these potential customers, and for incredulous Lowlanders to boot, William Laing wrote two pamphlets that used chemical analysis, personal testimony, and local case histories to substantiate the tonic power of the town's mineral well. The first, entitled *An Account of Peterhead* (1793), was printed by T. Evans in Pater-Noster-Row, London. It was sold in the capital, as well as in Peterhead, Aberdeen, and Edinburgh.[1] Over a decade later, in light of the commercial success of the well, Laing wrote a follow-up publication. Printed in Aberdeen by J. Chalmers, it was entitled *Account of the Cold and Warm Sea Baths at Peterhead* and went through two editions.[2]

Laing's publications bolstered the Lodge's attempt to make the water a material commodity. In particular, he identified several substances in it that were commonly known to have a therapeutic effect on the body. Indeed, his desire to connect both useful and commercial knowledge with scientific experimentation in this manner is similar to other material practices studied in this volume. In his case, he commodified the well's water by strategically appropriating practices and theories that were the domain of medical chemistry. As in the case of milk, dyestuffs, alcoholic spirits, and chemical remedies

1. Laing 1793b.
2. Laing 1804. Hereafter, *Sea Baths.*

(see Orland, Nieto-Galan, Spary, and Klein in this volume), his knowledge of the chemical composition of the well contributed significantly to its transformation into a commodity.[3] Studies that address this facet of spa water are few, and the present work seeks to draw attention to this fact by addressing the material basis of spa water therapy as understood by literate Scots who lived during the late eighteenth century.

One notable obstacle that often prevents historians from pursuing detailed studies on provincial spa publications is the obscurity of the authors who wrote them. For example, other than the brief biographical summaries devoted to Laing in *Scottish Episcopal Clergy* and the *Fasti Academiae Mariscallanae Aberdonensis,* the contours of his life remain murky.[4] Indeed, he does not even have an entry in the *Oxford Dictionary of National Biography.* The lack of such studies makes it difficult to explore how local practitioners became medical authorities, what kind of expertise they possessed, and how they used their knowledge to make mineral well water a therapeutic commodity. In Laing's case, the situation becomes even more intriguing when one considers that he was an ordained priest, practicing physician, and self-trained chemist. This being the case, this chapter uses Laing and several other frequenters of the well as examples of the complex world of pharmaceutical commodification. The first section unpacks why spa water was seen as a pharmaceutical commodity and how Laing acquired the necessary skills that allowed him to be a local medical authority. This is followed by a section that explains how chemistry was used to isolate substances that were widely known to have pharmacological value. The final section reveals how socioeconomic factors influenced the interpretation of the water's therapeutic efficacy and its popularization in Scotland's print culture.

## The Medical Context of Spa Water

### SPA WATER AND *MATERIA MEDICA*

The medical marketplace was on the rise in the eighteenth century.[5] Although the sale of naturally occurring "cures" was an ancient practice, the medical theories that governed the sale and use of drugs in Enlightenment

3. This approach to eighteenth-century pharmaceuticals is summarized in Curth 2006 and King 2006.

4. A basic outline of his vital dates and degrees appears in Bertie 2000 and P. Anderson 1889–1898.

5. Porter and Porter 1989. The standard work on the social and medical relevance of eighteenth-century spas is Porter 1990. For Scotland, see Durie 2003.

Scotland were significantly influenced by the chemical knowledge taught in schools and universities.[6] This practice conformed to a larger European phenomenon during the eighteenth century in which the therapeutic value of tonics became inextricably tied to the rise of spa towns.[7] With the rise of medical "chymistry" in the seventeenth century, it did not take long for physicians to combine chemical experimentation with pharmacology. By the 1700s, the pages of Europe's leading scientific journals carried a wide array of articles that sought to connect the material content of the mineral water with therapeutics.[8] As the century progressed, a plethora of books, monographs, articles, and chorographies were published on the chemical materials contained in mineral wells.[9] In mainland Europe, two sites that were examined frequently were Carlsbad in Bohemia (modern-day Karlovy Vary) and Pyrmont in Lower Saxony. The same trend continued in Britain, and by the middle of the century there were numerous spa towns that promoted the therapeutic value of the chemical substances contained in their water. In England, the most famous were Bath and Bristol, but there were many mineral wells that were frequented by regional, national, and international tourists.[10] In Scotland the relationship between therapeutics, chemical analysis, and spa water was commonly known among physicians and literate patients who lived in the last decades of the century. During this time, medically oriented articles or pamphlets were published on mineral wells such as Peterhead Spa, St Bernard's Well, Moffat Spa, and Dunse-Spaw.[11]

The common tactic of most chemically oriented spa literature was to isolate the material components of the water so that they could be identified and used to treat a disease. Although mineral waters were historically held to have medicinal properties, pharmacology in Enlightenment Scotland employed a chemical rationale to explain how the material composition of the water could be used therapeutically to restore health. This type of chemical pharmacology was in full swing by the 1750s, when the pharmacopoeias of

6. The chemical foundations of eighteenth-century pharmacology are also addressed in A. Simmons 2006; Klein 2007a, 2007b; Eddy 2005; Maehle 1999; Cook 1990.

7. See Porter 1990; Hembry 1990.

8. See Maehle 1999; Eddy 2001.

9. Coley 1982.

10. For the growth of eighteenth-century seaside resorts, including Brighton, see H. Robinson 1976, 3–13. The Georgian origins of modern spas and seaside resorts are summarized in Hassan 2003, 15–30, and in Corbin 1994, 57–96.

11. For a sample of these publications, see Taylor 1790 and Home 1751. For the experimental context that surrounded the interest in these wells, see Eddy 2001.

Edinburgh and London were translated from Latin into English.[12] The causative agent in a drug that instigated material change in the body was often called the "active principle." Physicians sometimes linked this "power" to the concept of "elective affinity," that is, the notion that certain kinds of chemical substances attract each other based upon a force (affinity) inherent in their material composition. For this reason, the concept of affinity was often explained in pharmacopoeias.[13] There were many types of drugs that contained such active principles, but some of the more common were expectorants, diuretics, and tonics. The causal therapeutic powers of these drugs were linked to experiments conducted on animals and patients in the country's leading universities and infirmaries. Most physicians, apothecaries, and surgeons who operated in these settings held that body tissue was made up of fibers that, depending on their composition, could be made to contract or relax. Contraction forced fluids out of the viscera and flesh, thereby making the body hard. Relaxation allowed fluids to seep in, thereby making it soft. Medical historians usually group the therapeutic theories that resulted from this physiological model under the term *neohumoralism* on account of their conceptual similarities to humoralism, the ancient and medieval medical theory which held that health was regulated through the use of fluids (black bile, yellow bile, phlegm, and blood). As intimated above, Scottish neohumoralism was closely linked to chemical powers attributed to different substances. If one wanted to maintain a healthy nervous system, one had to ensure that the tissues remained properly balanced between hardness and softness. When the body became imbalanced, drugs with specific types of active principles had to be used to restore the equilibrium.[14]

When it came to mineral wells, Scottish neohumoralism gave great weight to the therapeutic power of the tonics found in a spa's water. The link between the tonic and its power was usually established by the testimonies of those who had imbibed the water. In Peterhead, such evidence came from case histories of Aberdeen Infirmary patients and the personal testimonies of spa denizens from both the affluent and common classes who held that the water had cured their nervous disorders. Tonics were closely connected to cures that stimulated or invigorated the nervous system, which, at the time, included not only the brain, spinal column, and nerves, but also muscles and tissues that are now considered to be part of the circulatory and digestive systems. William Cullen, an influential professor of medicine at the University

12. Cowen 2001.

13. A good example occurs in Lewis 1770, 27–29.

14. Neohumoralism is treated throughout Risse 2005 and 1992.

of Edinburgh, for instance, promoted what historians have called "a single and ultimately indivisible neuro-muscular system."[15] It was within this intellectual context that most literate Scots learned about the tonics contained in Scotland's spa waters.

Laing's publications indicate that his perception of the nervous system was similar to Cullen's. In his first Peterhead pamphlet, he discussed nervous disorders that produced "lamentable affections both of body and mind, called *hypchondriacal* or *hysterical* disorders; which last complaint is usually nothing else than a debilitated stomach, in an irritable condition, disordered accidentally by passion, or some excess or irregularity."[16] Based upon his analysis of the well's material contents, he held that Peterhead's spa water could alleviate these nervous complaints. Overall, he was respectably acquainted with therapeutic theories used to calm the nervous system, and his comments on Peterhead's tonic power over dropsy provide a particularly good example: "Dropsy also is a disease of debility; wherein the vessels exhaling a fluid into the various cavities of the body for moistening them, from weakness, as is supposed, allow too great a quantity of the fluid to escape through them; while the vessels that take up that fluid, and convey it again into the mass of blood, fail to do their office from want of power. This points out the propriety of Peterhead water, as a tonic and stimulant."[17]

Because spa water contained naturally occurring therapeutic substances, there were many sources available to the reading public that explained and categorized their ameliorative contents. One of the most widely read books on the chemistry of spa water in Enlightenment Scotland was William Buchan's *Domestic Medicine.*[18] It was first published in 1769, and by 1802 it had gone through at least seventeen editions. As such a popular book, it had a considerable impact on the public's understanding of the medical and chemical relevance of spas. The book classified the therapeutic substances of spa water into four categories: *ferruginous, gaseous, saline,* and *sulfurous.* These categories were directly linked to several irreducible chemical components that guided experimentation during the 1760s: water, metals (ferruginous matter, that is, a substance containing iron), airs (gaseous matter), inflammables (sulfurous matter), and salts (saline matter).[19] This process of classifying the

15. The clearest explication of Cullen's nervous theory that I have encountered is Bynum 1993.

16. Laing 1793b, 35.

17. Ibid., 37–38.

18. Rosenberg 1983, 22–42. See also the quotation taken from Buchan's *Domestic Medicine* that appears at the beginning of this chapter.

19. These principles are defined in Holmes 1989b.

content of spa water into categories that corresponded with chemical principles mirrored the practices used to categorize minerals and drugs in the medical courses given at Scottish universities.[20] Based on experiments and patient histories, drugs were associated with the specific substances of chemistry, thereby making matter theory an issue of personal relevance to many people.

Spa-related articles and pamphlets, both popular and academic, often focused on "salts," that is, dissolvable material agents that bore acidic and alkaline properties.[21] Identifying salts in this manner stretched back to the seventeenth century, when saline active principles became more associated with the notions of alkalinity and acidity as well as to specific types of therapeutic cures. In this context, the isolation of a salt from a sample of spa water often contributed to the water's commodification. The practice of ingesting salts contained in spa water (and many other drugs) remained a dominant form of provincial therapeutics in Scotland until it was challenged by mid-eighteenth-century medical theories that placed more emphasis on the curative power of heat. This evolution stemmed from the fact that both chemistry and therapeutics were changing from the 1770s forward. In the years that followed the publication of Laing's first pamphlet, the new chemistry of Antoine-Laurent Lavoisier began to be accepted throughout Europe, and medical theories based on body temperature enjoyed increasing popularity. In the wake of such changes numerous provincial works, Laing's *Sea Baths* (1804) for example, gave more attention to temperature-based cures and the tonic power of hot and cold baths.

## WILLIAM LAING: THE AUTHORITY OF A CHEMICAL COMMODIFIER

Although well-known medical professors like the University of Edinburgh's Joseph Black demonstrated a passing interest in the chemical content of Peterhead's water, the main medical commodifiers of the spa were Laing, Rev. Dr. James Beattie, and Rev. Dr. George Moir. This means that the principal experts on its contents had not formally studied medicine in university. This might seem odd to modern eyes, but becoming a medical authority in

20. Eddy 2002, 2004, 2008.

21. Salts could also be isolated via simple distillation, which did not necessitate expensive chemical equipment. On a more academic scale, articles on salts appeared frequently in London's *Philosophical Transactions* and Edinburgh's *Essays Physical and Literary;* however, related articles also appeared in *The Scots Magazine.* For the role of acidity and alkalinity in spa water, see Monro 1770.

eighteenth-century Scotland was not a straightforward enterprise. In addition to studying in a university, one could be granted an MD on the basis of an outstanding medical publication, or through a plan of self-study that culminated in an oral examination in a university. Such a situation is notable, as it shows that the three local experts were not invested members of the medical establishment, which at the time was regulated by the Royal College of Physicians of Edinburgh and by those who held medically related professorships in the universities. Here I wish to explore this context in more detail by giving a focused account of Laing's eclectic medical education and by showing why his comments on the therapeutic value of the spa's substances was taken seriously by his contemporaries. I will return to the expertise of Beattie and Moir later in this chapter.

Laing was born in Fraserburgh, Scotland, on 29 March 1742. In 1762 he matriculated at Marischal College, Aberdeen. His studies were overseen by William Kennedy (Greek), Frances Skene (civil and natural history), George Skene (natural philosophy), and James Beattie (moral philosophy). In 1766 he was awarded an MA. While pursuing his studies, he developed a close relationship with Beattie, and they went on to become lifelong friends. Although the precise nature of Laing's initial exposure to medicine and chemistry is not known, it can be safely assumed that he initially learned about the subjects while in university. George Skene's lectures, for instance, promoted chemistry, especially the works of Herman Boerhaave, and its usefulness for mineralogy and the related fields of medicine, agriculture, and industry.[22] It is also quite likely that Laing studied medicine with John Gregory. In addition to being Beattie's close friend and personal physician, Gregory was also the professor of medicine at King's College, Aberdeen, until 1764.[23] Upon graduating, Laing was appointed preceptor to William Fraser of Kirktown in Philorth and, with the help of Beattie, he was made Under Master of Aberdeen Grammar School in 1770.[24] He was ordained the next year, on Beattie's recommendation, into the Episcopal Church of Scotland by the Bishop of Down and Connor.[25] In the same year, he was appointed as minister to the English Episcopal Chapel in Peterhead and he remained there for the rest of his career.[26]

22. Wood 1993, 91–94.

23. B. Ashworth 2003, 67–69.

24. P. Anderson 1897, 38–41.

25. Henderson 1907, 375–76.

26. Beattie 1771 (R. Robinson 2004, 248). I will make substantial use of the correspondence between Rev. Dr. William Laing and Rev. Dr. James Beattie. These letters are housed in several collections around Scotland, and many of them have been summarized and placed in a numerated list published in R. Robinson 2004. Whenever I cite a Laing or Beattie letter, I list the date

Over the next ten years, Laing continued to foster his interest in medicine by reading books and observing the ameliorative effects of Peterhead's waters.[27] In time he became proficient in medical theory and *materia medica*, and began to recommend remedies and cures to his parishioners and friends. As he did not hold an MD, he was not legally allowed to take payment for any of his medical advice. Beattie, however, visited Peterhead often, and was able to witness Laing's abilities first hand. He was so impressed with Laing's skills as an "adept in medicine" that he approached Alexander Donaldson, Marischal's professor of medicine, to see if his former student could be awarded an honorary "Doctor's degree in Physick." Although such a recommendation from a professor of moral philosophy might at first glance sound rather unusual, Beattie's knowledge of medical theory was by no means slight, and he was certainly qualified to assess Laing's abilities.[28] Based on Beattie's testimony, Donaldson agreed to support the measure and both professors presented Laing's case to the university. As a result of this petition, Laing was awarded an MD in October 1782.[29] The excited Beattie then wrote Laing of the good news: "I will *Mr Laing* you no more henceforth. You are now to all intents and purposes The Reverend Dr William Laing Doctor of Medicine—quod felix faustumque sit."[30] Laing took his MD seriously and thereafter continued to educate himself on medically related subjects. As I will explain below, his Peterhead work shows that he was keen on performing chemical experiments. He also was involved with community health efforts.[31]

Throughout the rest of his life, the wider Peterburgh community continued to hold Laing's medical abilities in high esteem. Beattie, for example, continued to affirm Laing's medical authority throughout the 1780s

---

that the letter was written and the number given to it in Robinson's list. All letters and their archival locations are listed at the end of the chapter.

27. Depending on a person's state of mind, reading books could be beneficial or harmful (see Johns 2000).

28. Beattie apparently did this as an act of benevolence and not on the direct request of Laing. Beattie's medical knowledge was clearly evinced in his correspondence, and his extensive reading on the nervous system undergirded much of his philosophy of mind—especially as expressed in his *Elements of Moral Science* (Beattie 1790–1793). He was also knowledgeable of chemistry, especially the works of Joseph Priestley (Beattie 1774, in R. Robinson 2004, 556).

29. For the context of such honorary, or irregular, degrees, see Johnston 1987.

30. This quotation and Beattie's involvement in obtaining Laing's MD are detailed in Beattie 1782 (R. Robinson 2004, 1081). The Latin phrase reads as follows in English: "May it bring you happiness and good fortune."

31. For example, Laing played a role in the introduction of smallpox inoculations (Beattie 1788a, in R. Robinson 2004, 1506).

and 1790s.[32] Laing eventually became the Beattie family's personal physician when they visited Peterhead and this appointment motivated Beattie to express his approbation of Laing's medical expertise in a letter to Robert Arbuthnot: "Yet Dr Laing (of whose medical skill I have a very high opinion) is under no apprehensions, and assures me there is nothing the matter with him [Beattie's son] but weakness, which, being the effect of relaxation merely, good weather, fresh air, strengthening medicines and moderate exercise, will in time remove."[33]

Laing's role as a local physician soon led him to become interested in the chemical composition of the town's well and he began to conduct his own experiments on the water. His pamphlets on the well, mentioned above, indicate that he was well-versed in medical chemistry, especially as promoted in the works of William Cullen, Torbern Bergman, Carl Linnaeus, and Thomas Beddoes. Near the end of the 1790s, he expanded his interests into the therapeutics of temperature and took care to read the works by Benjamin Thompson (Count Rumford), James Currie, and William Wright on the subject. These authors used recent developments in the chemistry of heat and coldness to reevaluate the therapeutic value of water. Laing's provincial interest in consulting such up-to-date chemical rationales emanated from the experimental mindset that permeated medical theory in Scotland. Indeed, even though histories of chemistry often portray the Chemical Revolution of the 1790s as a time of upheaval, the truth of the matter was that whenever a "new" substance was isolated or renamed, it was quickly appropriated for spa water analysis. Such a context explains why Laing's *Sea Baths* included a more serious discussion of temperature (that is, a substance directly relevant to Lavoisier's new notion "caloric"). This was the case not only for Laing's works, but for other local chemists throughout Scotland and Europe in general.[34]

Laing's knowledge of *materia medica* was also enhanced and validated by the reading public's interest in the subject. The number of health-related periodicals and books were increasing in Britain at this time, thereby creat-

32. Beattie corresponded often with Laing about medical treatments and a variety of other topics (including politics, music, and poetry). Beattie also encouraged Laing's artistic interests in various ways, such as sending him a cello from Edinburgh and placing Laing in charge of two organs being constructed in Peterhead (Beattie 1772, in R. Robinson 2004, 335; Beattie 1773, in R. Robinson 2004, 372).

33. Beattie 1790 (R. Robinson 2004, 1631).

34. Coley (1982) addresses the rising popularity of fixed air, a new substance from the 1760s onward, throughout his article.

ing a more medically literate audience.[35] Discussions concerning the material basis of drugs appeared often in late eighteenth-century Scottish periodicals, handbooks, and even polite correspondence. The pages of local newspapers and the gentlemanly columns of *The Scots Magazine,* for example, were filled with chemically based suggestions about health and agriculture.[36] It was also common in Lowland Scotland for students and members of the public to learn medical chemistry by attending university courses on *materia medica,* purchasing their own pharmacopoeias, and reading articles in the periodic press.[37] Even books for young children and adolescents contained apothecary weight tables and conversion charts.[38] Many of these sources cited medical authorities like Cullen or Bergman, but, when it came to discussing the therapeutic value of local spa water, authors tended to give priority to observations that they had made themselves or that had been made by another respected local source. Laing was no exception to this narrative trend, and, aside from summarizing several patient histories, the bulk of the empirical evidence in his two pamphlets originated from his own observations.

## Material Analysis and Commodification

### LAING'S CHEMICAL EXPERIMENTS

When Laing published *An Account of Peterhead* in 1793, he was keen to connect the chemical composition of Peterhead Well to contemporary therapeutics so that the water could be commodified alongside the other tourist attractions of the town.[39] To isolate the material substances in the water, he turned to various contemporary forms of chemical analysis. Because he did not have easy access to the metrologically uniform instruments being produced in

35. Porter 1985. A basic knowledge of medical theory was often encouraged among polite members of British society (see Golinski 1992, 11–49).

36. To my knowledge, there is no definitive work on this topic. The chemical interests of the publishers are addressed throughout Eddy 2007 and 2009. For Scotland, the most helpful list of periodicals for the period appears in Craig 1931.

37. *The Scots Magazine* published a number of articles pertaining to chemistry every year. For example, the 1787 volume reprinted *Edinburgh Magazine*'s summary of the positions of the antiphlogistonists and phlogistonists. Obituaries of chemists were also given, including that of Lavoisier. See "Abridgement of M. Metherie's Retrospective View of the State of Natural Science for the Year 1787," 1787; "Account of M. Lavoisier," 1798; and "The Life of Antoine Laurent Lavoisier," 1798. The wider readership of chemistry is addressed in Golinski 1992.

38. Eddy 2009.

39. The commodification of health in eighteenth-century England is addressed in McKendrick, Brewer, and Plumb 1982. For Scotland, see Durie 2003.

metropolitan settings, he performed experiments using whatever his locality could provide. His descriptions of his experiments mention an eclectic array of common instruments, some of which were wine glasses, china cups, tin plates "tubulated" into retorts, glass retorts, glass bottles, siphons, a tin kettle, stoneware cups, and tobacco pipes for stirring.[40] He also used specialized glass instruments like Florence flasks and a Nooth's apparatus that were employed by contemporary chemists of the time.[41] As most of these common and specialized items were used regularly by apothecaries, it points to the key role that they played in nurturing chemistry outside large cities.[42] Laing used his instruments in conjunction with heat and acids to study a variety of chemical processes, including distillation, filtration, crystallization, evaporation, deliquescence, calcination, and phlogistication.[43] With the help of gravimetric tabulation, he determined the number of grains of each substance contained in a twelve-pound avoirdupois sample.

Despite the ingenuity of his experiments, Laing denigrated the exactitude of his results. More specifically, he believed that time constraints and the "imperfection" of his apparatus had created some quantities that could not be replicated in future experiments on the water. Indeed, on the front page of his first Peterhead pamphlet, Laing honestly stated that "the analysis of mineral water is a matter of no little difficulty." But, even so, he felt that some of his results could not have been "more accurately performed by another."[44] As the other chapters in this volume demonstrate, this situation was not unique to Laing or to other rural-based physicians per se. The challenges arising from the complexity of material substances were quite common, and

40. It must be noted that this list is a bit more basic than types of apparatus that Peter Shaw's *Enquiry* (1734) used for mineral water analysis. Shaw's list is sometimes used by historians to discuss eighteenth-century chemical instruments. It is reprinted in Coley 1982, 128. For chemical experiments occurring outside university laboratory settings in Scotland, the Duke of Argyll's instruments have received the most detailed attention. Many of them are cataloged in Emerson 2002.

41. The most well-known instrument used to impregnate water with fixed air was Nooth's apparatus. It was made of glass and was fragile. The first one was designed in the mid-1770s and it soon became popular throughout Britain. At the start of his experiments, Laing had hoped to track the water's quantity of fixed air by using a Nooth's apparatus; however, to his consternation, it broke and he could not get it repaired. Coley (1982) also addresses instruments used to extract gases from mineral water. See also Brownrigg 1765.

42. Laing most probably obtained his instruments from James Arbuthnot, the local apothecary.

43. The experimental relevance of these processes is addressed in Multhauf 1966; Holmes 1989b; Newman and Principe 2002; and Klein's chapter in this volume.

44. Laing 1793b, 18.

it must not be forgotten that self-deprecation was a common rhetorical tool used in published works of this time period.[45] But setting rhetorical issues aside, most eighteenth-century chemists, Scottish or otherwise, dreamed of having better facilities and equipment. This was the case for many chemically minded improvement writers and for medical students.[46] Laing's humble protestations notwithstanding, he was able to do a great deal with the tools and the substances that he had in his possession. His inability to achieve the exactitude that he desired was less a matter of his skill and more of an issue of limited free time and the financial resources required to acquire and repair equipment.[47]

Laing devoted the first one-third of *An Account of Peterhead* to recounting the experiments that he had performed on the water. He argued that several substances in the spa's water had therapeutic powers. In making this assertion, he was able to connect the materials in the water to chemical theories of illness and health that were being used by physicians in Scotland and elsewhere.[48] More specifically, he wrote that: "The medical effects of the Peterhead water, as of all other mineral waters, depend on a combination of various causes: but in so far as they depend purely on the water, they are founded chiefly on these three parts of it, the IRON, especially that part of it which is united to the *muriatic acid*, the COLD WATER, and the FIXED AIR."[49] With this statement, it can be seen that Laing wanted to emphasize that the medical effects of the water derived explicitly from its chemical composition. In addition to these three chemical components, he held that there were several other notable ingredients. In the end, Laing concluded that it contained the substances shown in table 8.1.

All of these substances were seen as commodifiable pharmaceutical simples that had significant therapeutic value at the time.[50] Isolating the spa's

45. These strategic literary practices are addressed throughout Golinski 1992.

46. For more on chemistry and its presence in improvement literature, see Eddy 2007; for the Royal Society of Medicine, see Risse 2005, 83–88.

47. Laing's time was monopolized by house repairs and by political and ecclesiastical responsibilities. This explains why it took him at least a year to write the first pamphlet (Beattie 1792, in R. Robinson 2004, 1769). Additionally, when his Nooth's apparatus broke, he could find no one to fix it and purchasing a new one would have been too expensive.

48. Two studies that address the interaction between medical theory and chemistry in Scotland are Maehle 1999 and Risse 2005. For another comparative view between Britain and Germany on this topic, see Maehle 2007.

49. The uppercase lettering in this quotation occurs in the original text written by Laing.

50. These substances appear in the pages of any early modern pharmacopoeia. A good reference work that lists the names, contents, and therapeutic value of eighteenth-century drugs is Estes 1990.

TABLE 8.1. Analysis of the water of Peterhead Well

| *Substance* | *Quantity* |
|---|---|
| Aeriated iron | 3 grains |
| Muriated iron | 30 grains |
| Muriated lime | 7 grains |
| Siliceous earth | 2 grains |
| Gypsum | 2 grains |
| Glauber's salt | 13 grains |
| Common salt | 7 grains |
| Fixed air ($CO_2$) | 83 1/6 cu. in. |

components in this manner, moreover, was commonly practiced in the texts that Laing cites in his pamphlets. Iron, fixed air (carbon dioxide), and coldness were known as tonics in most spa literature. Notably, two of these three materials mapped on directly to two types of spa water outlined in Buchan's widely read *Domestic Medicine:* ferruginous waters, which included iron; and gaseous waters, which included fixed air. The third category, cold waters, appealed to the rising value assigned to the therapeutic uses of "cold," which some held to be a substance called "frigorific." By flagging the presence of these three types of tonics, Laing was effectively highlighting what he believed to be the most salable contents of the water. In order to understand his rationale on this subject, more needs to be said about their therapeutic value.

## THE THERAPEUTIC POWER OF MATERIALS

Although the medical relevance of metals has been generally overlooked by historians, they played a strong role in Enlightenment Scottish pharmacology.[51] Iron in particular was thought to "constringe" bodily fibers.[52] For those whose nerves needed strong stimulation, pure iron filings were administered in the form of a pill. But the easiest way to ingest iron was to drink it. Waters that contained iron were called "ferruginous" or "chalybeate" during the early modern period. Laing was not the first to realize that the Peterhead's water was chalybeate in nature; indeed, the presence of iron had been addressed by George Skene and Donald Monro during the 1770s.[53] All three of these men knew that, based on eighteenth-century chemistry, the only materials that could be fully dissolved in water were acidic or alkaline salts. If

51. The therapeutic use of metallic simples dates back to ancient times. See McCallum 1999 and Lewis 1770, 489–546.

52. Lewis 1770, 492.

53. Skene 1773; Monro 1770, 262, 278.

an acid was poured over pure iron, it produced a salt known as "vitriol of iron" (sometimes called "salt of iron"), that is, an acid which was combined with iron and which could be dissolved into water. Liquids that contained vitriol of iron could be made artificially or they could be obtained naturally via chalybeate spas. In his first pamphlet on Peterhead Well, Laing stated that the spa's water contained one whole grain of salt of iron to the pint.[54] Such a dosage was considered to be just right for someone whose nerves had become too tense and needed to be loosened via a slight jolt of stimulation. Such an assertion also fit well with the mid-eighteenth-century belief in the tonic power of acidic mineral waters.

Gaseous water was an extremely popular experimental topic during the eighteenth century and it attracted the attention of leading chemists.[55] The tonic power of aerated waters was used for many diseases, two of the most common being a weakened nervous system (which caused sluggishness and inattention) and bladder stones.[56] In Edinburgh, Francis Home and Robert Whytt were particularly interested in them. Their early studies, however, assumed that the gas contained in the water was common air. After Joseph Black demonstrated the existence of "fixed air" in the 1750s,[57] William Cullen suggested that this new gas had tonic powers. Over the next three decades, mineral wells that contained fixed air were prescribed based upon the chemical impact they were thought to have upon the body. If aerated water could not be obtained naturally from mineral wells, it could be made artificially by mixing salts into distilled water or by using instruments to imbue water with fixed air. Some of Cullen's students went so far as to suggest that the levels of fixed air in the body needed to be balanced like any other fluid if health was to be maintained.[58] Laing, on the other hand, was more conservative and stated, "I make no doubt that the tonic effect of fixed air lies much more in its rendering cold waters pleasant to taste, agreeable to the stomach, and easy to pass off, than in any strengthening power inherent in itself."[59]

Cold water was often treated as a tonic by Laing's contemporaries. He does not reveal whether he thinks "cold" is a material or a property of a substance, an ambiguity that could very well be linked to the contested nomenclatural

54. The English Imperial troy grain at this time was 64.8 mg (Connor and Simpson 2004, 758).

55. Including Joseph Priestley. See Priestley 1772; Golinski 1992, chap.4.

56. Maehle 1999, 55–125.

57. Black 1756.

58. For the therapeutic relevance of fixed air, see Macbride 1764. This book went through multiple editions.

59. Laing 1793b, 22–23.

status of "heat" and "cold" at the time.[60] Even so, he held its tonic power to be similar to that expressed by iron and fixed air. In Scotland, the therapeutic power of coldness had been investigated by Cullen during the 1750s, especially in an essay that showed how evaporation facilitated a drop in temperature upon the surface of thermometer bulb.[61] From this experiment, Cullen reasoned that "cold" could possibly invigorate the body, which meant that it was a tonic. Drawing from this work, his pupils pursued further physiological experiments which convinced them that perspiration cooled the body. Thus, sitting in a heated room or taking a walk in cool air, ironically, created a coldness that had a tonic power.[62] Because most tonic remedies were taken orally, temperature-based cures were sometimes seen as being less invasive, and this contributed to their rise in popularity during the 1780s and 1790s.[63] This therapeutic shift engendered modifications in the way that mineral water was administered and, consequently, commodified. Although people continued to drink it, hot houses and cold baths were built on spa sites so that visitors could soak themselves in vaporous heat or cold water. Additionally another readily available medium for tonic coldness was the sea's water and air. Laing's first pamphlet on Peterhead Spa noted this trend, and his discussion of coldness focused more upon the town's sea baths. He held that their extreme cold (48°F) contributed to health by "gently irritating [the body's] nerves and blood vessels, and thereby producing that glow and sensible perspiration on the skin, which are so agreeable after cold bathing."[64] For those unable or disinclined to bathe in seawater, Laing recommended the benefits of the cold sea air.[65] Laing noted that Peterhead's cold baths ranged between 46°F and 48°F (the normal temperature of the sea) and the warm bath was heated to 96°F.

As mentioned above, Laing took a closer look at temperature-based therapeutics after he published *An Account of Peterhead.* His subsequent thoughts on this subject were influenced by James Currie's *Medical Reports on the Ef-*

60. Chang 2002. See also the various sections on "cold" and "heat" in Chang 2006.

61. W. Cullen 1756. Cullen's views on temperature (especially heat) are contextualized in Taylor 2006.

62. The experimental evidence for the tonic power of cold water, as investigated by Cullen's students during the 1780s, is addressed briefly in Risse 2005, 84. Donald Monro, also a student of Cullen, addresses the "coldness" of Peterhead throughout Monro 1770. Theories of heat and health are addressed in Forrester 2000.

63. The recalibration of hot and cold cures also paralleled the rise of portable, accurate, and cheap thermometers. See Estes 1991.

64. Laing 1793b, 45.

65. Rustock 2002 addresses the therapeutic aspects of airs (meteorology) in chapter 5. Notably, Laing's assessment of cold waters and airs is more optimistic than the strictures against extreme temperatures present in Buchan 1786 and 1788, 716–18.

*fects of Water, Cold and Warm* (1797).[66] This was one of the most authoritative texts during the late 1790s on the tonic and relaxatory powers of high and low temperatures. He also had read the works of Count Rumford, William Wright, and Thomas Beddoes.[67] Throughout both Peterhead pamphlets, Laing cites all four of these authorities. Though the temperature of aerial and aqueous fluids had a long therapeutic history, these works were written during the transition from the older principle-based classification system of chemistry to that proposed by Antoine Lavoisier and his colleagues. Like cold, heat (or caloric) was still treated as a therapeutic material during the 1790s. Laing competently discusses all of the foregoing sources in *Sea Baths* and relates their therapeutic conclusions to Peterhead's waters and airs. In doing so, he introduced names taken from the French nomenclature, thereby popularizing new chemical terms to locals and visitors who may not have previously encountered them.[68]

## The Socioeconomic Context of Material Authority

### THE ECONOMICS OF PETERHEAD'S WATER

Like many spas in Enlightenment Europe, the mineral water of Peterhead had attracted local attention for several centuries. During the 1760s and 1770s, the city started to receive more visitors on account of improvements made to the country's roads and ports by the crown in response to the 1745 Rebellion. Positive tourist reports of its spa soon followed. For instance, in 1778 David Loch's *A Tour through Most of the Trading Towns and Villages of Scotland* stated that "[Peterhead] is the Scarborough of North Britain, and has excellent accommodation for bathing. The mineral waters are much in vogue, and the inhabitants obliging and industrious."[69] Laing's role as a local medical authority and his interest in Peterhead's spa water, therefore, occurred at a time when the well's contents could be accessed more easily by a rising number of visitors. As he relates in his pamphlets, the town of Peter-

66. Currie 1797. Currie's work on the therapeutics of heat and cold is addressed in Forrester 2000.

67. B. Thompson 1798, 1804; Wright 1786; Beddoes 1799. Notably, Currie, Beddoes, and Wright studied medicine in Edinburgh under Cullen and Black. For more on their thoughts on temperature, see W. Cullen 1756 and Black 1770—although it must be kept in mind that the latter source was a pirated edition of Black's lectures.

68. Two such terms were "carbonate of iron" and "sulfurated hydrogen" (Laing 1804, 10–11).

69. Loch 1778–1779, 120. This quotation is also contained in Loch 1778, 61.

head had rapidly expanded in recent years. From the 1770s to the 1790s it had doubled its size from a few hundred residents to a population of well over five hundred. Although the well had been a popular attraction for northern aristocrats, gentry, and wealthy professionals, the economic success of the town was largely dependent on its role as a base for fishing vessels and as a port for cargo ships arriving from Europe (Norway, Sweden, and Holland), England (Newcastle, Sunderland, and London), and Scotland (Leith and Inverness). With the rise in trade also came a much higher profile for Peterhead in Britain, and from the 1780s onward Edinburgh newspapers and the London *Times* ran stories about the port and its inhabitants. Like many such accounts, the brief reports praised the well's ameliorative effects but gave no clear indication as to how its contents connected with the medical theories that guided therapeutics; the same could be said for advertisements for the well in the local press.[70] Thus, Laing's unique contribution to the popularization effort was the fact that his research identified specific therapeutic substances that were relevant to the rising tide of medically literate readers of the late eighteenth century.

Peterhead's rising status within the shipping industry made it more convenient to visit and, correspondingly, more appealing to tourists. One of the leading local groups that sought to exploit this potential market was the town's Freemasons.[71] In 1793 the society bought a plot of land adjacent to the well. Under the leadership of James Arbuthnot (junior), the town's apothecary, the society improved the facilities available to visitors. Because this investment was predicated upon the expectation of financial success, it should come as no surprise to see that Laing's first pamphlet appeared during the same year that the Freemasons began to seriously improve the well's facilities. His decision to commodify the well's contents was particularly well timed and followed a trend in which pamphlets and metropolitan newspaper reports served to popularize provincial spas and their facilities. While such publications were sometimes vague on dosage, Laing's pamphlet offered specific quantities. If someone simply wanted to be reinvigorated, he recommended bathing in the water and drinking one gill of it per day for six

70. See the Peterhead Spa articles in *Aberdeen Journal* for 22 July 1771 and 17 July 1775.

71. I have not been able to ascertain whether the Freemason's Lodge employed any other local physician to test the well's waters, nor whether they stocked the pump room's reading area with other types of medically relevant literature. For the interaction between Freemasonry and natural philosophy, see Elliot and Daniels 2006. The reading rooms of spas and the presence of spa pamphlets in reading libraries is briefly addressed in Grenby 2002, see especially pp. 24 and 31.

weeks.[72] Because summer admission to the pump room cost one guinea and each bath cost a shilling, it is clear that the Freemason's Lodge benefited from the attention that Laing's pamphlet attracted to the well. Based on the salability of the work, it would seem that he succeeded in popularizing the site. After the initial order of one hundred copies sent to Peterhead sold rapidly, Laing's gout-stricken publisher was keen to have him publish more on the subject.[73] However, as mentioned earlier, *Sea Baths,* the supplement, did not appear until 1804 (mainly because Laing was too busy with his parish duties).

Because most of the paying visitors came from the middle and upper ranks of society, Laing was quick to mention that the well's clientele included "ladies and gentlemen," including "senators, judges, philosophers, military officers, clergy, [and] merchants."[74] Aside from the entrance fees, visitors had to pay for accommodation, meals, and entertainment, all of which benefited the local economy. Thus, while the spa's water was known to those in the north of Scotland, the circulated copies of Laing's pamphlet communicated a much more specific idea of the water's contents and the environment in which they could be enjoyed, thereby justifying the prices advertised in his essay. Such a situation benefited both Laing and the townspeople. Beattie summarized this viewpoint in a letter that he sent to Sir William Forbes in 1793: "Our friend Laing at Peterhead has, at Mr Arbuthnot's desire and mind, written an Account of Peterhead, its mineral water, air, and neighbourhood: it is a pamphlet of 80 pages closely printed; and I think will do honour to him and good to the publick; I hope too he shall get a little money by it. It is dedicated to the Merchant Maiden Hospital, who certainly ought to make him a handsome present, as his work can hardly fail to draw strangers to Peterhead, and so raise the value of their property."[75] Beattie's prediction soon came true. Two years later Peterhead's summer spa business was booming, and this motivated the Freemason's Lodge to add cold and warm baths, contiguous with the well, that used seawater on account of "the gentle stimulus of the saline particles applied to the nerves and vessels of the skin."[76] This use of hot

72. Laing 1793b, 49–50. A gill was a "measure of liquid capacity, being 1/16 of a pint, 1/4 of a mutchkin" (Connor and Simpson 2004, 758).

73. Despite the Aberdeen printer (Chalmers) of the pamphlet being crippled by gout, Beattie also mentioned that it was popular in the town (Beattie n.d., in R. Robinson 2004, 1804; Beattie 1793b, in R. Robinson 2004, 1820b). Once Chalmers recovered later in the year, he was keen to get Laing to write a supplement (Laing 1793a, in R. Robinson 2004, 1834).

74. Laing 1793b, 75.

75. Beattie 1793a (R. Robinson 2004, 1796).

76. Laing 1804, 18.

and cold baths drew on the temperature-based therapeutic theories outlined above and followed similar additions to other European spa towns (Bath and Carlsbad, for instance). By 1799 James Arbuthnot had installed a forty-foot by twenty-foot gentleman's bath and turned the Lodge into a pump room where visitors' nerves could be stimulated by drinking coffee or relaxed by listening to organ music. It was these additions that were discussed by Laing in the *Sea Baths* sections that addressed the temperature of the town's well, seawater, and air.

## VARIETIES OF MATERIAL KNOWLEDGE

I have found no evidence of any major disputes that took place over the types of substances that the water actually contained. There was one point of contention, however, that exercised the pens of Peterhead's local experts. Although they all agreed on the value of the chemical materials contained in the water, they differed on the therapeutic power that such substances had over the human body. A few of these disagreements are worth noting, as they shed light on the social context that affected the way in which Laing's contemporaries, especially those who were not part of the medical establishment, interpreted the impact the well's material substances had upon the physiology of their own bodies.[77] In what remains of this chapter, therefore, I shall treat the varieties of material knowledge evinced in the writings of Laing, Professor James Beattie, and Rev. Dr. George Moir, with a view to show the individual interpretations that were attributed to the effects of the substances contained in the water, thereby highlighting three forms of therapeutic expertise that were guided by personal experience.

Beattie's knowledge of the well stemmed directly from the tonic power that water had over the many maladies that he and his family experienced.[78] Indeed, his oldest son, the poet James Hay Beattie, expressed the family's affection for the water in a poem that he wrote about the well in 1793. Entitled "Invitation to Peterhead," it extols the healthful virtues of the well's location and its "sparkling nectar."[79] Such a positive perception perhaps explains why

77. Late eighteenth-century academic disputes over the therapeutic effects of different drugs are addressed throughout Maehle 1999. The constant revaluation of the drugs contained in the Edinburgh pharmacopoeia is treated in David L. Cowen 1957, 1974, 1982, and 1985.

78. Beattie believed in the tonic power of Peterhead's water until the end of his life (Laing 1803, in R. Robinson 2004, 2031).

79. Laing 1793b, 78–79. An excerpt from this poem is quoted at the beginning of this chapter.

the senior Beattie drank much of the "nectar" over the course of his adult life. He was plagued by a weak and overworked nervous system that engendered headaches, a "windy" stomach, weak bowels, "depression of spirits," and vertigo. During the Enlightenment, such disorders were often associated with intellectuals, as the process of intense thinking was thought to weaken the nerves.[80] Additionally, his youngest son, Montagu, was sickly as a child and his wife, Mary, suffered from a nervous disorder that eventually led her to be committed to an asylum.[81] The entire family, then, suffered from nervous disorders that justified the use of the water.

The Beattie physicians were John Gregory, George Skene, Thomas Livingston (physician to the Aberdeen Infirmary), and Laing. As was common, the Beatties were prescribed spa water for their nervous conditions. Laing even used James Beattie (whom he called the "benevolent Dr B***") as one of his "nervous" case histories in the first Peterhead pamphlet.[82] Although he was familiar with the waters of Tunbridge, Pitcaithly, and Pannoninch, Beattie's letters repeatedly emphasized that he preferred the salubrious effects of Peterhead's water and the air of its surrounding environment. Based on the tonic power that the iron and fixed air in the water had on his viscera and flesh, Beattie drank the spa's "nectar" and occasionally bathed in it as well.[83] After a few comments about his past use of the water, he outlined his personal therapeutic approach in a letter written to Sir William Forbes in 1788: "Instead of taking it with an empty stomach and exercise after it, which is the established method, I drink it at six in the evening, at going to bed, and between three and four in the morning, sleeping after it. It has produced one symptom [flatulence], which every body tells me is a favourable one . . . and I am upon the whole much better."[84] Here we can see that he had confidence in his own testimony over the water's ameliorative powers—so much so that he felt that he understood the *effects* of the water well enough to medicate himself. The medical theory behind this decision was undoubtedly taken from what he had read and from conversations with his physicians. The ultimate authority, however, rested with Beattie, and his case serves as an informative example of the way in which patients

80. The context of this intellectual condition is treated in Porter 1991; Rousseau 2004; and Böhme and Böhme 1983.

81. R. Robinson 1996.

82. Laing 1793b, 72.

83. References to his drinking and bathing in the water are made throughout his correspondence from the 1760s up until his death. An outline of his health (and that of Mary Beattie) is given in the introduction to R. Robinson 2004.

84. Beattie 1788b (R. Robinson 2004, 1499).

played a respected role in therapeutics on the eve of the onset of clinical medicine.

Whereas Beattie is representative of those polite spa denizens familiar with medical theory who drank the water according to their own interpretation, Laing is more representative of eighteenth-century physicians whose livelihood required that they at least mention the relationship between the contents of spa water and the therapeutic theories promoted by the medical establishment.[85] This being the case, Laing gave a select handful of examples of instances in which the contents of the water had ameliorated some of his own nervous conditions as well as those of a few townspeople.[86] But he did not comb the town's streets and infirmary wards to compile an exhaustive list of case histories. There were two main reasons for this, and both shed light on the intended readership of his pamphlets. First, his retelling of the physiological effects of the water's contents followed the familiar format of the brief patient histories used by physicians to relate the therapeutic power of pharmaceutical simples and compounds in academic journals like Edinburgh's *Essays and Observations Literary and Physical,* popular periodicals like *The Scots Magazine,* and successful self-help medical handbooks like Buchan's *Domestic Medicine.* The readers of these publications were usually satisfied with one or two clear examples. That is to say, Laing's testimonial format would have been familiar to both academic and common readers. Second, Laing clearly believed that the mere identification of the individual tonics, especially iron, was sufficient evidence for an audience already familiar with the power that contemporary medical theory attributed to such substances. This explains why he did not offer a detailed account of the theories that underlay their physiological impact (as an introductory medical text might do). When it came to pharmaceuticals, Laing's readers did not need to be given detailed patient histories, because they were already familiar with the medicalized presentation of the effects of drugs (and their material contents) that permeated Scottish print culture.

Though Laing's analysis of the water's content was not seriously challenged, his interpretation of the effect of the tonics attracted some criticism. In particular, Rev. Dr. George Moir took issue with Laing's comments on the spa.[87] As well as being the Presbyterian minister of the town, Moir also held

85. The nature and efficacy of the therapeutic treatments in Edinburgh's Royal Infirmary are addressed in Risse 1986 and Tröhler 2000.

86. Laing 1793b, 4. Elsewhere, in his section "Of disorders that required strengthening," he states that the water had cured a woman who suffered from the "sinking of the spirits" and convulsions (that is, two nervous disorders) (Laing 1793b, 35).

87. Moir 1791–1799.

an MD from Aberdeen. His interest in the chemistry of the well stretched back at least to 1773, when he had corresponded with Joseph Black about the chemical composition of the water.[88] Moir accepted Laing's chemical analysis of the well's composition but disagreed with the therapeutic effect that Laing had assigned to it: "This water has long been deservedly in repute, for general debility, disorders of the stomach and bowels, flatulencies and indigestion, nervous complaints which flow from these causes, and diseases peculiar to the fair sex; and in all these disorders, I can from 30 years of observation and experience affirm, I know of no remedy no more efficacious."[89] Having established himself as a local authority, Moir unabashedly criticized the therapeutic value that Laing had attributed to the metallic and gaseous tonics in the water. Moir opposed the stimulatory power of tonics as being too harsh, favoring instead the relaxatory effect of diuretics, diluents, fomentations, and opiates. That is to say, he was not attacking Laing's views per se; rather, he was challenging the stimulatory theory of tonics upon which they were based. To challenge Laing, Moir used evidence gathered from local patient histories (especially on dropsy) and from Francis Home's *Clinical Experiments.* Home taught *materia medica* at the University of Edinburgh from the 1760s to the 1790s and preferred the use of laxatives, diuretics, and deobstruents.[90] In challenging the use of the tonics found in Peterhead's well, Moir was voicing a turn against stimulants that was moving through the Scottish medical community during the 1780s and 1790s (especially after Cullen's death in 1790).[91] In addition to these relaxatory alternatives, Moir also asserted that the therapeutic value of bathing was inhibited in Peterhead because it lacked a "warm bath, and a bathing machine on the sea-beach."[92]

Although his criticism of tonics was well grounded in medical theory, Moir gave little attention to Laing's comments on cold water sea bathing. This oversight might possibly be linked to differing political opinions and not the specifics of medical theory.[93] Laing, however, took care to respond to Moir's criticisms when he published *Sea Baths* in 1804. There he used the chemis-

88. Moir 1773. He also sent Black a sample of the water.

89. Moir 1791–1799.

90. Home 1780. Home thought that stimulants were too harsh on the viscera of the body.

91. Cullen's therapeutic practices were under attack by the followers of John Brown ("Burnonians") at this time. They challenged the analogical and inductive links that had been made between chemical substances (including heat and cold) and the cures engendered by drugs (Bynum and Porter 1988).

92. For more on the social placement of the Moir-Laing "debate," see Durie 2003, 212–13.

93. Moir engaged in a heated war of words with Laing and Beattie during the 1780s in a disagreement over nonjurors (Laing 1788, in R. Robinson 2004, 1540).

try of heat, especially observations based on Currie's work on the subject, to counter Moir's accusation that he did not properly understand the therapeutic value of using heated or cooled water. Although he did not specifically state Moir's name, Laing strongly criticized the "wrong direction" given by other local "advisers" who offered poor advice on account of their ill knowledge of the medical theory behind temperature-based therapeutics.[94] Thus, again, the dispute was not over the contents of the water. Nor was it over the types of instruments used to ascertain temperature. The real disagreement concerned the question of how the tonic affected the body—something that could not be easily resolved merely through the use of a test tube.

## Conclusion

In this chapter I have addressed how Rev. Dr. William Laing and other local authorities contributed to the commodification of the water of Peterhead Spa by linking its chemical analysis to medical theory. This process was guided by Laing's heterogeneous expertise as a physician and chemist, as well as by the economic rise of the city of Peterhead in the last third of the eighteenth century. In particular, I showed that Laing's role as a medical authority also qualified him to be a local expert who used chemistry to isolate and measure the iron, fixed air, and "cold" that was present in the well's water. All of the substances were tonics that, according to late eighteenth-century medical theory, were held to have an ameliorative effect on the viscera and flesh of the body. Thus, in using chemistry to identify these materials—the "sparkling nectar," so to speak—Laing was effectively participating in a wider story of commodification that gripped many provincial European spas at the time. But whereas his conclusions over the content of the well were generally accepted by his provincial contemporaries, his views concerning the therapeutic value of the materials were subject to debate. Such a situation mirrored the pharmacological disagreements on this subject that were occurring at the same time in academic medicine. As mentioned throughout the chapter, leading physicians like Cullen and Home concurred that chemically isolated materials were therapeutically viable, but they could not agree on how such cures were best put to use. While their theories were applied in different ways by local authorities like Laing, Beattie, and Moir, all of the authorities under discussion in this chapter operated under the assumption that the neohumoralist framework was essentially correct, thereby creating a shared canon of sources and ideas by which they could discuss the material contents of the

94. Laing 1804, 15–16.

well. No matter which stance a local expert took on the therapeutic power of the substances in the water, however, the fact that Laing's pamphlet was just one of a plethora of contemporary publications that addressed the chemical composition of spa water shows that their commodification was linked not only to the material knowledge of the local authors, but also to that of a medically literate audience that was at least familiar with the chemical nomenclature of the day.

## Primary Sources

Secondary sources can be found in the cumulative bibliography at the end of the book.

### MANUSCRIPT SOURCES

*Abbreviations*
AUL: Aberdeen University Library
EUL: Edinburgh University Library
NLS: National Library of Scotland

Beattie, James. 1771. Letter to William Laing, 25 June 1771, NLS Acc.4796, Fettercairn Box 91 (cf. R. Robinson 2004, 248).

———. 1772. Letter to William Laing, 13 October 1772, AUL MS 30/2/89 (cf. R. Robinson 2004, 335).

———. 1773. Letter to William Creech, 17 April 1773, AUL MS 30/1/40 (cf. R. Robinson 2004, 372).

———. 1774. Letter to Elizabeth Montagu, 27 May 1774, AUL MS 30/1/78 (cf. R. Robinson 2004, 556).

———. 1782. Letter to William Laing, 28 October 1782, AUL MS 30/1/210 (cf. R. Robinson 2004, 1081).

———. 1788a. Letter to James Hay Beattie, 3 August 1788, AUL MS 30/1/288 (cf. R. Robinson 2004, 1506).

———. 1788b. Letter to Sir William Forbes, 10 July 1788, NLS Acc.4796, Fettercairn Box 92 (cf. R. Robinson 2004, 1499).

———. 1790. Letter to Robert Arbuthnot, 25 April 1790, NLS Acc 4796, Fettercairn Box 92 (cf. R. Robinson 2004, 1631).

———. 1792. Letter to Robert Arbuthnot, 9 July 1792, AUL MS 30/1/332 (cf. R. Robinson 2004, 1769).

———. 1793a. Letter to Sir William Forbes, 23 February 1793, NLS Acc. 4796, Fettercairn Box 94 (cf. R. Robinson 2004, 1796).

———. 1793b. Letter to William Laing, 10 October 1793, NLS MS 3648 f. 169 (cf. R. Robinson 2004, 1820b).

———. n.d. Letter to Robert Arbuthnot, AUL MS 30/1/33 (cf. R. Robinson 2004, 1804).

Laing, William. 1788. Letter to James Beattie, 13 December 1788, AUL MS 30/2/579 (cf. R. Robinson 2004, 1540).

———. 1790. Letter to James Beattie, 19 December 1790, AUL MS 30/2/636 (cf. R. Robinson 2004, 1680).

———. 1793a. Letter to James Beattie, 6 November 1793, AUL MS 30/2/696 (cf. R. Robinson 2004, 1834).

———. 1803. Letter to Sir William Forbes, 1 November 1803, NLS Acc. 4796, Fettercairn Box 95 (cf. R. Robinson 2004, 2031).

Moir, George. 1773. Letter to Joseph Black, 19 March 1773, EUL GB 0237 Joseph Black Gen. 873/I/53–54.

Skene, George. 1773. Letter to Joseph Black, 25 March 1773, EUL GB 0237 Joseph Black Gen. 873/I/55–56.

### PRINTED PRIMARY SOURCES

Abridgement of M. Metherie's Retrospective View of the State of Natural Science for the Year 1787. 1787. *The Scots Magazine* 50:269–72.

Account of M. Lavoisier. 1798. *The Scots Magazine* 60:584–85.

Beattie, James. 1790–1793. *Elements of Moral Science.* 3 vols. Edinburgh: Creech.

Beddoes, Thomas. 1799. *Essay on the Causes, Early Signs, and Prevention of Pulmonary Consumption for the Use of Parents and Preceptors.* Bristol: Biggs & Cottle.

Black, Joseph. 1756. Experiments upon Magnesia Alba, Quick-lime, and some other Alkaline Substances. In *Essays Physical and Literary*, ed. Philosophical Society, 157–225. Edinburgh: G. Hamilton & J. Balfour.

———. 1770. *An Enquiry into the General Effects of Heat; with Observations on the Theories of Mixture.* London: J. Nourse.

Brownrigg, William. 1765. An Experimental Enquiry into the Mineral Elastic Spirit, or Air, Contained in Spa Water; as well as into Mephitic Qualities of this Spirit. *Philosophical Transactions* 55:218–43.

Buchan, William. 1786. *Cautions Concerning Cold Bathing, and Drinking the Mineral Waters.* Edinburgh: Balfour & Creech.

———. 1788. *Domestic Medicine: or, a Treatise on the Prevention and Cure of Diseases by Regimen and Simple Medicines.* Edinburgh: Balfour & Creech.

Cullen, William. 1756. Of the Cold Produced by Evaporating Fluids, and some other Means of Producing Cold. In *Essays Physical and Literary*, ed. Philosophical Society, 145–56. Edinburgh: G. Hamilton & J. Balfour.

Currie, James. 1797. *Medical Reports on the Effects of Water, Cold and Warm, as a Remedy in Fever and Febrile Diseases Whether Applied to the Surface of the Body or Used as a Drink: with Observations on the Nature of Fever and on the Effects of Opium, Alcohol, and Inanition.* London: Cadell & Davies.

Home, Francis. 1751. *An Essay on the Contents and Virtues of Dunse-Spaw.* Edinburgh: A. Kincaid.

———. 1780. *Clinical Experiments: Histories, and Dissections.* Edinburgh: Creech.

Laing, William. 1793b. *An Account of Peterhead, Its Mineral Well, Air, and Neighbourhood.* Edinburgh: Creech.

———. 1804. *An Account of the New Cold and Warm Sea Baths at Peterhead: a New and Excellent Mineral Spring; Various Advantages from the Warm Sea Baths; and Disorders that have been Remedied by it.* Aberdeen: Chalmers.

Lewis, William. 1770. *The New Dispensatory: Containing, I. The Elements of Pharmacy. II. The Materia Medica.* 3rd ed. London: J. Nourse.

The Life of Antoine Laurent Lavoisier. 1798. *The Scots Magazine* 59:147–50.

Loch, David. 1778. *A Tour through Most of the Trading Towns and Villages of Scotland; Containing Notes and Observations Concerning the Trade, Manufactures.* 3 vols. Edinburgh: Ruddiman.

———. 1778–1779. *Essays on the Trade, Commerce, Manufactures, and Fisheries of Scotland.* 3 vols. Edinburgh: Ruddiman.

Macbride, David. 1764. *Experimental Essays on the Following Subjects: On the Nature and Properties of Fixed Air.* London: A. Millar.

Moir, George. 1791–1799. Number XXVII. Parish of Peterhead. In *Statistical Account of Scotland,* ed. J. Sinclair. Edinburgh: Creech.

Monro, Donald. 1770. *A Treatise on Mineral Waters.* Vol. 1. London: Wilson.

Priestley, Joseph. 1772. *Directions for Impregnating Water with Fixed Air.* London: J. Johnson.

Shaw, Peter. 1734. *An Enquiry into the Contents, Virtues and Uses of the Scarborough Spaw-Waters: with the Method of Examining any other Mineral-Water.* London: for the author.

Sinclair, John, ed. 1791–1799. *The Statistical Account of Scotland. Drawn up from the Communications of the Ministers of the Different Parishes.* Edinburgh: Creech.

Taylor, John. 1790. *A Medical Treatise on the Virtues of St Bernard's Well.* Edinburgh: Creech and J. Ainslie.

Thompson, Benjamin [Count Rumford]. 1798. An Inquiry concerning the Source of the Heat which is excited by Friction. *Philosophical Transactions* 88:80–102.

———. 1804. An Account of a curious Phenomenon observed on the Glaciers of Chamouny; together with some occasional Observations concerning the Propagation of Heat in Fluids. *Philosophical Transactions* 94:23–29.

Wright, William. 1786. Remarks on Malignant Fevers; and their Cure by Cold Water and Fresh Air. *London Medical Journal* 7:109–15.

9

# Liqueurs and the Luxury Marketplace in Eighteenth-Century Paris

E. C. SPARY

Well-to-do Parisians in the 1750s and 1760s could take their pick from a remarkable range of fashionable luxury products. Mustard, lemonade powder, rum, candy, cake, and game pie jostled with a host of other goods, from fabrics, wigs, snuffboxes, natural history specimens, watches, medicaments, and fireworks to perfumes, prints, and an oyster fork, specially designed, that administered a measured volume of pepper to the shellfish in the act of prizing it from its shell.[1] By the mid-eighteenth century, food shops had spread into the newly built-up wealthy areas of the city, and this expansion was coupled with the spread of luxury and semiluxury goods to a broader social spectrum of buyers.

Traders in such goods faced fierce rivalry from commercial competitors, but their clients were ones whose every gesture, item of dress, or habit was determined by courtly standards of politeness, even if they themselves would never attend at court. As one silk merchant complained, even the Parisian *petite bourgeoisie* copied, down to the last degree, every innovation espoused by members of the court. And courtiers, he added, were extremely discriminating: they had "such an exact attention to and . . . such a perfect knowledge of" the mode that nothing smacking of last year's fashion could ever be sold to them.[2] A complex system of knowledge went into the construction of fashionable identity, and its experts were the connoisseurs: persons of taste

1. *Avantcoureur,* 1762, 186–87.

2. Poni 1998; see also Roche 1994. Poulot 1997 offers an excellent summary of the debate over emulation. The most trenchant critique has come from economic historians (e.g., Weatherill 1988, 20, 194–96; Fine and Leopold 1990; Vickery 1993), while historians of consumption are more favorably inclined toward emulation (see, e.g., Coquery 2003; Fairchilds 1993).

and high social rank. This polite science was gleefully satirized by contemporaries, even as they practiced it. The Italian-born marquis Louis-Antoine de Caraccioli parodied the excessive variety of everything fashionable in Paris: "In one decade, how many ways of styling one's hair, bedizening oneself, applying beauty spots, gaudifying, perfuming oneself, dressing, swelling with conceit, introducing oneself, greeting, speaking, carving, eating, dancing, walking, blowing one's nose?"[3] Every detail of dress and gesture, every accessory, every table manner was part of the polite science; only true experts—both sellers and buyers—could master the vast amount of information and detail required to remain ahead of the emulators. In this sense the marketplace was the meeting-place for at least two kinds of knowledgeable experts, artisanal vendors and polite clients; and the community that judged fashion goods, including prepared foods, was a community of practitioners of polite science, led by connoisseurs.

It is now a commonplace of anthropologically informed histories of food to assert that foods cannot be treated merely as material substances serving nutritive purposes: they must be viewed as part of broader systems of communication, social distinction and self-fashioning. It behooves us to ask just how meaning can be said to inhere in material objects, specifically in foods, and what happens to meaning during consumption, when food disappears and becomes ourselves. Although it is not new to assert that foods have symbolic meanings, I mean to take this claim in a stronger sense and systematically explore how meaning and foodstuff were interrelated at the level of production and commerce. My chapter will argue that the meaning of foodstuffs is a product of the repertoires of manipulations, uses, and technologies within which foods as objects exist and circulate. The same object can be at one and the same time the property of different groups of experts and users. If meaning arises out of negotiations, conflicts, and appropriations over material objects, consensus then appears an exceptional and fragile state maintained only by labor. These issues have been explored within the sociology and anthropology of science, most notably in the debate over boundary objects and in a slightly earlier literature on actor-network theory. However, in the former case, objects have an independent existence; in the latter, the objects themselves are transformed in passing between actors. My interpretation tends toward the latter rendering of liqueurs and other prepared foods; indeed, it will become clear that there was ultimately no way of defining a ma-

3. Caraccioli 1757, vi–vii. The expression "polite science" is a translation of the "science du monde" critically characterized by Helvétius (1988, 100; 1988, discourse 2, chaps. 4–10). Caraccioli's satire was partly aimed at this work.

terial substrate for these goods that achieved universal assent. Their chemical and alimentary identity and physical and social effects were all differently defined by the individual constituencies of actors who in one way or another interested themselves in the production and consumption of liqueurs in mid- to late eighteenth-century Paris. I will seek to show that, outside certain very limited spaces, no epistemological superiority can be assumed for scientific objects over everyday ones in the eighteenth-century city.[4]

Among some forty prepared-food manufacturers advertising in one weekly newspaper, the *Avantcoureur,* between 1761 and 1773, were apothecaries and grocers, distillers and *limonadiers, traiteurs,* pastry cooks, roasters, mercers, and vinegar makers. These vendors were united more by the range of goods they had on offer than by guild affiliation. The largest single category was the liqueurs, whose urban manufacture was legally restricted to the distillers and limonadiers' guild. However, they were also imported, manufactured, and sold by members of other guilds. Other well-represented categories included confectionery, pastry goods, vinegars, and mustards.[5] Common to all these goods was their artistry, or, negatively viewed, their artificiality; to many contemporaries they were, as will become clear, allied with the products of chemical operations.[6] These food products were often the subject of commercial rivalry. Luxury merchants from different guilds vied with one another to present the latest novelty or fashionable item, and prepared foods, which had associations with exoticism, artificiality, and pleasure, attracted criticism from writers of health manuals.

The struggle to capture the meaning of liqueurs, to produce them as objects of consumption and knowledge, went on at commercial, scientific and medical levels. Because of the fine line between foods, remedies, and toiletries, and the fact that all three were purchased by the same clientele, prepared-food entrepreneurs needed to address not only consumers' views of their products, but also other meanings for those goods supported by members of rival corporations and organizations, including the scientific institutions and the medical guilds. I will explore in detail some treatments of distillation penned by members of the apothecaries' and distillers' guilds and show how

4. Bachelard 1991; Pickering 2005; Star and Griesemer 1989, taken up by Fujimura 1992, esp. 170–76; Latour and Woolgar 1979; Latour 1986, 2000; Daston 2004; Turner 1996, esp. 37–43. Haraway (1991, 200–201) argues that the boundaries of scientific objects "materialise in social interaction."

5. The advertising practices for such items resembled those used for cosmetics or medicaments. See Martin 1999, chap. 2; C. Jones 1996; also the important discussions in Berg and Eger 2003a, 2003c; Perrot 1995; Coquery 1998.

6. Flandrin 1996a, 1996b.

they defined and utilized chemistry and how they juggled the world of scientific institutions and polite print culture to further their commercial ends. As I will show, liqueurs were the focus for debates about the relationships between matter, bodies, and expertise in the age of Enlightenment.

## The Compound Liqueurs

After its union with the limonadiers in early 1676, the distillers' guild encompassed merchants operating at different levels, including large-scale manufacturers of brandy for consumption and the arts, entrepreneurial liquorists who specialized in the production of the so-called compound liqueurs, combinations of brandy or water and fruit, spice, or floral flavorings, and café owners who sold brandy and liqueurs alongside the exotic beverages of coffee, chocolate, and tea and sorbet, ices, and preserved fruits and nuts.[7] The boundaries between these categories were not clear cut. The limonadiers, the *vinaigriers,* or vinegar makers, and the grocers and apothecaries were all licensed to distill wine into brandy, and any member of the distillers' guild could make and sell a range of liqueurs. There were also street vendors who offered a tot of brandy to passing pedestrians. Brandies and liqueurs were thus marketed at various social levels, from the elite to the city's poor.[8]

Liqueurs were variously evaluated according to the terms of different alimentary knowledge systems. Some contemporaries argued that liqueurs fed the nervous system and enhanced mental abilities, a particularly important function if eating and drinking were to be allied to mental self-cultivation and self-presentation.[9] For an audience dominated by entrepreneurs and fashionable consumers, liqueurs denoted pleasure and tastefulness. I shall call this the connoisseurial interpretation. According to others, liqueurs were ingesta without being nutriment. Thus, debates about their status amounted to debates about the function of "empty" foodstuffs, ones that were consumed solely for their pleasurable qualities and for their mental effects. Liqueurs were the ultimate hedonistic food, and for those who argued that everything nonnutritious should be treated as a medicament, they played a particularly problematic cultural role in their guise as the signifiers of fashionable life and of self-pleasuring. They reduced alimentary intake to pure symbolism and to commercial transaction; moreover, they were substances explicitly con-

7. Demachy 1775, 109.

8. In the latter case, they would come to be associated with public disorder and drunkenness (Brennan 1988, 88). However, my concern here is with the production of luxury liqueurs.

9. Poncelet 1755.

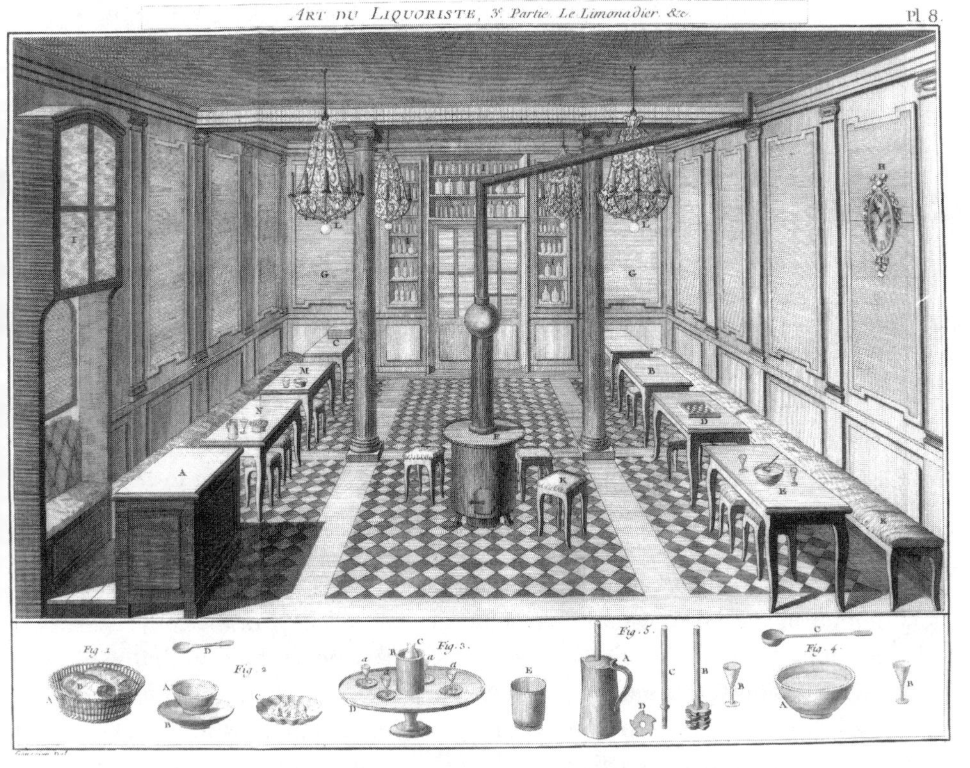


FIGURE 9.1. Interior of a café. (Demachy 1775, plate 8; copperplate engraving after Goussier. Reproduced by kind permission of the Syndics of Cambridge University Library.)

sumed for the purposes of altering the healthy individual's moral state. For this audience, dominated by medical reformers, pharmacists, and physicians, liqueurs denoted degeneracy and addiction—the health interpretation. No wonder, then, that liqueurs were often cited as evidence of the moral excesses of the modern age.

In the eighteenth century, the term *liqueurs* was applied to all the drinks produced and sold by the *limonadier*, including coffee, tea, chocolate, and lemonade, as well as to cosmetics, perfumes, and medicaments. For eighteenth-century consumers, liqueurs epitomized the cornucopian, the embrace of Nature's treasury of comestibles. Déjean, author of a didactic treatise on distillation, included recipes using orange blossom, pinks, jasmine, violet, marjoram, lemon balm, aspic, thyme, basil, sage, rosemary, citrus, clove, anise, cinnamon, nutmeg, coffee, and cocoa.[10] According to one description, a

10. Déjean 1759, viii, 3.

typical liquorist's laboratory was supplied with some thirty aromatic spirits, twenty to twenty-four essential oils, thirty tinctures, and three types of basic brandy. Using these substrates, the liquorist could compose any liqueur quickly, creating a judicious balance of flavors and aromas.[11] Compound liqueurs were produced in large households for domestic consumption, but by the early eighteenth century, they were already a highly competitive commercial terrain, a showpiece of Paris entrepreneurialism. As one city guide noted in 1715: "It is well-established that Paris is a very delicious abode; here one finds everything which might awaken the mouth's pleasure, & flatter the most sensual. Pastry, Liqueurs, Confectionery & other things of that nature are not lacking here, everything prepared in the last. degree of perfection."[12]

Paris distillers like Déjean or his successor Sauvel were not large-scale brandy manufacturers. They specialized, rather, in the compound liqueurs, as being "preparations, or mixtures," of brandy that, according to Déjean, "facilitate or accelerate its sale."[13] The name of *eaux* was given to a class of liqueurs made by distilling water or brandy with aromatic herbs, spices, or fruits; thus, there was cherry-water, cinnamon-water, and so on. Two important new classes of spirituous liqueurs were invented in the early eighteenth century: the ratafias and oils. Ratafias resembled the *eaux*, but used brandy exclusively, combining fruit juices with exotic spices.[14] The oils were loaded with sugar syrup at levels that produced an oily consistency. These too included exotic ingredients, such as pineapple, cinnamon, or star anise, becoming more readily available thanks to the growth of global and colonial trade.[15]

## Innovation and Invention

To recruit customers, liqueur makers presented their wares as modish, pleasurable status symbols, whose manufacture taxed and displayed the entrepreneur's ingenuity. Such agendas are particularly evident in an eighteenth-century print innovation, newspaper advertisements. By the later eighteenth century, merchants of fashion items, including food entrepreneurs, used the medium to engage with a well-to-do, literate, health-conscious urban population. Food advertisements created a market ostensibly built around

11. Demachy 1775, 84.

12. Liger 1715, 352.

13. Déjean 1759, ix.

14. Macquer 1766, "Eaux distillées," 1:373–76; Delamare 1705, 3:797.

15. Mintz 1985, 79–96; Flandrin 1996b.

the consumer's social distinction, in practice increasingly permitting anyone with enough money to own taste. They concealed corporate rivalries behind appeals to the civilizing power of commerce.[16] Thus, they united two of the themes prominent in Habermas's *The Structural Transformation of the Public Sphere* (1989), the expanding world of printing and reading and the world of social and financial commerce.[17]

Only a fraction of Parisian food and drink entrepreneurs exploited the new opportunities for marketing and self-promotion offered by print.[18] In producing a luxury item as an object of knowledge in published descriptions, artisans coproduced their own persona and that of their client. Food producers constructed their clients as tasteful and discriminating, setting standards for health, fashion, and beauty that merchants merely emulated. Innovation and progress were built into entrepreneurs' marketing practice.[19] Advertisers changed or improved their products regularly—to meet "the increase in desires and luxury appetites," according to Sauvel, a leading Parisian liqueur seller.[20] This commerce rested on claims that innovation was a characteristically French phenomenon. It fitted well with the professed agenda of the *Avantcoureur,* whose editors declared in January 1767 that the newspaper's role was to make known new inventions, experiments, and objects. Sgard (1991, 1:155) describes the *Avantcoureur* as having difficulty "in finding its identity, cultural or commercial"; however, it addressed a certain conjuncture of consumers, entrepreneurial merchants, and savants, a certain mode of Enlightenment mediated through print culture and centered on commerce, taste, and innovation.[21]

Nonetheless, the status of advertising and invention was evidently far from stable, particularly with regard to food products. In one advertisement for pasties from Amiens, the *Avantcoureur*'s editor adopted a defensive tone: "We will not neglect any object of industry. Is it any more ridiculous to announce a good dish than a beautiful fabric? We exhort the scoffers to

16. Martin 1999, chap. 2; C. Jones 1996, 24–27.

17. For relevant discussions of this work, see Goodman 1992; Nathans 1990; C. Jones 1991; "Public Sphere" 1992; Broman 1998.

18. Coquery 2003, 193.

19. See analogous claims concerning Wedgwood in McKendrick, Brewer, and Plumb 1982, 100–145; also C. Jones 1996. On fashion and its critics in eighteenth-century Paris, see, among others, Ribeiro 1995; Roche 1994; Kennedy 1989. On novelty and innovation, see Sargentson 1996; Pennell 1999; Hilaire-Pérez and Garçon 2003; Bruland 2004.

20. *Avantcoureur,* 1762, 102; 1763, 86.

21. As Smeaton (1957) noted, the *Avantcoureur* was the first to publish some of Lavoisier's earliest experiments and was also a leading forum for other chemical controversies.

make use of all these things, if they can, rather than laughing at the advertisement."[22] Such concerns were well-founded, for contemporaries portrayed the restless pursuit of novelty as a moral problem particular to the French, the fickleness and shallowness of fashion a national failing.[23] Innovation in the luxury arts was morally dubious. As early as 1685, the limonadiers were attacked for unbridled alimentary innovation. On "the pretext that they have the right . . . to retail brandy, lemonade & other well-known and widely used liqueurs in their shops," limonadiers were taking the original legislation as a license "to invent, compose & sell, & serve to drink . . . other extraordinary and unknown liqueurs, which they compose as seems good to them, of rectified brandy, spices & other, more violent drugs, which render furious those who use them frequently, & which have been judged to be bad & very dangerous, after an infinity of unfortunate accidents."[24] Such accidents included the death of one competitor in a drinking competition held in a Paris café in 1756, which involved consuming one pint each of the liqueurs Cinnamomum, white Escubac, and Crême des Barbades.[25] The dangerous business was thus the production of *new* liqueurs.

From their earliest origins, the limonadiers had thus been professional innovators. Their wares were unknown to earlier generations. More than many food products, liqueurs exemplified the dietary transformations Parisians had undergone. Debates over these new forms of consumption centered less on the formation of social spaces for new stimulants than on the significance of novelty as evidence for disturbing social change. According to Mercier (1995, 1:187–88), "Our ancestors went to the cabaret, and it is said that they preserved their good humour: we hardly dare go to the café; and the black water that one drinks there, is more harmful than the generous wine on which our fathers got drunk: sadness and caustic humour reign in these parlours of glass, and a despondent tone is manifest on all sides: is it the new drink which has produced this change?" Mercier singled out liqueurs and spices in particular as instrumental in altering modern Parisian minds:

> The air of Paris, if I am not mistaken, must be a unique air. What substances flow together in such a small space! Paris can be compared to a large saucepan, in which meats, fruits, oils, wines, pepper, cinnamon, sugar, coffee, the most distant productions come to mingle. . . . From all these juices, assembled and concentrated in the liqueurs which flow into every household in great streams,

22. *Avantcoureur*, 1766, 10.

23. Caraccioli 1757, 41.

24. Delamare 1705, 3: 810.

25. *Annonces, Affiches et Avis Divers*, 1756, no. 48 (1 December): 191.

> which fill whole streets . . . , how should there not result attenuated parts in the atmosphere, which compress the mental fibre more than anywhere else? Hence perhaps that lively, light sentiment which distinguishes the Parisian, that carelessness, that flowering of spirit which is particular to him.[26]

Though manufactured in Paris, compound liqueurs and the limonadiers' products in general, like the products of fashionable cuisine, produced an exotic environment in which Parisian bodies developed. The very possibility of constituting Paris as a center of learning was potentially compromised by liqueurs. Liqueurs, especially commercially produced liqueurs, thus stood at the nexus of concerns about innovation, artificiality, consumption, and Enlightenment. As such, their polite status was open to question. The apothecary Jacques-François Demachy remarked in the 1770s: "However agreeable these Liqueurs may be, at the time of writing they have lost somewhat of their value. I do not know what coarseness in the palates of persons who were formerly the most delicate, led them to be no more difficult than the common people. They take from sensuality that which the latter only use from need." Only neat brandy or Andaye brandy, scented with fennel, were "served informally on our best tables, & . . . drunk there without blushing."[27] As objects of polite knowledge, compound liqueurs were unstable; they failed to achieve universality, exemplifying for critics the local, contingent, and variable status of fashion objects and luxury consumption. It is in this light that we must read both historical accounts of liqueurs and contemporary critiques and defenses of them.

Compound liqueurs usually contained brandy, which in the eighteenth century was a by-product of wine manufacture. It was also a measure of the failure of wine making, because only poor-quality wine, unacceptable to the consumer market, was distilled. Fashions for compound spirituous liqueurs were linked to the availability of the brandy that was their central ingredient: Oil of Venus, the first of the class of oils, appeared in the 1720s, as brandy commerce began to develop on a large scale, and the proliferation of advertising in contemporary periodicals concerning new liqueurs corresponds with a second commercial peak in the 1760s.[28] Although the compound liqueurs would appear to be useful meters of the flourishing of French commerce and invention in the eighteenth century, they have received scant and usually negative assessments from historians. Delamain (1935, 26–33), in his history

26. Mercier 1995, 1:24–25.

27. Demachy 1775, 98; see also Le Maître de Claville 1740, cited in Dornier 1997, 177.

28. Lachiver 1988; L. Cullen 1998, 1, 21, 47.

of cognac, treats them as a temporary aberration from the rise of the cognac industry. Lachiver (1988, 255–71) portrays the rise of a French market for brandies and liqueurs as a perversion of the natural French taste for wines, and it is thus important to him to present the market as one founded entirely by outsiders, in this case, the Dutch.[29]

For such commentators, liqueurs are the very opposite of Frenchness; they were foreign marketing ploys that distracted from the truly French drink of wine and perverted tastes through the addition of substances foreign both to France and to wine. These modern formulations recapitulate contemporary aspersions on liqueurs as adulterated products. They largely efface liqueurs as a historical subject, despite the central role these goods played in debates over luxury, French character, and innovation in the midcentury. In an advertisement for a new peach wine, Lecouvreur, owner of the Grand Café on the Pont Saint-Michel, commented that the art of distillation "seems now to have been brought to its highest degree. Every day sees it making some new discovery."[30] A practice indispensable to attracting and retaining fashionable clienteles—the production of novelty—was here presented as the epitome of perfection in the mechanical arts, the symbol of the progress of civilization. Liqueurs, in fact, were the very stuff of modernity, and the French excelled in their production. By 1773 an advertisement in the *Avantcoureur* for the liqueurs of the Lorraine distiller Cheneval described liqueurs as being "of nearly universal usage" and added that "the torch of Chemistry has contributed not a little to enlightening the theory & practice of distillation, and to accelerating the progress of that art."[31] Such readings of the art of distillation seemed, on the face of it, to credit distillers with considerable abilities in the practice of chemistry. Chemistry was the basis for distillers' claims to innovate, the source for their claims to possess enlightenment, and the resource that enabled them to develop a market for this particular luxury food. The next part of the chapter will be concerned with the chemical production of liqueurs and brandies.

## The Delights of Distillation

Idle browsers through Antoine Baumé's *Élémens de Pharmacie theorique et pratique* (1762) could be forgiven for believing they were reading a cookbook or a domestic economy manual. Perhaps half of the receipts it contained

29. See also Dion 1959, chap. 13; Sournia 1990, chap. 2; L. Cullen 1998.

30. *Avantcoureur*, 1769, 6.

31. *Avantcoureur*, 1773, 147–48.

were as much alimentary as medicinal: chocolate, quince jelly, cherry wine, and barley sugar jostled with pills, plasters, and electuaries. Baumé defined confectionery and distillation as arts dependent upon pharmacy.[32] An innkeeper's son from Senlis, he became a master apothecary in 1751, eventually establishing a wholesale trade supplying other apothecaries' shops and hospitals with medicaments and laboratory apparatus.[33] This astute pharmaceutical businessman also sought to participate in the learned life, in a typical combination for chemists in mid-eighteenth-century Paris. Baumé taught courses and pursued extensive research in chemistry over many years, achieving election to the Académie Royale des Sciences in 1772. He collaborated with Pierre-Joseph Macquer on a chemistry course advertised in newspapers, including the *Avantcoureur*.[34] Here, too, Baumé publicized his own innovations alongside the food advertisers: procedures for the manufacture of plant starch and borax, the improvement of earthenware, the preservation of wheat, and a new product, "Crystals of Venus."[35] An active reformer of many arts, especially those concerned with alimentation, he spanned the divide between commercial pharmacy and academic chemistry. His memoir on distillation procedure was reprinted several times, and his views on distillation were cited by others writing on chemistry, pharmacy, and the arts.[36] Accounts of his experiments on distillation and confectionery making appeared, again, in the *Avantcoureur*. Within the chemical community, before judicial authority, and among the literate urban readership, Baumé counted as an expert in distillation.

The concerns of distinct types of practitioners thus intersected in distillation. Baumé represented big pharmaceutical business in the eighteenth century and was also firmly located within the chemical networks of old regime France, thanks to his ties to Macquer and the Académie Royale des Sciences. The likes of Sauvel or Déjean were also successful urban merchants. Both sides were innovators of new products in three domains, the medicinal, alimentary, and cosmetic. Expertise and inventiveness were common to distillers

32. Baumé 1762, 446–49, 531–32, 547–55. Similar claims were made in Venel 1753, 420. W. Anderson (1984, 19–34) argues that Macquer's chemical project was legitimated as a philosophical enterprise through claims about chemistry's benefits for the arts; see also Klein 2005c.

33. See Bouvet 1937, 128, 190.

34. Sturdy 1995. On Baumé's relations with Macquer, see Julien 1992; on his pursuit of academic status, Poirier 1996, 28. Macquer and Baumé's course was taught annually for sixteen years, beginning in 1762, and was the basis for Macquer 1766 (see Bouvet 1937, 88–89).

35. *Avantcoureur*, 1761, 675; 1763, 178; 1766, 296–98; 1768, 24–27, 309–12, 391–94.

36. Baumé 1778; Dujardin 1900, 159–61; Macquer 1766, 1:373–76; Cossigny de Palma 1782. Baumé wrote the entry "Distillateur" for Macquer 1766–1767, 1:x–xi, 488–500.

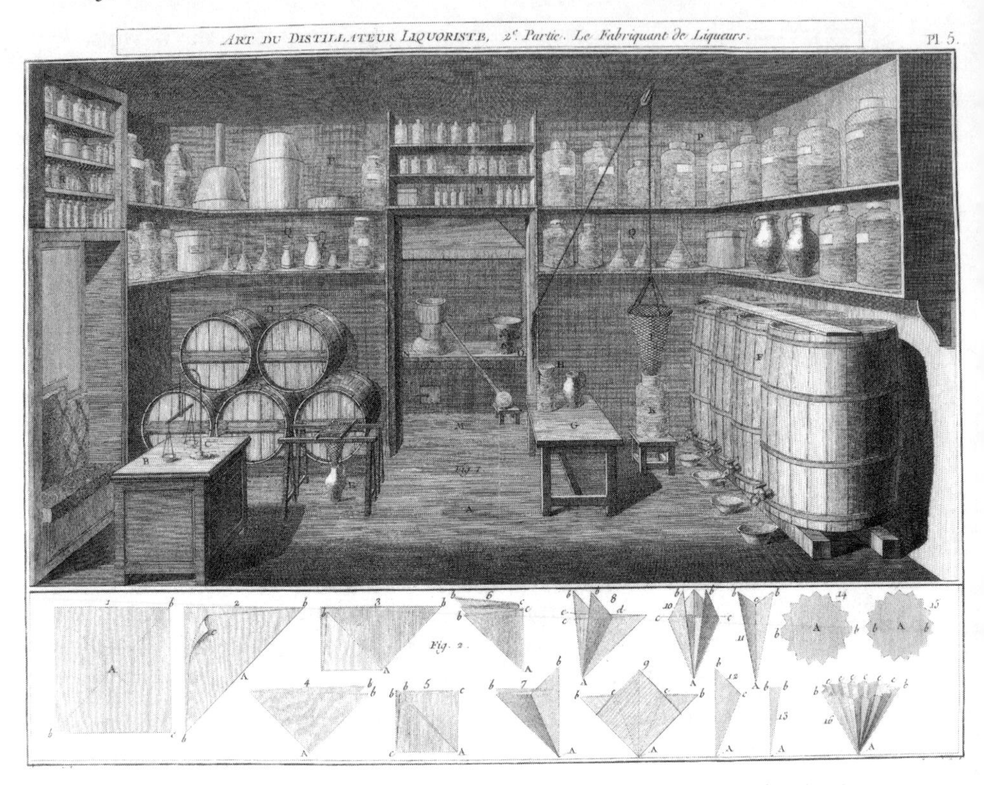


FIGURE 9.2. Liquorist's workshop. (Demachy 1775, plate 5; copperplate engraving after Goussier. Reproduced by kind permission of the Syndics of Cambridge University Library.)

and pharmacists; both shared an artisanal origin and a place in the developing business world of eighteenth-century France. Like apothecaries, distillers worked in a laboratory, often a room behind the public area of the café, where they prepared liqueurs of all kinds, from coffee to Cinnamomum.[37] Scientific practice in academies and other scientific institutions of the old regime has been well studied; but it was also conducted outside these establishments, in just such private laboratories (see figure 9.2).

At the time when Baumé began his interventions in the distillers' trade, he was still a decade away from being an academician. In institutional, practical, social, and entrepreneurial terms, he was comparable with the liquorists in their laboratories. The debate over distillation, seen from this standpoint, defies classification into a simple opposition between "artisans" and "scientific practitioners," and the overlap creates problems for tacit definitions of the

37. Demachy 1775, 114.

scientific sometimes deployed in history of science. Experimental protocol for distillation was discussed at length in chemical or pharmaceutical treatises of the mid-eighteenth century, as well as in didactic works written by distillers or about distillation as a mechanical art. Within this world of print, many seemingly self-evident distinctions between polite science and natural science vanish.

Distillation involved heating the materials to be distilled in an alembic and allowing the vapor to rise up and escape the vessel by means of a spiral tube, the serpentine, which was cooled along its length before passing into a receptacle at the other end.[38] The Paris distiller Déjean defined distillation as the "art of extracting spirits from bodies." By means of heat, the insensible parts of a body were put into motion and the trapped spirits freed of phlegm and earthy parts, thus enabled to volatilize. Déjean drew heavily on technical chemical terminology, referring to primary principles such as phlegm, fire, and earth. With his chemical explanation of distillation, he sought to confer epistemological nobility on the practice, which "may be worthy of the care & attention even of Savants."[39] While many chemical treatises represented no more than three different arrangements of distillatory apparatus, Déjean described thirteen, suited to various different products and employed by different artisanal groups. Each had its own name, taken from the particular alembic shape involved. Twenty years of distilled practical knowledge allowed Déjean to recommend maneuvers for every product. Unlike chemistry textbooks for polite audiences, his account acknowledged the technical complexity and multiple variants of distillation as used by the corporate distiller.[40] Baumé's memoir on distillation also systematically investigated all the apparatus used in distillation, but did so with a view to eliminating many variants. As a man who made his living as a bulk producer of chemicals, Baumé's aim was to speed up both the distillation process and the manufacture of the apparatus, and he placed the cost of production and purity of the product ahead of the manipulation of flavors and aromas.[41]

Spirit of wine was the twice-distilled product of vinous fermentation. The first distillation produced the more strongly flavored brandy. In the manufacture of compound liqueurs, distillers used distillation to extract and concentrate flavoring substances from individual herbs, spices, and fruits

38. Expositions of the history of distillation include Dujardin 1900, esp. chap. 9; R. Forbes 1948; Brock 1992, 23–26.

39. Déjean 1759, 3.

40. Déjean 1759, chap. 2. For some standard chemical formulations of distillation, see, for example, Macquer 1766, 1:354–57; Baumé 1773, 1:83–84.

41. Baumé 1778; for a critique, see Cossigny de Palma 1782, 10–13.

in combination with brandy; skill was located both in this procedure and in the combination of different flavors that generated commercial liqueurs. Such techniques were attacked by Baumé. Ordinary brandy was not adequate for making up liqueurs, nor ordinary spirit of wine for making remedies, he claimed. He called on pharmacist distillers to compose their liqueurs using spirit of wine manufactured in their own shops and not to buy product from other guilds. "It is on the purity of spirit of wine, & on the separation of its coarse essential oil, that the perfection of compound spirituous waters, & table liqueurs, depends in large part."[42] Baumé's was one of numerous contemporary suggestions for improving liqueurs, but it is typical of a trend running through his treatise: the denunciation of rival corporations as concerned with profit over quality. Claiming the epistemological high ground by promoting standardization was itself a competitive strategy. Baumé was entering the hotly contested arena of liqueur innovation, asserting that his techniques yielded better results for connoisseurs.

Macquer's *Dictionnaire de Chymie* (1766, 1:396–97) was keyed to Baumé's material technology; the book collapsed brandy into spirit of wine or "burning spirit," so that brandy, a substance in the production and manipulation of which the distillers' guild specialized, ceased to exist in its own right. Instead, brandy was merely spirit of wine, imbued with impurities: "plenty of superfluous water, & oil of wine, substances totally foreign to spirit of wine, properly speaking." This ontology was not commercially neutral. In writing of the process of redistilling or rectifying spirits to remove these impurities, Macquer referred to Baumé's technique for rectifying large quantities at a time.[43] The call for pure spirit of wine as the substrate for chemical operations as well as medicinal, alimentary, and cosmetic production served as an advertisement for Baumé's enterprise. Macquer's term for spirit of wine in his dictionary, "burning spirit," reflects his concern to represent spirit as a general substance produced by all forms of spirituous fermentation, not just wine.[44] Again, we are apparently getting away from the commodity here toward a general chemical class of substances. There were, in fact, two distinct accounts of spirit of wine current in the European chemical community. In Macquer's *Elémens de chymie-pratique* (1751), spirit of wine was portrayed as a compound of oily, acidic, and aqueous principles.[45] Stripping the spirit of its

42. Baumé 1762, 332–35; see also C.-J. Geoffroy 1741; Montigny 1770, 436. Déjean limited spirit of wine to cordials and perfumes (1759, 75).

43. Macquer 1766, 2:359–62.

44. Macquer 1766, 1:407–14.

45. Quoted in Baumé 1757, 76.

water rendered both its acidic and oily properties more evident. In this, Macquer followed the German chemist Georg Stahl's account of spirit as a very attenuated and light oil, combined with water by the mediation of an acid. This oil, according to Stahl, was phlogiston itself.[46] By 1766, Macquer's view had altered. Although not ruling out the possibility of discovering an oily principle, he now portrayed most oils as compound substances that emitted phlogiston in decomposition. Experiments by Boerhaave and Johann Juncker had suggested that spirit was a simpler compound of phlogiston and water, the latter being the only residue left after combustion. For them, phlogiston and the fire principle were one and the same thing, and this claim was now strongly supported by Macquer.[47]

Macquer also described, in some detail, Baumé's well-known experiments in distilling burning spirit with a concentrated acid. The aim was to strip the spirit of its water, thus getting closer, perhaps, to a substance that was pure fire. The acid first removed water in which the spirit was dissolved, then the water that was, as Macquer put it, "essential" to the spirit itself, in other words, that which, held in combination with the fiery substance, produced the characteristic properties of spirit.[48] The result was ether, a fluid more volatile and flammable than rectified spirit of wine, which was completely consumed in burning, leaving no residue. In a treatise on ether of 1757, Baumé claimed that the properties of ether placed it midway between oil and spirit of wine. These experiments on spirit of wine were conducted as part of a wider inquiry into the nature of fire, prompted by Boerhaave's identification of phlogiston in its purest state with "alkool," the most rectified possible form of spirit of wine, or in other words, with ether. This was the view toward which Macquer moved between the 1750s and the 1760s. Although familiar with this explanation, which had its origins in the ether experimentation of German chemists during the 1730s, Baumé continued to defend Stahl's account of phlogiston as a compound between fire and vitrifiable earth.[49] Famously, Baumé claimed that during calcination, phlogiston could combine with metals, while fire was evolved.

46. Macquer 1766, 1:407–8; 1:574.

47. Macquer 1766, 1:413.

48. Macquer 1766, 1:409; 1:455–63; also discussed in "Pêche," 1765, 288. Ether was the subject of Baumé's first memoir before the Académie Royale des Sciences in 1755 (see Baumé 1760).

49. Both Macquer and Baumé referred to Boerhaave's experiments (Macquer 1766, 1:408; Baumé 1773, 1:128). On Boerhaave's view of ether as identical to the fire principle, see Heimann 1981, 70; however, "Vin, & Fermentation vineuse" (1765, 286–87) indicated that Boerhaave's view was contested. Chemical inquiries into ethers are discussed by Klein (2005a, 12–14), Kopp (1966, 3:299–321), and Priesner (1986, 130–32); see also Klein's chapter in this volume.

Baumé and Macquer agreed on phlogiston's properties, if not its chemical nature. Ether and spirit of wine resembled phlogistical substances in their extreme volatility and flammability, as well as their ability to affect the nerves, causing dramatic effects: "drunkenness, dizziness, suffocation, unconsciousness & death."[50] In 1773, for the first time, Baumé cited Boerhaave's definition of alcohol as the most highly rectified form of spirit of wine. Although he did not pass judgment upon Boerhaave's claim, it is likely that Baumé, like Macquer, had come to regard the fire principle as corresponding to the volatile part of spirit of wine and ether.[51] Despite the broader scientific claims underlying Baumé and Macquer's commentaries on ether, it should still be viewed as a simultaneously chemical and commercial object. Ether was the subject of Baumé's first memoir to the Académie Royale des Sciences in 1755, prompted by the interest of Académie chemists a decade earlier in developing methods of producing ether on a large scale.[52] Baumé not only became an acknowledged expert in the complex technical maneuvers involved in producing ether, but also a manufacturer and leading merchant of the substance. Ether thus linked Baumé's bid for academic recognition and his market entrepreneurialism.

None of these transactions concerning purity, procedural improvement, and displays of technical expertise were truly distinct from distillers' own strategies for self-promotion. Even the matter theory that underpinned experimentation on spirit was shared by Déjean, who drew upon Macquer's *Elémens de chymie-pratique* (1751) in defining spirit as the lightest part of bodies, by nature disposed to great mobility and possessing an igneous substance.[53] All three thus shared a single model of spirit. What differed was Baumé's and Macquer's experimental pursuit of a general, systematic, and interconnected cosmology of principles. Phlogiston, ether, and spirits were not only central to Baumé's cosmology, they were part of his self-presentation as experimenter and discoverer, which, like those of the distillers, sustained his commercial reputation. This goes some way toward explaining Baumé's defense of phlogiston against Antoine Lavoisier; not only was he perpetually

50. Macquer 1766, 2:222; 1:414, 463.

51. This has been almost the only aspect of Baumé's work that has entered histories of chemistry, where it has figured as an attempt to preserve phlogiston against Lavoisier's reforms. See, for example, Legrand 1973; Flahaut 1995. But the picture is far more complex; see, among others, Allchin 1992; Musgrave 1976; Poirier 1996, 67, 113; Donovan 1993, 107–8, 295; Bensaude-Vincent 1993, 50, 268.

52. Baumé 1760; 1757, 2–20.

53. Déjean 1759, 45, chap. 15.

at work with the fiery matter in the laboratory, he had, moreover, built his scientific reputation upon manipulating it.

## The Experimental Body

Ontology was not the only domain over which distillers and pharmacists diverged. Déjean's suggested methods for judging brandy involved the distiller's own body as meter. Good brandy should be neither too white nor too cloudy; in a glass, it should bubble and foam, and each bubble should last for a long time. A small amount rubbed between the palms should quickly evaporate, and it should not soak into paper. The "best manner of judging it" was, however, by means of taste. The distiller should learn "to be a little gourmet in this respect: taste decides even better than all the above-mentioned proofs. Those who do not want to rely on themselves alone, or who mistrust their taste, can make use of the proofs above." For Déjean, noncorporeal proofs clearly took second place in a commercial world in which connoisseurship was central to the distiller's art, and not to possess connoisseurial skills was "shameful." According to the *Dictionnaire de Trévoux*, "gourmets" were "those who taste wine, and judge of its goodness and keeping qualities"; the gourmets numbered among the wider group of skilled connoisseurs.[54] The gourmet was thus in himself an accurate measure of the goodness and composition of distilled liqueurs. This was the model Voltaire (1764) would use to define the "man of taste": "The gourmet promptly senses and recognizes the mixing of two liqueurs. The man of taste, the connoisseur, will promptly notice the mixing of two styles." The polite connoisseur could be a liqueur expert, and vice versa.[55] Accordingly, Déjean advised the trainee distiller to emulate the connoisseur's palate, learning how to distinguish between the different natural substances used in liqueur manufacture.[56] Baumé worked, rather, to efface connoisseurialism. High-quality spirit of wine should be "perfectly pure," having "no foreign odour"; the "spirituous part should promptly evaporate, & leave neither humidity nor an odour approaching that of the phlegm of brandy." So far, this is not too distant from Déjean's second-order criteria. But Baumé added one more definition: "Perfectly rectified spirit of wine should not weigh more than six gross forty-eight grains in a bottle which holds an ounce of water, the temperature being ten degrees above freezing."[57] This was typical of his redef-

54. Quoted in Giaufret Colombani and Mascarello 1997, 60.
55. Quoted in Flandrin 1989, 300.
56. Déjean 1759, 77–78, 252; see also Dubuisson 1779, 1:196–97.
57. Baumé 1762, 332.

inition of quality in terms of chemical purity and quantifiable criteria rather than corporeal skill, innate or acquired. By such methods, he sought to pin down the elusive property of chemical identity and create a single category of flammable spirit. Of fermented liqueurs, including wine, beer, and cider, Baumé noted: "All these flammable spirits are of the same nature, they have the same properties; they only differ among one another in the flavours & odours particular to each, & which cannot be entirely removed by reiterated rectifications; however, the thing may not be impossible. For example, I did everything possible to spirit of wine extracted from Spanish wine, to remove its odour & flavour, without succeeding."[58] In fact, Baumé was *unable* to demonstrate experimentally that odor and flavor were accidental to flammable spirit, in other words, to show that all flammable spirits were the same. Nevertheless, he continued to maintain that separating gustatory properties from spirit of wine was theoretically possible, thus upholding the possibility of dissociating connoisseurship from quality.

Baumé's search for chemical ways of measuring liqueur quality replaced the body of the gourmet or connoisseur with that of the calibrated laboratory instrument, transferring expertise from distillers and their polite clientele to chemists.[59] Toward the end of 1768 he announced his invention of an areometer, a device for measuring the spirit content of liqueurs by specific weight, in the *Avantcoureur.*[60] However, Baumé's instrument worked only for his own product, purified spirit of wine: the sugar in flavored beverages made up with brandy or spirit of wine, such as the composed liqueurs, invalidated its readings.[61] The ideal of an instrumental control over the spirituous content of drinks remained elusive. The year after his final article on the areometer appeared in the *Avantcoureur,* Baumé achieved election to the Académie Royale des Sciences.

## Distillation, Publication, and the Academy

Most individuals termed "chemists" before 1789, notably excepting Lavoisier, were apothecaries.[62] Distillers began to enter print culture from the 1750s

58. Baumé 1762, 326–27.

59. For the British case, see W. J. Ashworth 2001; Sumner 2001. I am grateful to Simon Schaffer for these references.

60. *Avantcoureur,* 1768, 712–16; 1768, 793; 1768, 806–11. On Baumé's areometer, see Bensaude-Vincent 2000.

61. Cossigny de Palma 1782, 46.

62. Meinel 1983. Apothecary chemists included Baumé, Rouelle, Vauquelin, Cadet de Vaux, Demachy, and Déyeux. See also Court and Smeaton 1979; J. Simon 2005; Klein 2005c.

onward to defend their own chemical expertise, central both to commercial production and public self-promotion. This is evident in the readership for scientific and technical works, the only area where merchants' and nobles' reading overlapped.[63] Growing public interest in the improvement of the mechanical arts is evident in the many periodicals, dictionaries, encyclopedias, and manuals appearing in the second half of the century. As Darnton remarked of the *Encyclopédie*, "For every remark undercutting traditional orthodoxies, it contains thousands of words about grinding grain, manufacturing pins, and declining verbs."[64] Déjean's publication of artisanal techniques, an unusual move, was part of an explicit attempt to reform and perfect the art of distillation before a polite audience familiar with chemistry. It inserted him within the program of useful, rational public knowledge pursued by the Académie Royale des Sciences. "Distillers had failed to destroy people's false opinion concerning the danger of using Liqueurs. The simple Exposition I give of composition and the several combinations, will undeceive many people who have believed up to now that their consumption is pernicious; & they will consume them with that much more confidence."[65] In using chemical terminology and theory, Déjean drew upon a literature familiar to his scientifically informed clientele. Fashionable academic chemists like Nicolas Lémery had offered lecture courses before polite audiences since the 1700s, and most chemical textbooks covered distillation. Distillers were well advised to keep abreast of chemical publications; and within print culture, a distiller like Déjean might engage on equal terms even with academic chemists like Macquer and Lémery.[66]

Within the academy itself, however, the balance was altered. Apothecaries had been represented in the Académie Royale des Sciences since the late seventeenth century; distillers remained excluded.[67] Membership of the academy conferred considerable social and political prestige. The academicians' joint publishing enterprise, the *Description des Arts et Métiers,* aimed to reform the mechanical arts throughout the kingdom by setting out the principles of their practice in print. The series has generally been portrayed as a neutral manifestation of Enlightenment enthusiasm for practical and utilitarian matters. Yet such publications too might be interventions in corporative tussles, as is evident from the two *Arts et Métiers* treatises on distillation, penned by

63. Roche 1978a, esp. 391
64. Darnton 1984, 191.
65. Déjean 1759, V.
66. Déjean 1759, chaps. 11, 12, and 15. See Klein 1996.
67. Hahn 1971.

Jacques-François Demachy, apothecary, satirist, and spy.[68] Demachy, who became a master apothecary in 1761, rose to prominence in the apothecaries' guild but never achieved election to the academy.[69] His contributions to the *Arts et Métiers* series must be read as a product of the tangled interrelations between guild and academy as sites for the production of authoritative knowledge.

Demachy began his treatise on liqueurs by lamenting the resistance offered him by guild distillers asked to disclose the secrets of their production. His account played on a well-worn contemporary theme: the greed of private artisanal knowledge opposed by the civic benefits of openness and publication. To support this dichotomy and suppress distillers' claims to epistemological equivalence, Demachy marginalized and localized the distillers' technical language by characterizing it as a "jargon" that needed translation into the "Chemists' terms that are generally known." He also suggested that the distillers' tinkerings with alembics and receptacles did not constitute chemistry: only in fermentation—the domain of vintners and brewers, not distillers—did "the Liquorist become the imitator of the Chemist." Read in one light, Demachy's treatise was nothing other than a sustained satire on the claims to expertise and good taste advanced by the liquorists, the newest of the apothecaries' rivals in the art of distillation and the most epistemologically insecure. Among other things, he claimed that the class of compound liqueurs known as the oils had been invented by an apprentice tanner, a snide reference to contemporary fears that spirits would harden and desensitize the palate and stomach.[70] The composition of liqueurs was nothing more than the skill of disguising revolting or flavorless ingredients through artifice:

> Star anise on its own smells of lice, but a little green anise saves it from this awkwardness; amber alone produces no odour, a little musk gives it the necessary relief; quince on its own is detestable, a little clove picks up & corrects its perfume. Could one believe, if Authors worthy of a certain confidence had not assured it to be so, that cow manure, combined with aromatics, could transform its disgusting denomination into the pompous title of Water of mille-fleurs? All this resembles in every point the Art of the Cook, seasoning saves all; & the great Lord who devoured his slippers made into a ragoût by his industrious Cook, proves that the Liquorist equally is the master of creating whatever

68. Demachy 1773, 1775.

69. On Demachy, see Toraude 1907. On the *Arts et Métiers* series, see Torlais 1961, 46–47, 253–54; Hahn 1971, 68; Gillispie 1980, 340–60.

70. Demachy 1775, 9, 88, 56, 96–97. The purported name of this inventor, "Bouillerot," was itself a pejorative reference to a bulk distiller, or *bouilleur,* of cheap brandy.

illusion he wishes, as long as a fine palate, a discernment which is skilful in its mixtures, perhaps also a wise discretion, preside over his operations.[71]

Such attacks would not be taken unresistingly by the distillers. François-René-André Dubuisson, a distiller since the early 1730s, represented the peak of corporate commercial prowess. A retired café proprietor, he published on distillation largely as a corrective to Demachy's presumptions to distillatory authority.[72] His strategy for discrediting Demachy challenges our usual assumptions about the history of chemistry. For Dubuisson, Demachy was a representative of the apothecaries' guild and not of the Académie Royale des Sciences. Indeed, he warned the academy that it was at risk of being swindled by Demachy's account of distillation. Dubuisson thus placed himself on an epistemological par with Demachy, leaving the reader in no doubt that Demachy's "official" text was, in fact, just another tactic in the ongoing war between the apothecaries and the distillers. Like Déjean, Dubuisson laid claim to chemical expertise: he termed the *limonadier* a "Chemist-Distiller" and claimed to have instructed himself, early in his career, by consulting leading chemical and medical publications. His authority of first resort had been Nicolas Lémery, "because in the 1730s that Author still enjoyed the titles of Restorer of Chemistry in France, & Reformer of Pharmacy in the Capital."[73]

Lémery's chemical project had coupled a concern with the manipulation of materials with a Paracelsian heritage emphasizing the extraction and purification of essences from natural bodies. Iatrochemists argued that the removal of coarse and earthy parts from essences or spirits allowed them to penetrate the body's fabric more readily and there work their specific effects. Dubuisson adopted this model to legitimate his claims for the healthfulness of his products. Distillers were skilled in extracting the spiritus rector, the most volatile part of aromatic substances, from the harmful "acrid & bitter salts" that coexisted with it in plant materials.[74] Dubuisson also pursued experimental research into rectification to improve standards of liqueur manufacture. In one set of experiments, he replicated procedures described by medieval iatrochemists, in which brandy was distilled six times before being set to circulate in a closed still for many weeks. The result was a brandy that

71. Demachy 1775, 68.

72. Dubuisson 1779, "Avant-Propos," vol. 1.

73. Dubuisson 1779, "Avant-Propos," 1:ix.

74. Dubuisson 1779, 1:50, 175–76. The term was also used by chemists; see Macquer 1766, "Esprit recteur," 1:420–22; Baumé 1762, 302. Aroma was still identified by many French chemists with the spirituous or volatile principle, the Paracelsian sulfur (see Déjean 1759, 140–41; Brock 1992, 45–46; Moran 2005, 11–22).

became stronger but also purer, losing the "heterogeneous parts" of wine, its salts, acid, and oil. Dubuisson agreed with Baumé that once-distilled brandy was inadequate for making good quality liqueurs but insisted that highly rectified spirit, as produced by big distillers (see figure 9.3), was a harsh product suitable only for "the fabrication of varnish." His lengthy procedure generated a stronger brandy that was more flavorsome, though very penetrating, and "more spirituous than it had been before having undergone these different operations."[75] Dubuisson's definition of a pure, high-quality brandy or spirit of wine was quite different from that of Baumé; it did not depend on spirit content as measurable by an instrument. Nor did he view a tasteless and odorless spirit as the ultimate outcome of rectification, claiming instead that flavor intensified with rectification. His experimentation was entirely in keeping with the program of enhancing and managing flavors that typified the activities of guild liquorists in the period.

The agendas, procedures, and authority claims of apothecary chemists were freely contested by members of rival guilds, who also laid claim to the chemical heritage of the academy and who did not accord innate epistemological superiority to the apothecaries. The issues that were most controversial in commercial chemistry touched on matters that fitted well within Lémery's program for analysis and extraction, still a primary published evidence of academic interest in distillation. In this sense the program of extraction of essences remained alive outside the academy, even though academic chemists had switched as early as the 1710s to a more systematic inquiry into the general principles of chemical activity and the relations between different categories of chemical substances.[76] Dubuisson was representative of distillation's development into an increasingly technically and theoretically complex activity "in which, beside the old-fashioned distiller, more sophisticated distillers worked consciously to a higher standard."[77] But he simultaneously opposed the conversion of distilling into a bulk industry, acknowledged by a new entry in the 1798 edition of the *Dictionnaire de l'Académie Française:* "Distillery," a "place where one carries out distillations on the large scale."[78]

Only the names of Demachy and Baumé figure in histories of chemistry, even though Déjean and Dubuisson counted themselves chemical experts.[79] In fact, neither group had a legal prerogative where chemical preparations

75. Dubuisson 1779, 1:126, 97.

76. See especially Kim 2003, chap. 2; Klein 1996, 275–76; Holmes 1989a, 61–83.

77. L. Cullen 1998, 100.

78. Académie Françoise 1798, 1:432.

79. Holmes 1989a, 3–20.

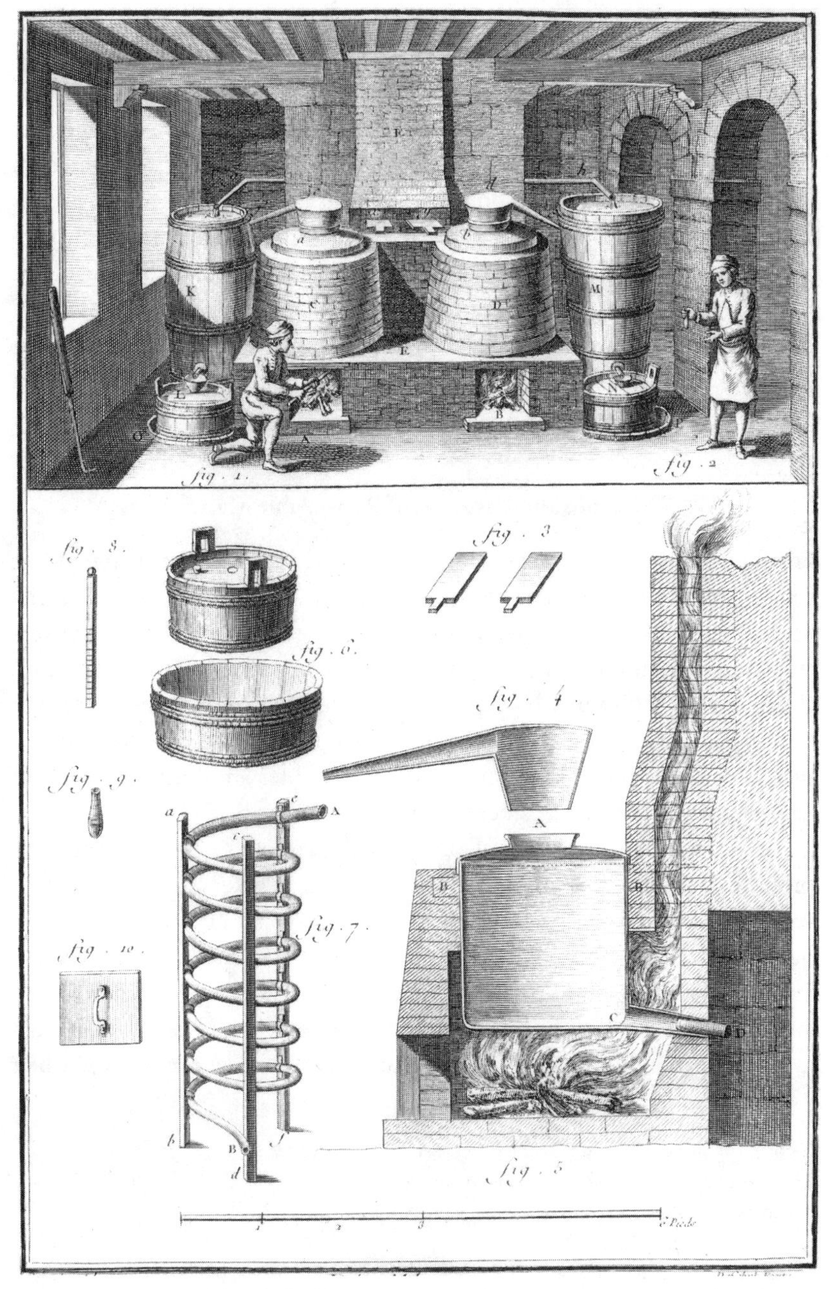


FIGURE 9.3. Large-scale brandy distillation (Diderot and d'Alembert 1763, 20:22; copperplate engraving by Defehrt after Goussier).

were concerned; these were the corporate property of a tiny guild, the *distillateurs d'eaux-fortes*, by an *arrêt du Conseil d'Etat* of 23 May 1746. By this ruling, the limonadiers were banned from "meddling in any of the operations belonging to the art of chemistry," as Baumé put it.[80] The struggle over the nature and purpose of distillation in the second half of the eighteenth century can be viewed as a product of attempts by different artisanal groups to redefine their domain of expertise and the practices and products that accompanied it. This is true of chemists and pharmacists as well as distillers. Increasingly, the history of eighteenth-century chemistry is being rewritten in terms of its practices, materials, and public uses, and here Paris distillers of this sort deserve a place, because they explicitly attached themselves to a chemical project.[81]

## Conclusion: Histories of Pleasurable Matter

The final aspect of this account of liqueurs and spirits concerns the public debate over their status. The contemporary term *liqueur* did not necessarily denote what we would understand as an alcoholic drink, a matter often overlooked in commodity histories.[82] In his well-known account of the history of stimulants, Schivelbusch proposes that coffee rose to prominence among the European elite because it denoted industriousness. For this reason, he argues, it was more important in nations where a Protestant work ethic reigned; Catholic, absolutist, and aristocratic cultures such as Spain or Austria favored chocolate. Apart from the fact that this makes an anomaly of France, perhaps the most famous nation of European coffee drinkers, it also leaves unexplained the fact that tea dominated the English market for much of the eighteenth century. An appeal to a category of "stimulants" is too coarse-grained an instrument to permit us to understand why coffee predominated over tea in France.[83] Additionally, anachronistic categories such as "alcohol" and "stimulants" beg the question of what contemporaries experienced when

80. Scagliola 1943, 72–94; Baumé, "Distillateur," in Macquer 1766–1767, 1:500. On the subsequent divergence of pharmacy from chemistry, see Court and Smeaton 1979; J. Simon 2005, esp. chap. 5; Levere 1994, chap. 4.

81. See especially Klein 2005c.

82. Authors such as Blocker (1994), Valverde (1998), Dornier (1997), and Warner (1994) present a largely teleological account in which there is a direct filiation between discussions of liqueurs in the eighteenth century and the invention of alcoholism in the nineteenth. However, see Nahoum-Grappe 1987.

83. Schivelbusch 1993, 38. W. D. Smith 1992–1993 offers an economic explanation for tea's popularity in England; see also Mintz 1985.

consuming such substances, how they explained those effects, and how they themselves classified ingesta. Notwithstanding these basic difficulties, the principle of Schivelbusch's inquiry—that one ingestum can stand for a single worldview, emerging out of the concerns of a homogeneous culture—has lately assumed prominence as a model for writing the history of food, including subsequent histories of stimulants and the substantial literature on the temperance movement.[84] Such accounts, however, rest on two unexamined assumptions: first, that accounts of any foodstuff can be found to be univocal to the extent that Schivelbusch's analysis demands; second, that the composition and effects of foods and drinks, as understood in the eighteenth century, can readily be translated into today's categories. Neither of these premises bears close scrutiny.

Histories of the temperance movement often construe their subject in terms of alcohol consumption in the modern sense, attending little to eighteenth-century chemical and consumer categories. The familiar Anglocentric model, in which temperance was a religious movement serving to prevent unruly behavior among the poor, provides an inadequate explanation for French debates over the polite consumption of liqueurs and spirits.[85] There were three ways in which compound liqueurs drew criticism from contemporaries. First, they were often made with brandy, and thus featured in a critique of drunkenness that gained ground in the later eighteenth century. Second, they lent themselves to an attack, deriving from humoral medicine, upon the inflammatory effect of certain foodstuffs, primarily the productions of hotter climates, such as spices. Third, as commercial inventions by experts in the manipulation of tastes, liqueurs were a target of critics opposing the commercialization of food production. All three critiques against liqueurs were manifest in eighteenth-century writings, prohibiting the construction of a single overarching discourse of "temperance" or "antialcoholism." Indeed, according to the *Encyclopédie*, "intemperance" was any form of alimentary overindulgence. It had no specific restriction even to beverages, let alone to alcoholic ones.[86]

84. See, for example, the discussion of tea in Hobhouse 1985, part 3; also Walvin 1997. On "stimulants," see Schivelbusch 1993; Matthee 1985; Camporesi 1994; Rudgley 1993.

85. Most studies concern Britain, for example, Halimi 1988; Clark 1988. Sournia (1990, chap. 3) explains the comparatively low-key French temperance movement by claiming that spirit drinking was uncommon, an impression no doubt gleaned from French-language histories of winemaking. This chapter, together with L. Cullen 1998, chap. 5, and Brennan 1988, contradicts that assumption.

86. See Venel 1757. Sournia notes this different usage, but explains it as a product of contemporary ignorance (1990, 37).

"Alcohol" was thus not a category of ingestum throughout the early modern period, as historians of temperance have implicitly assumed. As late as the 1760s, the primary meaning of the word *alcohol* in French publications was still a chemical procedure for purification, or else the quality of extreme subtlety; it was not a type of beverage.[87] Even if one identifies spirit of wine with "alcohol," as most historians have, it is important to recognize that "spirit of wine" was an equivocal experimental object among chemical practitioners. There could be no general agreement about classifications of individual foods based on their chemical composition; even for Macquer or Baumé, spirits were one manifestation of a larger category of substances troubling the nerves and brain. At this time, therefore, there was a gap between the chemical account of spirits and the social critique that has implicitly been taken as founded on it.

Nor did all critiques of liqueurs relate to their content of spirit of wine. In many early eighteenth-century writings, liqueurs were allied not with wine but with a range of foods and beverages especially associated with the expertise of cooks and limonadiers. In this formulation, the limonadiers' sphere of expertise was primarily the manipulation of flavors and aromas, not of spirit of wine. Distillation using brandy or spirit of wine was merely one of several techniques employed by limonadiers for liberating the spiritus rector from aromatics. Other methods, specific to the substance being worked upon, relied upon water, heat, maceration, infusion, or grinding. This formulation created a category for liqueurs that encompassed coffee, tea, and chocolate, spirituous and nonspirituous beverages.[88] It is reflected in the fact that all the fashionable drinks made and sold by distillers, and not merely the "alcoholic" ones, came in for attack. Thus, the Paris medical faculty physician François Thierry classed tea, coffee, and chocolate among the liqueurs in the category of "slow, but universal poisons."[89] His erstwhile colleague Etienne-Louis Geoffroy allied liqueurs as often with coffee or with spiced and salty dishes as with wine, all being heating substances engendering fever and inflammation.[90] In Galenic medical terms, liqueurs were not distinct from other

87. Hecquet 1710, 1: "Explication de quelques termes de médecine," unpaginated; Macquer 1766, 1:10; Dujardin 1900, chap. 2; Giaufret Colombani and Mascarello 1997, 57. The author of the article "Alkool" (1751) in Diderot's *Encyclopédie* noted, however, that the term was almost exclusively applied to highly rectified spirit of wine.

88. Dubuisson 1779, vol. 1, chap. 1.

89. Thierry 1755, chap. 7; see also, for example, Review of Adrien Richer 1760.

90. É-L. Geoffroy 1997, for example, 25, 41, 181, 197, 206, 227, 263, 273, 277, 292, 352, 356, 365, 386, 424.

heating foods, and the term *spirit* denoted both volatile and aromatic parts of foods. Furthermore, foods and drinks had individual effects on the constitution deriving from their unique composition, and medical practitioners rarely agreed about what these were. Physicians and apothecaries themselves participated in the invention of new liqueurs, in technical reforms of distillation, and, not least, in the tasteful consumption of luxury goods. The well-established medical literature attacking the unhealthiness of liqueurs is thus misinterpreted when read exclusively as an attack on "alcohol" drinking.

Lastly, liqueurs could be critiqued in their role as symbols of artistic progress and artisanal expertise. In his health manual, the abbé Armand-Pierre Jacquin identified two kinds of spirituous liqueurs, "bourgeois" and "distilled." The former, less harmful, were composed by infusion, while the latter were fermented in wine or brandy then extracted with an alembic. Jacquin poured equal scorn upon distilled liqueurs and nonfermented liqueurs such as barley water or lemonade; the latter upset the digestion by cooling the stomach excessively.[91] Only the bourgeois liqueurs were spared, though they too contained spirits. His criticism was thus principally directed against commercial production, singling out the cluster of goods made and sold by the limonadiers, together with their expert practice of distillation. As the alembic became the symbol of artificial intervention in drinks, so liqueurs epitomized contemporary fears about contamination. Ending his didactic treatise with an account of infusion liqueurs, Déjean noted that "the foremost" reason for studying their preparation was "that it is necessary to satisfy many people who are very convinced that anything which has passed through an Alembic is prejudicial to health, & who will never be disabused of their views."[92]

The work of chemists like Baumé was central to the invention of alcohol as a social and chemical category. Baumé's writings foregrounded and unified the class of spirituous liqueurs as fluids containing a chemically analyzable and measurable spirit. This move was mediated by importing Boerhaave's redefinition of alcohol as "the purest form of the flammable principle reduced to its greatest degree of simplicity."[93] To replace a myriad spirituous drinks with a single category of spirit, identical in all fermented substances, which varied in (measurable) degree of purity, was to make an important epistemological assertion. In effect, spirit, a chemical substance, was elided with alcohol, formerly a value (purity) and a procedure (rectification). Such tau-

91. Jacquin 1762, 194–97. See also, for example, "Pêche," 1765.

92. Déjean 1759, 458–59.

93. Baumé 1773, 1: cxlviii: "Vocabulaire."

tologous definitions explained alcohol as a substance that could be identified through the chemical procedures by which it was produced. Formulations of this sort aligned with literature that joined wine and liqueurs in critiques of drunkenness and blamed their consumption for numerous diseases.[94] Here is the temperance debate presented in familiar form and identifiable with modern concepts of "alcohol."

We are thus faced with several overlapping frameworks for classifying and ascribing value to liqueurs, considered as objects of knowledge. In describing how material objects such as liqueurs and spirits circulated within and between domains of epistemological ownership, I am perhaps approaching Michel Serres's account of the "quasi-object," also borrowed by Bernadette Bensaude-Vincent for her discussion of the balance. To reprise Bruno Latour: "If choosing words for the network-tracing activity has to be done, quasi-objects or tokens might be the best candidate so far. . . . As a rule a quasi-object should be thought of as a moving actant that transforms those which do the moving because they transform the moving object. When the token remains stable or when the movers are kept intact, these are exceptional circumstances which have to be accounted for."[95] We cannot speak of "a meaning" for liqueurs but rather of parallel deployments of liqueurs in distinct epistemological projects relating to luxury and taste, health, chemical analysis, and so on. To single out chemical content or physiological effects as the explanatory principle for historical accounts of liqueurs or coffee, tea, and chocolate is to subscribe to a winner's history. As such substances were bartered between the worlds of the shop and the laboratory, they were effectively reinvented, because, in different settings, their modes of production and consumption, their definition, and their social, physiological, and chemical effects were also reconfigured. Luxury, social change, methods of preparation, and spaces and manners of consumption were at stake in the debate over liqueurs; in Old Regime Paris, they would prevail over chemical or governmental attempts to redefine these objects of knowledge.

This was true, at any rate, before 1780, the time when this chapter ends. Were I to extend the study beyond 1800, it would become clear how well the dual operations of analysis and classification served chemists in their attempts to reform the mechanical arts. The problematic category of alcohol was adopted by Lavoisier, by Parisian chemists and pharmacists, and eventually by most literate groups in the capital. It would become a point of agreement

94. Plagnol-Diéval 1997, 248. Macquer (1766, 2:642–43) also allied wine with brandy in attacking excess consumption.

95. Serres 1989, 1982; Bensaude-Vincent 1992, 234; Latour 1993; quotation from Latour 1997.

on which practitioners could ground the reform of arts and trades, including distillation and sugar manufacture, as well as new etiological categories and new regimens of chemical analysis. Lavoisier (1789, 1:139–52) showed just how to bootstrap an explanation in chemistry: in the very example he used to explain the conservation of matter in chemical reactions, alcohol was defined as one product of fermenting sugar; sugar was defined as that which, fermenting, produced alcohol and carbonic acid.[96] I have endeavored to avoid historical teleology by beginning this chapter from the "wrong" place, namely, the marketplace and not the academy. It is only thus that we can see that there was nothing inevitable about the stabilization of such a category. In different social spaces of learning, the ontological map into which liqueurs and brandies fitted had distinctive contours. Not until the settlement of the conditions under which explanations invoking "alcohol" had priority, could chemists' various instrumental, analytical, and quantitative criteria for demonstrating its existence become authoritative. Liqueurs were troublesome precisely because they eluded such categorizations for so long; where scientific explanation was concerned, they were matter in the wrong place.

## Primary Sources

Secondary sources can be found in the cumulative bibliography at the end of the book.

Académie Françoise, ed. 1798. *Dictionnaire de l'Académie Françoise, revu, corrigé et augmenté par l'Académie elle-même.* 2 vols. Paris: J. J. Smits et Cie.

"Alkool." 1751. In *Encyclopédie, ou Dictionnaire raisonné des sciences, des arts et des métiers, par une société de gens de letters,* ed. D. Diderot and J. d'Alembert. 17 vols. Paris: Briasson et al., 1:277.

Baumé, Antoine. 1757. *Dissertation sur l'Æther, dans laquelle on examine les différens produits du mêlange de l'Esprit de Vin avec les Acides Minéraux.* Paris: Jean-Thomas Hérissant.

———. 1760. Mémoire sur l'Éther vitriolique. *Mémoires de Mathématique et de Physique, Présentés à l'Académie Royale des Sciences, par divers Savans, & lûs dans ses Assemblées* 3:209–32.

———. 1762. *Élémens de Pharmacie theorique et pratique: contenant toutes les Opérations fondamentales de cet Art, avec leur définition, & une Explication de ces Opérations, par les Principes de la Chymie; la maniere de bien choisir, de préparer, & de mêler les Médicamens, avec des Remarques & des Réflexions sur chaque procédé; les Moyens de reconnoître les Médicamens falsifiés ou altérés; les Recettes des Médicamens nouvellement mis en usage; les Principes fondamentaux de plusieurs Arts dépendans de la Pharmacie: tels que l'Art du Confiseur, & ceux de la préparation des Eaux de Senteur & des Liqueurs de Table. Avec une Table des Vertus & Doses des Médicamens.* Paris: Veuve Damonneville et al.

———. 1773. *Chymie expérimentale et raisonnée.* 3 vols. Paris: P. Franç. Didot le jeune.

96. Brilliantly explicated in Poirier 2007, chap. 2. On the wider question of the fate of alcohol, see Spary, forthcoming.

———. 1778. *Mémoire de M. Baumé. sur cette question : quelle est la meilleure manière de construire les fourneaux et les alambics propres à la distillation des vins.* Paris: Clousier.

Caraccioli, Louis-Antoine, marquis de. 1757. *Le Livre de quatre couleurs.* Aux Quatre-Éléments (Paris): De l'Imprimerie des Quatre-Saisons.

Cossigny de Palma, Joseph-François Charpentier de. 1782. *Supplément au Mémoire sur la fabrication des Brandies de sucre.* Isle de France: Imprimerie Royale.

Déjean, M. 1759. Traité raisonné de la Distillation, ou la Distillation réduite en Principes, avec un Traité des Odeurs. 2nd ed. Paris: Nyon & Guillyn.

Delamare, Nicolas. 1705. *Traité de la police, Où l'on trouvera l'histoire de son etablissement, les fonctions et les prerogatives de ses magistrats, toutes les loix et tous les reglemens qui la concernent: On y a joint une description historique et topographique de Paris & huit Plants gravez, qui representent son ancien Etat, & ses divers Accroissemens. avec un recueil de tous les statuts et reglemens des six corps des marchands & de toutes les Communautés des Arts & Métiers.* 3 vols. Paris: Jean & Pierre Cot.

Demachy, Jacques-François. 1773. *L'Art du distillateur d'eaux-fortes.* Paris: Jean Desaint & Charles Saillant.

———. 1775. *L'Art du distillateur liquoriste; contenant le bruleur d'eaux-de-vie, le fabriquant de liqueurs, le débitant, ou le cafetier-limonnadier.* Paris: Jean Desaint & Charles Saillant.

Diderot, Denis, and Jean Le Rond d'Alembert, eds. 1763. *Encyclopédie, ou Dictionnaire raisonné des sciences, des arts et des métiers, par une société de gens de lettres.* Paris: Briasson et al.

Dubuisson, François-René-André. 1779. *L'Art du Distillateur et Marchand de Liqueurs. Considérées comme Alimens médicamenteux.* 2 vols. Paris: Dubuisson.

Geoffroy, Claude-Joseph. 1741. Méthode pour connoître & déterminer au juste la qualité des Liqueurs Spiritueuses qui portent le nom d'Eau de Vie & d'Esprit de Vin. *Histoire et Mémoires de l'Académie Royale des Sciences. . . .* 1718:37–50.

Geoffroy, Étienne-Louis. 1997. *Manuel de médecine pratique: ouvrage élémentaire auquel on a joint quelques formules.* Electronic document. Paris: INALF.

Hecquet, Philippe. 1710. *Traité des dispenses du carême, dans lequel on decouvre la fausseté des prétextes qu'on apporte pour les obtenir, en faisant voir par la mecanique du corps, les rapports naturels des alimens maigres, avec la nature de l'homme: et par l'histoire, par l'analyse & par l'observation, leur convenance avec la santé. . . . Revûe, corrigée, & augmentée par l'Auteur, de deux Dissertations, l'une sur les macreuses, & l'autre sur le tabac.* 2nd ed. 2 vols. Paris: François Fournier.

Helvétius, Claude-Adrien. 1988 *De l'Esprit.* Paris: Fayard.

Jacquin, Armand-Pierre. 1762. *De la santé. Ouvrage utile a tout le monde.* Paris: Durand.

Lavoisier, Antoine-Laurent de. 1789 *Traité élémentaire de chimie: présenté dans un ordre nouveau et d'après les découvertes modernes.* 2 vols. Paris: Cuchet.

Le Maître de Claville, Charles-François-Nicolas. 1740. *Traité du vrai mérite de l'homme, considéré dans tous les âges & dans toutes les conditions.* 4th ed. 2 vols. Paris: Saugrain père.

Liger, Louis. 1715. *Le Voyageur fidele, ou le Guide des Etrangers dans la Ville de Paris, qui enseigne tout ce qu'il y a de plus curieux à voir: les noms des Ruës, des Fauxbourgs, Eglises, Monasteres, Chapelles, Places, Colleges, & autres particularitez qüe cette Ville renferme; les Adresses pour aller de quartiers en quartiers, & y trouver tout ce qu'on souhaite, tant pour les besoins de la vie, que pour autres choses.* Paris: Pierre Ridou.

Macquer, Pierre-Joseph. 1751. *Elémens de chymie-pratique, contenant la description des opérations fondamentales de la chymie, avec des explications et des remarques sur chaque opération.* 2 vols. Paris: n.p.

———. 1766. *Dictionnaire de Chymie, Contenant la Théorie & la Pratique de cette Science, son application à la Physique, à l'Histoire Naturelle, à la Médecine & à l'Economie animale; avec l'explication détaillée de la vertu & de la maniere d'agir des Médicamens Chymiques. Et les principaux fondamentaux des Arts, Manufactures & Métiers dépendans de la Chymie.* 2 vols. Paris: Lacombe.

Macquer, Philippe, ed. 1766–1767. *Dictionnaire portatif des arts & métiers, contenant en abrégé l'histoire, la description & la police des arts et metiers, des fabriques et manufactures de France & des Pays étrangers.* 3 vols. Yverdon: n.p.

Mercier, Louis-Sébastien. 1995. *Tableau de Paris. Edition établie sous la direction de Jean-Claude Bonnet.* 2 vols. Paris: Mercure de France.

Montigny, Etienne Mignot de. 1770. Mémoire sur la construction des Aréomètres de comparaison, applicables au commerce des Liqueurs spiritueuses, & à la perception des Droits imposés sur ces Liqueurs. *Histoire et Mémoires de l'Académie Royale des Sciences. . . .* 1768:435–59.

"Pêche." 1765. In *Encyclopédie, ou Dictionnaire raisonné des sciences, des arts et des métiers, par une société de gens de letters,* ed. D. Diderot and J. d'Alembert. 17 vols. Paris: Briasson et al., 12:230–31.

Poncelet, Polycarpe. 1755. *Chymie du Goût et de l'odorat.* Paris: P. G. Le Mercier.

Review of Adrien Richer "L'Histoire moderne des Chinois, des Japonais, des Indiens, des Persans, des Turcs, des Russes." 1760. *Année Littéraire* 1760 (3): 202–4.

Thierry, François. 1755. *Médecine Expérimentale, ou Résultat de nouvelles Observations Pratiques & Anatomiques.* Paris: Duchesne.

Venel, Gabriel. 1753. Chymie. In *Encyclopédie, ou Dictionnaire raisonné des sciences, des arts et des métiers, par une société de gens de letters,* edited by D. Diderot and J. d'Alembert. 17 vols. Paris: Briasson et al., 3:408–37

———. 1757. Intempérance. In *Encyclopédie, ou Dictionnaire raisonné des sciences, des arts et des métiers, par une société de gens de letters,* ed. D. Diderot and J. d'Alembert. 17 vols. Paris: Briasson et al., vol. 7.

"Vin, & Fermentation vineuse." 1765. In *Encyclopédie, ou Dictionnaire raisonné des sciences, des arts et des métiers, par une société de gens de letters,* ed. D. Diderot and J. d'Alembert. 17 vols. Paris: Briasson et al., 17:283–89.

PART THREE

# State Interventions

## Introduction to Part 3

The essays in this third part, by Marcus Popplow, Agustí Nieto-Galan, and Seymour H. Mauskopf, highlight institutional efforts of learned societies and the early modern state to promote hybrid experts and the useful ventures that combined artisanal and higher natural knowledge. They study social groups, institutions, and administrations that sought to improve the production of agricultural materials, dyestuffs, and gunpowder for the benefit of the public. These organizations fulfilled significant functions as mediators between different sectors of production or between the spheres of production and consumption.

The chapters in the previous two parts have studied experts who found individual ways to fuse local hands-on knowledge about production processes with more systematic, often text-based, experiential knowledge. In the sixteenth and seventeenth centuries it was mainly individual expertise and princely patronage that made technical innovations and new constellations of useful knowledge possible. The chapters in this part shift attention away from the court and individual experts and toward new social institutions of expertise, innovation, and economic mediation that emerged in the course of the eighteenth century. The economic societies studied by Popplow are a good case in point. These institutions, which spread throughout Europe during the second half of the eighteenth century, undertook systematic efforts to find sustained ways to improve agriculture, involving peasants, landowners, and administrations in a country. In this way, they sought to prevent hunger crises and to meet the needs and expectations of artisans using agricultural raw materials. Economic societies thus fulfilled the function of mediators between different social groups and sectors of the economic system.

Popplow scrutinizes strategies of German economic societies to create advanced bodies of useful knowledge for agriculture as well as the obstacles they encountered and their eventual failure. "Viewing themselves as catalysts for the acceleration of the production of raw materials," he argues, "the economic improvers strove systematically to assemble a publicly accessible body of knowledge." Members of economic societies collected data about agricultural practices, performed experiments, and proposed technical measures for agricultural innovation as well as methods for the dissemination of knowledge, including the education of peasants. State authorities approved these activities and often directly financed them. By around 1800, Popplow points out, these innovators saw themselves faced with "insurmountable obstacles." Such failures of the economic societies shed light on the complexity of the ventures at stake in this volume and the difficulty of attempts to amalgamate "scientific" methods and forms of knowledge with the traditional and local know-how of peasants and artisans.

The complexity of material production, the "thick" of materials, is also highlighted in Seymour H. Mauskopf's study of the crisis of English gunpowder in the eighteenth century. State administrations intervened to guarantee sufficient supply and quality of a material that was of crucial importance for the state. The forms of intervention and mediation were certainly no less complex than in the economic enlightenment at stake in Popplow's chapter. A whole national organization of gunpowder production was created in eighteenth-century Britain, which involved administrations such as the Ordnance Office as well as its different types of expert-officers. Another important element of this organization was the Royal Laboratory at Woolwich along with its comptroller, who was entrusted with the supervision of gunpowder proof. Furthermore, there were also the Royal Military Academy at Woolwich, which trained cadets in artillery and engineering; the private owners of gunpowder mills, where gunpowder was produced from the raw materials saltpeter, sulfur, and charcoal; the different producers of or traders in these raw materials; and the military, which continually complained about the quantity and quality of British gunpowder. As in other European countries, Mauskopf observes, the British authorities tried "to bring scientific expertise to bear on the amelioration of gunpowder." In their attempts to rationalize British gunpowder production and to standardize the quality of gunpowder, officials performed systematic experiments and measurement in the late eighteenth century. To these belonged, in particular, ballistic testing of gunpowder carried out under the supervision of the proof master at Greenwich and experiments performed at the Royal Laboratory at Woolwich by comptrollers such as George Napier and William Congreve. As Mauskopf demonstrates, these two men corresponded exactly with the type of hybrid experts also highlighted in the other chapters in this volume: They "were military officers, held positions of authority in military establishments, and carried out research and publications in artillery and ballistics."

Agustí Nieto-Galan's chapter moves on to a similarly complex system of material production, social cooperation, and expertise in eighteenth-century France organized around the manufacture and quality control of natural dyestuffs in the Manufacture royale des Gobelins. This system included the workshops of the manufactory (*manufacture*), the laboratory of the Gobelins, as well as a dyeing school (*école de teinture*), where students became familiar with published treatises on dyestuffs and methods of bleaching. It also meant a social hierarchy of ordinary dyers, the *chef d'atelier,* the general inspectors of dyestuffs appointed by the Paris Academy, and the royal administration. As Nieto-Galan observes, historical evidence shows that in this system "the formation and development of the art of dyeing depended on strategic appropriations of academic knowledge." The quality tests for dyestuffs, in particular, introduced and further elaborated by academicians such as Charles-François du Fay, Jean Hellot, Pierre-Joseph Macquer, and Claude-Louis Berthollet, played a significant role in the rationalization of the art of dyeing. The laboratory of the Gobelins, where such quality tests were performed, thus was a "crucial intermediate space between academia and the workshop." But not always, Nieto-Galan points out, did the workshop and the academic world cooperate smoothly, as several debates about expertise and authority between the *chef d'atelier* and the academic inspectors demonstrate.

10

# Economizing Agricultural Resources in the German Economic Enlightenment

MARCUS POPPLOW

Investigations into the properties of materials, which brought "artisanal" and "learned" knowledge into contact in the early modern period, concerned not only the arts and crafts, as the domain where materials were turned into finished products, but also agriculture, as the domain where many such materials, including foods, were produced.[1] This chapter investigates strategies for increasing the production of domestic agrarian resources by creating bodies of advanced knowledge, envisioned in the second half of the eighteenth century in many European states in the so-called industrial or economic enlightenment.[2] The term "economic enlightenment" is preferred in what follows, as it appropriately echoes the broad notion of "economy" shared by the eighteenth-century protagonists, which addressed matters pertaining to both agriculture and the arts and crafts. With a view to avoiding hunger crises, as well as shortages of plant resources for the arts and crafts, govern-

My thanks go to the participants of the two workshops at which earlier versions of this paper were presented for their critical commentaries, an anonymous referee for most valuable advice, and especially to Ursula Klein and E. C. Spary for their extensive comments, extraordinary patience, and support.

1. This chapter forms part of a wider project, "On the 'Oeconomization' of Nature in the 18th Century—a Turning Point in the History of the Environment?" funded by the Deutsche Forschungsgemeinschaft (DFG) at the Brandenburg University of Technology (BTU) in Cottbus, under the direction of Günter Bayerl, chair for the history of technology. For earlier project publications, see, for example, Bayerl 1994, 2001; Meyer and Popplow 2004.

2. Mokyr argues for the crucial role the "industrial enlightenment" has played in transforming the "scientific revolution" into the later "industrial revolution" (2002). For the term "industrial enlightenment," see Mokyr 2002, 34–35; Mokyr 2005.

ment officials, clergymen, and members of the Republic of Letters proposed a broad set of measures for agriculture, forestry, and animal breeding. Such interests were pursued in informal contexts in the Republic of Letters, but also in institutions like academies and, in particular, in the countless—if often short-lived—economic and patriotic societies founded across Europe in the second half of the eighteenth century. Viewing themselves as catalysts for the acceleration of the production of raw materials, the economic improvers strove systematically to assemble a publicly accessible body of knowledge. They set themselves the tasks of collecting knowledge concerning promising innovations, of determining whether such proposals could be successfully turned into practice, and of distributing their results to the actors concerned. Economic and patriotic societies thus acted as sites where diverse forms of knowledge and practice intersected around the issue of agricultural materials.

One of the goals of this chapter is to offer a broad account of the problems and challenges faced by systematic eighteenth-century inquiries into the methods of increasing supplies of agricultural materials. In analyzing this "economization" of natural resources—with a focus not on their economic exploitation as such, but on the knowledge systems and practices related to the investigation of their material qualities—this chapter is thus not limited to a case study of one among the many "new and useful" plants introduced to Europe in the early modern period, such as the potato or sugar beet. Instead, it will investigate the general program of the economic enlightenment, which, in modern terms, attempted to encompass all the stages of production up to the nationwide implementation of successful improvements. Unlike England, the German lands studied here did not spearhead agricultural innovation, something that strengthened their efforts to foster innovation by generating a practice-oriented knowledge base concerning agricultural materials.

Another important issue of this chapter is mediation. In spite of their focus on the material dimensions of agriculture, the activities of the economic enlighteners were often carried out at a considerable distance from the materials as such. Members of economic societies were certainly unlikely to have worked the fields themselves, even if they grew plants of interest on their own estates or studied them in experimental gardens or on their journeys. Their extensive publications suggest that much of their time must have been spent not working on plants, but with a pen and paper in hand. In addition, economic societies, like any other type of eighteenth-century society, were also attended on account of the opportunities they offered for sociability and for the enhancement of personal status. The improvers saw these extended knowledge networks, which brought them close to administrations and to political power, as a crucial advantage over peasants and landowners, who in their view were limited to local expertise.

Economic societies thus fulfilled the important function of mediations between farmers, the state, and consumers. Comparable to the experts on milk studied in this volume by Barbara Orland, they collected data and experiential knowledge from many different practical areas and at the same time sought to systematically improve this knowledge in order to bring it back to agricultural practice. But they did so in an institutional and thus comparatively sustained way. Hence, the eighteenth-century economic societies took up a goal that had been associated with the newly created academies in the seventeenth century, namely, to render science useful to the common weal. From a modern standpoint, the apparent divergence of the economic enlightenment from agricultural materials as such fits well with the slow and winding paths by which technical knowledge was appropriated in administrative and scientific contexts during the early modern period. In many other fields of technical activity, similarly, this process was characterized by the development of media-based strategies for assembling, testing, and distributing technical knowledge in specialized, mediating institutional contexts, backed by administrative and political measures. Seen from this perspective, the investigations of the economic enlightenment into the production of agricultural materials seem representative of an important aspect of long-term processes of modernization in Europe.

Such a statement does not contradict the fact that eighteenth-century economic societies in fact involved a significant lack of integration of different communities, especially as concerned the farmers themselves. By around 1800, such obstacles would come to be seen as insurmountable in many respects. A logical consequence of this conclusion was the institutional differentiation of tasks such as knowledge production, administration, and education during the nineteenth century, the complexity of which had been underestimated by the members of the economic societies. The quick dissolution of economic societies after 1800 might be read as a sign that some of their major goals had not been achieved. However, the discourses and practices they had initiated were integrated with only slight changes into the much more differentiated nineteenth-century institutions concerned with agricultural knowledge production and distribution. Even if these later developments lie outside the focus of this chapter, one might read this case study as an indication that processes of institutionalization do not represent linear paths toward modernization. Nevertheless, the suspension of a certain institutional form does not mean that new approaches pursued in such a context, as, in this case, the study of the material qualities of natural resources, vanished along with their institutional "containers."

In spite of the European dimension of the economic enlightenment, its activities have not received much scholarly attention, aside from a number of descriptive and often well-researched case studies on particular economic societies and Lowood's excellent overview for the German lands.[3] Economic and agricultural historians usually mention the economic enlightenment only in passing, as a futile attempt by overoptimistic intellectuals to administer eighteenth-century agriculture that was rebutted by the realities of everyday practice. Assessments that emphasize the importance of such activities within long-term processes of innovation are rare.[4] The effect of their activities upon administration practice—many members doubled as administrators—also remains underresearched to date.[5] Conversely, the recent studies of early modern academies conducted by historians of science have thus far not included economic societies, presumably because of their orientation toward agricultural practice.[6]

The first part of this chapter sketches the historical context of the economic enlightenment in the German lands to illustrate the circumstances under which this new knowledge-based approach to agricultural resources developed in the second half of the eighteenth century. The second and main part describes in detail what the improvers understood to be their practice-oriented, "scientific" approach to agricultural reform, namely, techniques of compiling, validating, and disseminating knowledge about natural resources.

3. For England, see Schaffer 1997a, Goddard 1998; for France, Justin 1935, Roche 1978b, Steiner 1996; for Spain, Shafer 1958; for Venice, Simonetto 2001; for the German lands, Vierhaus 1980, Lowood 1991, Schlögl 1993, Popplow, in press. For well-researched case studies of the Leipzig society, see Eichler 1978; Schöne 2001.

4. "A more positive evaluation, however, emerges from understanding these societies not as a means to turn technological innovations directly into practice, but as institutions concerned with the handling of such innovations, . . . from a social historian's point of view, their decisive function was education and training, and the replication of new structures of action and interaction, which allowed a means-end orientation toward both internal and external nature" (Schindler and Bonß 1980, 256; my translation). "The patriotic societies clearly had a catalytic effect on economic progress, social change, and contemporary changes in mentality" (Bödeker 1999, 301; my translation).

5. For a case study, see Holenstein 2003.

6. In one otherwise impressive guide to eighteenth-century science, for example, a mere few pages are devoted to agriculture. As part of a chapter on "Science and the Popular," the emphasis is on the plurality of different forms and sites of agricultural knowledge and on well-known agricultural innovators such as Jethro Tull, William Cullen, and Arthur Young (see Fissell and Cooter 2003, 139–46).

The third part briefly sketches the problems faced by the mediatory approach of the economic societies and the reasons why their specific institutional model lost its attraction in the early nineteenth century.

## The Public-Knowledge Approach to Agricultural Resources in the Eighteenth Century

Focusing on the ideal conditions for growing certain plants in order to increase the supply of raw materials, the economic improvers considered plants only in terms of their useful properties for human needs. In the period investigated here, the second half of the eighteenth century, their strategies for dealing with agricultural materials still far exceeded, and sometimes even dismissed as inadequate, "scientific" approaches in the modern sense, such as the study of plant physiology or soil composition. Their program of studies included an investigation of the conditions of the local environment by means of field trials, as well as strategies for encouraging the acceptance of proposals for reform by large numbers of peasants. However, it was hardly possible to reduce the complex interaction of plants, soils, and farmers, under highly variable local conditions, to a mere handful of variables. Under such conditions, the economic improvers struggled to generate public knowledge of how to increase a state's supplies of agrarian materials and to supplement or even replace traditional agrarian knowledge, which was rooted in daily practice and usually inaccessible to outsiders.

Apart from these efforts in the context of the economic enlightenment, in the early modern period, individuals in numerous European states sought to increase supplies of agricultural raw materials through a variety of different enterprises, including the cultivation of new plants and experimentation with breeding techniques. The motivation for such unsystematic measures is still a subject of dispute, but the advantages offered by new market opportunities at the local or regional level often prompted activities of this sort.[7] Travelers from all over Europe admired the large-scale transformations of agrarian production realized in the Netherlands and, especially, early modern England.[8] The eighteenth-century economic enlightenment broke with such "traditional" processes of innovation in one critical respect. Peasants' and landowners' agricultural innovations were transmitted informally, by means of personal relations and instruction. Such processes now became the subject of macroeconomic considerations, and attempts were made to inte-

7. See Kopsidis 2006.

8. With regard to England, see, for example, Overton 1996; Allen 2004.

grate them into learned knowledge. This approach is thus comparable to that of the eighteenth-century French encyclopedists and others who collected "secret" knowledge in artisanal workshops in order to publicize it so as to encourage innovation. However, the economic improvers' attitude toward agricultural materials went further. In contrast to the authors of earlier agricultural manuals, the members of eighteenth-century economic societies treated publication as just one aspect of a wide range of activities. They saw themselves as not mere "reporters," but as active testers and disseminators of advanced knowledge about natural resources of a given territory. Their attempts might thus be characterized as the first top-down approach to encourage agricultural production not by political means, be they peaceful or violent, or through large-scale measures like land reclamation, but above all by creating a sophisticated knowledge base, certain elements of which were ideally to be applied by all agrarian actors, from simple farmers through estate owners to state officials.

## THE "TECHNICAL" CHARACTER OF REFORM MEASUREMENTS

The attempts of eighteenth-century economic improvers focused on the possibilities for increasing the production of agricultural materials by technical means. Here I use *technical* in the broad sense, to denote all kinds of measures aimed at increasing supplies of agricultural materials, provided such moves did not entail political and social measures that risked overthrowing the existing order. Even if members of economic societies discussed social reform behind the scenes, and such proposals occasionally found their way into some of their publications, conflict had to be avoided with the authorities, who approved the founding of such societies and often financed their activities to a large extent. One major focus was on proposals for intensified land-use by means of the classical cycle of intensified agricultural production, which entailed increased fodder production, allowing more cattle to be raised; stall feeding; the systematic distribution of dung; crop rotation; and other elements.[9] Other recurring topics included apiculture, fishing, viticulture, brewing, horticulture, and forestry. The issues of animal diseases and measures to destroy "harmful" animals also figured frequently in such writings and should be understood in the same way, as part of an attempt to increase production by reducing losses.[10]

9. See, for example, Prass 1997, 36.

10. See Meyer 2003 and Herrmann 2006; on cattle diseases, see Hünemörder, in press.

As regards the growing of commercial plants, which is the focus of this chapter, countless experiments were devoted to surrogate plants for dyeing, tanning, and paper making. The marsh lily was to be used to produce a yellow pigment, tormentil root for tanning, willows or nettles for paper production. Proposals for generating domestic sources of raw materials for high-quality textiles recommended merino sheep, as well as the floss of the poplar seed as a surrogate for cotton, or the planting of mulberry trees for silkworm cultivation. American plants, such as corn and especially the potato, had been known in many European regions for a long time, and reports on the qualities of an ever-increasing number of foreign plants and their acclimatization potential were commonplace. In a single year, 1771, the Leipzig economic society published announcements concerning "Siberian buckwheat," "Canadian cabbage," "sweet potatoes," the "East Indian olive," "Siberian cress," and "Pennsylvanian tobacco."[11]

## MOTIVATIONS AND PRECEDENTS

German economic improvers did not interest themselves in agricultural materials purely in response to perceptions about actual shortage. To be sure, famines prevailed in the German lands because of crop failure from 1739–1741, after the Seven Years' War (1756–1763), and again from 1770–1772. The population growth in the second half of the eighteenth century imposed additional demands on the provisioning of staple foods. However, such recurrent supply crises had long been typical of preindustrial societies. It was, rather, the interplay of a number of economic and cultural factors that, in the end, induced improvers to attempt to systematize knowledge about advanced techniques of agricultural production. Of crucial relevance was the perceived threat of future shortages in the supply of raw materials. Thus far, this topic has been researched in most detail for early modern wood shortages; comparative studies of other plant and animal resources, shortages of which were the subject of eighteenth-century public discourse, are still lacking. Studies comparing contemporary discourses on wood shortages with reconstructions of actual wood supplies have on the one hand brought to light actual local or regional shortages; on the other, they show that discourses about wood shortages often apparently were motivated by the particular economic interests of forest owners rather than by actual shortages.[12]

11. *Göttingische Anzeigen* 1771, no. 135, 1160.

12. See the recent references in Freytag and Piereth 2002; Warde 2006, 7–8 and *passim*.

The economic enlightenment's goal of tackling the supply of domestic agricultural materials in a comprehensive way was related to three central factors: first, traditions of investigating non-European plants within the context of European expansion; second, the establishment of institutions like academies and learned societies that collected knowledge for public benefit; and third, a bevy of economic theories arguing that a sufficient supply of natural resources was an indispensable condition for the well-being of the emerging states.

A major stimulus for gathering knowledge about new agricultural materials in the early modern period was European expansion, which has been interpreted, among other things, as a strategy by European powers to increase the space available to them for producing raw materials.[13] Starting in the sixteenth century, the growth of European colonies facilitated plant transfer between different parts of the globe, while Europe itself experienced a "colonization and commodification" of natural resources in the context of imperial trade, as leading colonial powers increasingly sought to introduce raw materials for the crafts and trades at home.[14] The accumulation of systematic knowledge about ways of introducing plants from the New World to European climatic and soil conditions was facilitated by the early modern academies; by state initiatives, including the financing of research expeditions and the founding of botanical gardens; and, finally, in less formal contexts of the Republic of Letters, such as private collections and extended networks of correspondence.[15] The activities of Linnaeus, who sought to stimulate the Swedish economy through the introduction of new commercial plants, have been thoroughly analyzed in this regard.[16] The transformation of natural history into a "science of resources," beginning before the mid-eighteenth century, has recently been highlighted by pioneering studies.[17] In this context, attempts to adapt non-European flora to European conditions gradually merged with comparable attempts to encourage domestic agricultural growth by setting up a comprehensive knowledge base. The Georgics Committee, founded by the Royal Society in the 1660s, can in many respects be seen as a forerunner of the eighteenth-century economic societies in its aims and its methods, as, for example, in its attempt to investigate English farming practices by means of a detailed questionnaire.[18]

13. See, for example, Pomeranz 2000, 264–78.

14. Stewart 1989, 836.

15. See Lowood 1995; Spary 2000, esp. 99–154; Drayton 2000; Smith and Findlen 2002.

16. Koerner 1999; Müller-Wille 2004.

17. Spary 2003, 31.

18. Lowry 2004, 93.

The Royal Society of Arts, founded in 1754, pursued similar activities on a larger scale.[19]

For non-European flora, as I have said, strategies for generating advanced knowledge about natural resources were pursued in European academies from the seventeenth century onward. Both in their techniques of knowledge production and in their character as social enterprises, academies served as role models for the economic societies, most of which were founded after the mid-eighteenth century: "They adopted the same research methods, the same reliance upon verifiable evidence, the same methods of stimulating research by setting examples and offering prizes, and the same means of publishing their findings and recommendations."[20] With their emphasis on encouraging agricultural improvement in a given territory, economic societies singled out one of many topics of interest to early modern academies. Their aim was to create a less exclusive space, one in which, for example, vernacular contributions and publications were the norm right from the start. This reflected the distinctly regional focus of economic societies, which primarily pursued local results. Often backed—like most academies—by funding from the sovereign's purse, and cooperating with state agencies, economic societies brought university scholars together with officials in the upper administrative echelons, burghers, noble landowners, and clergymen.[21] Around 1800, their membership in the German lands has been roughly estimated at several thousand (though, usually, only about a dozen active members coordinated the activities of a given society in practice). At a conservative estimate, there were around one hundred economic societies, some of which were only short-lived. Among the most important were those at Bern, Erfurt, Celle, Burghausen, Kaiserslautern, Breslau, Halle, and Leipzig.[22] The Göttingen and Berlin academies of science also stand out for their exceptional degree of interest in economic matters. Among early modern scientific institutions, economic societies occupied a special niche, uniting learned knowledge with administrative action and economic theory with practical local experience.

In this context, eighteenth-century economic concepts provided a decisive stimulus for addressing agricultural materials in the systematic man-

19. Hudson and Luckhurst 1954, 57–100.

20. Hahn 1963.

21. For example, the Royal Economic Society of Brunswick-Lüneburg in Celle received funding from the Elector of Hanover, King George III of Great Britain; the Burghausen Agricultural Society from the Elector of Bavaria, and the Kaiserslautern Physical-Economic Society from the Elector Karl Theodor.

22. For the figures cited here, see Lowood 1991, 26–27.

ner adopted by the members of economic societies. While sixteenth- and seventeenth-century mercantilist concepts had not greatly emphasized the production of domestic agricultural materials, apart from general advice to avoid exportation, physiocratism and cameralism, regardless of their specific theoretical outlines, both put considerable emphasis on increasing such production.[23] In the German context, the authors of the earliest cameralistic texts, writing in the late seventeenth century, saw knowledge about endemic natural resources as a central prerequisite for the promotion of the arts and crafts by the state, a line of argument developed in the mid-seventeenth century by a number of authors who had evaluated and described the "natural riches" of German lands, often focusing first and foremost on mineral resources.[24] In the wake of the economically disastrous Seven Years' War, the cameralists began to question the mercantilist dogma of trade as the main source of a territory's wealth. In its stead came the insight that production was of greater economic importance. This theoretical shift was closely linked to the state goal of population growth (*Peuplierung*) and the novel concept of a worldly "public felicity" (*Glückseligkeit*). The authors saw a rise in productivity in the manufacturing arts as indispensable, and a precondition for this was an increased supply of raw materials. They thus saw advances in both realms—agriculture and the arts and crafts—as prerequisites for a flourishing economy.

From the mid-eighteenth century onward, the institutionalization of German cameralism with the foundation of professorial chairs at universities provided a supply of trained civil servants to administer the increasingly complex economic interests of the early modern state. By the end of the Seven Years' War, cameralists increasingly called for practical measures. As early as 1755, the prominent cameralist Johann Heinrich Gottlob von Justi formulated a program for producing and distributing advanced knowledge, which gave ample coverage to the supply of agricultural resources. Academies of science, Justi argued, should require their members to conduct practical research, "just as in the Paris academy a member is obliged to engage in experiments with dyes . . . for the sake of the manufactures."[25] Colleges (*Hohe Schulen*) should establish a faculty of economics, while "private persons should tell the world about their insights." Economic societies were to be founded in keeping with the English model. In the country, economic supervisors (*Oeconomieinspectores*) were to instruct "the farmer about new

23. See Dittrich 1974, 56ff; Tribe 1988a; Meyer 1999; Sandl 1999; T. Simon 2004.

24. See Cooper 2004.

25. See Nieto-Galan's chapter in this volume.

advantages in husbandry." Justi saw periodicals as of special importance: "Administrative journals (*Intelligenzblätter*) should report only on matters of trade, manufactures, the arts and crafts, agriculture, stock-breeding and similar issues concerning the urban and rural economy."[26] In this context, natural resources were perceived as the foundation for producing all sorts of goods, and measures to encourage their availability were, at least implicitly, elevated to one of the state's aims. This intention was clearly summarized toward the end of the period considered here, in a submission to a prize contest held by the Bavarian academy of sciences in 1803: "Which goods of nature can be found in Bavaria and the Upper Palatinate . . . deserving more attention than has been devoted to them so far? And which of these products would be suited to employ many hands in manufactories for their modification and perfection? The academy wishes that a theoretically underpinned procedure should be disclosed, by means of which these materials might be worked so that in the end the artifacts might be sold not only in our home country, but abroad as well."[27] The seemingly arbitrary manner in which the economic enlightenment had considered virtually all plants and animals for possible economic uses by this time thus derived from this view—in today's terms, a macroeconomic perspective—on domestic production, as part of contemporary economic thinking. However, it is beyond the scope of this study to investigate the extent to which differences between the two major schools of physiocratism and cameralism remained evident in the practical manipulations of agricultural materials practiced during the economic enlightenment. At any rate, the economic societies were the decisive link between economic theory and agricultural practice, much more so than cameralist teaching at universities. On the other hand, it is clear that even if the economic enlightenment adhered to the general framework of ideas promoted by contemporary economic writing, its protagonists did not aspire to develop economic theory as such.

## Elements of the Economic Societies' "Scientific" Approach to Agriculture

The search for "useful" exotic plants in the context of European expansion, the role model provided by the scientific academies, and cameralistic thought

26. *Göttingische Anzeigen* 1755, no. 95, 881–82. This article refers to Justi's work *Abhandlung von den Mitteln die Erkenntniß in den ökonomischen und Cameral=Wissenschaften recht nützlich zu machen,* which had been published in the same year.

27. Quoted in Müller 1975, 76; my translation.

thus formed three traditions to which patriotic and economic societies adhered in their production of knowledge about agricultural resources during the second half of the eighteenth century. In many cases, their methods of producing knowledge differed from those of the academies, though their unpublished records still require further study to confirm the extent of such differences.[28] The economic society of the Palatinate explained in detail in its first publication how the society aimed "to investigate all useful or new discoveries, suggestions and experiences in different branches of agriculture, to perform actual trials [*wirkliche Versuche*] on most of them, as far as circumstances allow, and to represent successes and execution methods faithfully to all our citizens."[29]

A brief sketch of a typical example may provide an idea of the fundamental problems confronting improvers once the study of plants included attempts to cultivate them *in situ* in a given region. Between 1794 and 1803, toward the end of the period considered here, Friedrich Casimir Medicus, a prominent member of the Palatinate's economic society and the administrator of the prince elector's botanical garden at Mannheim, edited a journal devoted exclusively to the fast-growing *Robinia*, or bastard acacia. *Robinia* had been imported from North America to France early in the seventeenth century and, acquiring some renown as an exotic tree, was planted in a number of parks made in the fashionable English style in the early eighteenth century. Medicus propagated the fast-growing *Robinia* as a solution to the wood shortages, which were of intense concern throughout Europe at the time. In his journal, he repeatedly stressed the durability of *Robinia* wood, its advantages as a fuel, and its success even on meager soils. But *Robinia* saplings were relatively demanding during their first year of growth in a European climate. Medicus collected all information that might help to overcome this deficiency in a plant he saw as a possible remedy for severe wood shortages. The main aim of his publication was thus to provide data on suitable conditions for *Robinia* in the German lands. Characteristic of Medicus's approach was an explicit marginalization of inquiries into plant physiology, which, he was convinced, would not yield the information needed to realize large-scale plantation projects. Medicus tried growing about 170 *Robinia* in a test plot in Mannheim's botanical garden under different conditions of light and fertility. He also actively solicited reports on *Robinia* grown under varying condi-

28. Currently under investigation in a project "Nützliche Wissenschaft. Naturaneignung und Politik. Die Ökonomische Gesellschaft Bern im europäischen Kontext (1750–1850)," in progress at the Department of History, University of Bern.

29. *Bemerkungen* 1769, 14–15 (my translation). This volume was reprinted in 1771.

tions elsewhere. Twice he presented his readers with detailed questionnaires concerning *Robinia* specimens in their locality. By means of this standardized data collection, he sought detailed information on the age and size of *Robinia* plantations, the effects of fertilizers on the trees, and the soil structure, altitude, and orientation of the locations where they grew. His multivolume work of some 2,500 pages incorporated 128 letters by 84 individuals from all over Germany, half of whom—mostly forest administrators and higher level administrative officials—reported their own trials of growing *Robinia.* Apart from the multitude of information submitted, Medicus's efforts represent the typical intention of the economic enlightenment to provide technical information concerning the best conditions for growing promising plants. While repeatedly complaining that too few citizens engaged in cultivating *Robinia,* Medicus only occasionally mentioned structural obstacles like the fact that foresters customarily raised tree plantations from seed, while *Robinia* had to be grown from saplings raised by gardeners, which then required careful planting by the forester.[30]

Tree cultivation presented even greater problems than other agricultural resources, such as cereals, on account of the extended timeframe involved. But in all such cases, the economic improvers' primary aim was not to generate universal truths. Instead, they directed their efforts toward universal applicability within a given territory. They understood their endeavors as scientific in their own terms, though they sometimes remarked that investigations in contemporary sciences were not usually directly applicable to their concrete practical aims. As Medicus had done in the case of *Robinia,* economic societies primarily tried to generate "proven" knowledge, ready for economic application. Their general aim was appropriately outlined by Lowood: "to investigate local and regional conditions, collect examples of useful inventions and natural products, and disseminate information with the ultimate goal of improving regional industry and agriculture by building up an inventory of useful knowledge."[31]

Such an approach differed from the study of plants in natural history or chemistry. Certainly, improvers were interested in the chemical aspects of soil composition, the chemical analysis of plants, and the chemical and mechanical explanation of physiological processes, programs begun in the seventeenth century at the Academy of Sciences in Paris or by Johannes Baptista van Helmont.[32] James Hutton in Edinburgh studied the effects of growing

30. Hess 1987, 73–81; Knoll 2003; Popplow 2006.

31. Lowood 1991, 206.

32. See Stroup 1990, 65–166.

carrots and radishes in the dark and of cultivating seeds in poor versus rich soil, tying the results to theoretical concepts like the phlogiston theory.[33] In general, however, it was presumed that such endeavors were not directly applicable to local agricultural practice, where the aim was to yield viable economic results quickly. Thus, economic societies did not undertake detailed laboratory analyses of plants, for the most part, but concentrated instead on administering knowledge about growing them *in situ*. The most fruitful results were expected from field trials, where promising plants were cultivated under varying circumstances and related experiences collected and distributed. It was thus "descriptive natural history, topography, and the mapping of natural resources" that were of interest for these societies in their aim of achieving economic improvements in the particular conditions of their own region.[34] Knowledge in these fields primarily consisted in reliable information about what nutrition a plant or an animal needed or what climatic conditions would be appropriate for its cultivation or raising. Meteorology was of great interest, and the effects of weather phenomena on the growth of plants or the behavior of bees were intensively studied.[35]

The "scientific" status of the economic societies' activities was thus a flexible concept, the precise meaning of which depended on the other groups to which their members were comparing themselves. While the improvers proclaimed the scientific status of their own activities, in order to distinguish themselves from the mere probing and testing of "common" artisans or the "stubborn" adherence of the peasant to traditional methods of agrarian production, it was clear that they were not interested in theoretical reflection as such, unless this promised concrete results in the sense of rules to be followed. "Agricultural science" was mostly defined as the long-term investigation of the success or failure of new methods, and thus as an advanced form of experience: "if a member observed agricultural practice of the lay of the land carefully and often enough, he had raised *Oeconomie* to a *Wissenschaft*."[36] In this sense, contemporary authors explicitly spoke of the "science of economy" and of "ancillary sciences" to economy like natural history, natural theory (*Naturlehre*), and mathematics.[37]

Sometimes this approach generated programmatic statements that pointed out tensions between a strategy of applying contemporary science to practical

33. See J. Jones 1985.

34. Lowood 1991, 210; on the "eclectic quality" of eighteenth-century natural history, see Spary 2000, 5–6.

35. See Lüdecke, in press.

36. Lowood 1991, 149.

37. Beckmann 1769, preface.

needs and a more self-conscious strategy of converting practical experience into general rules. As the secretary of the Hamburg Patriotic Society stated explicitly in 1790: "The society . . . had decided from the beginning 'not to work directly from the sciences'; but instead to 'apply every useful result of human knowledge, discovery, and invention' to appropriate realms of 'practical, bourgeois life.'"[38] In more academic contexts, the lack of theoretical engagement in economic matters was nevertheless perceived to be a problem. In 1769 the Göttingen academy of sciences offered a prize on the question "How may the fact that *Oeconomie* has hitherto gained so little advantage from physics and mathematics be explained?" That the prize-winning essay again consisted, seemingly, solely of the usual set of proposals for agricultural reform might be read as evidence of the gaps perceived by contemporaries when it came to uniting practical and learned knowledge.[39]

In what follows, I will outline the broad range of learned, or "scientific," knowledge, as understood during the economic enlightenment, as it was intended to supplement or even replace traditional modes of agrarian knowledge. I will describe the individual stages of generating, validating, and communicating advanced knowledge about agricultural materials. Compared to eighteenth-century attempts to create such advanced knowledge in the arts and crafts, two major differences stand out. First, the process of validating knowledge was more complex because of the factors of space and time involved in field trials in contrast to laboratory experiments. Second, a much greater emphasis was laid on the impersonal, mediated communication of advanced knowledge, because the aim was to reach a broader audience than was addressed by inquiries into the specialized processes of a given mechanical art. A more detailed description of these three steps shows how broad this approach was compared to later definitions of scientific knowledge and also shows why, in the end, it always remained unclear to the historical actors themselves how these two worlds of local, personal knowledge on the one hand, and general, public knowledge on the other could ever be merged.

## PRACTICES OF COMPILING KNOWLEDGE

As I have already suggested, economic societies understood themselves for the most part as catalysts for increasing agricultural production. Prior to the foundation of most economic societies in the 1760s and 1770s, far-reaching

38. Lowood 1991, 187–88.

39. As reported in *Göttingische Anzeigen* 1771, no. 16, supplement, cxxxiv–cxxxvi.

agrarian transformation had occurred in the first half of the eighteenth century, accompanied by public discussions on agrarian innovation. In their self-descriptions, the societies underlined the value of the communal efforts made by their members, which surpassed those of the lone scholar. In keeping with the sociable character of economic societies, a large proportion of their activity was devoted to discussing measures for promoting agricultural growth. Meetings were usually annual or, in the more active societies, up to four times a year. These events usually included lectures and, where possible, demonstrations. In the interim, the executive committees met on a more frequent basis, sometimes even weekly.[40] Libraries and collections played an increasing role in the life of these institutions. In the collections, plant and animal specimens, models of mechanical contrivances for agricultural purposes, and physical instruments featured most prominently. Toward the end of the century, these collections and libraries, as well as the societies' publications, became the most visible result of the economic societies' activities and might arguably be termed "storage devices" for the knowledge they had assembled.[41]

There were several routes for acquiring this innovative knowledge. Despite the recurring emphasis on practical experience, farmers themselves took part in the economic societies' activities only exceptionally. Direct experience thus came mostly from administrative officials or clergymen who doubled as enlightened landowners. In addition, traveling, an extended correspondence network within the Republic of Letters, and the reading of economic literature produced in other regions and abroad were seen as vital sources of ideas for projects that might be successfully introduced in a society's own territory. It seems that the improvers saw their particular expertise as lying in the investigation of foreign plants, which represented a skill that by definition could not be matched by local expertise. At the same time, a major aim was to get hold of up-to-date knowledge of the home territory. The collection of "statistical" data and prize contests were the most common ways of procuring such information.

Investigations into the material qualities of certain plants, which aimed to increase agricultural production throughout a given territory, were often preceded by efforts to gain an overview of the state of the art of agricultural practice. In this context, large-scale data collection represented an important step in systematic efforts to produce an enlightened economy. Detailed knowledge on the availability of raw materials, topographic and climatic con-

40. See Willer 1967.
41. Mokyr 2005, 300.

ditions, and the distribution of the workforce, the arts and crafts, manufacturing, and trade was perceived to be a precondition for the proper administration of natural and human resources. In part, these efforts involved taking stock of the natural resources of certain regions in order to determine their economic uses. Such "statistical and topographical descriptions" were highly valued in the economic enlightenment.[42] To consider a regional example, in 1779 Johann Philipp von Carosi published his first treatise on the natural history of Lower Lusatia, evaluating the soils and minerals to be found in the region. In 1787 a "historical, geographical and statistical description" of the Cottbus region, published in the *Lausitzisches Magazin*, furnished detailed data on the species of plants, animals, and minerals to be found there. The description included details of agricultural and commercial plants grown there, as well as of wood usage in the different arts and crafts, wild animals in the area, cattle reared, and local deposits of minerals and peat. Tables to be filled out by members of societies or interested citizens were frequently used as a way of achieving standardized data collection; a typical example is tables of weather observations.

Generating expertise about the increased use of natural resources was also encouraged by prize contests, common in all major European academies during the eighteenth century and often concerned with economic issues.[43] The economic societies hoped that the rewards offered in prize contests would bring to light the knowledge of individuals who were not in the habit of publicizing their experiments. In awarding prizes, preference was given to those contributions that had already proven their practicability.[44] These aims can be discerned, for example, in a prize question submitted by the Göttingen academy of sciences for November 1776: "Which plants are still growing wild in the territory of Hannover that might be used to considerable advantage, especially by the farmer, without neglecting his other business, and which for that reason are judged worthy to be made known to him?"[45] The Göttingen academy was an especially active initiator of such economic prize contests, and the range of subjects is revealed in a survey of the period between 1757 and 1761. The prize questions included the advantages of Swedish iron over that of German origin; a pigment resembling indigo made from dyer's weed; recipes for preservable potato flour and bread; methods for achieving better

42. For a detailed case study, see Gerber-Visser, in press.

43. See Müller 1975. The project "Preisfragen als Institution der Wissenschaftsgeschichte im Europa der Aufklärung" at the Forschungszentrum Europäische Aufklärung, Potsdam, is currently compiling an extensive database on eighteenth-century European prize contests.

44. Müller 1975, 80.

45. *Göttingische Anzeigen* 1775, no. 138, 1179; my translation.

soil fertility; savings on fertilizer produced by soaking grain; ways of improving the quality of German wool to match the Spanish, or at least English, product; the nature and means of preventing grain rust; a fire-resistant wood paint; a comparison of German and English field divisions; and the best way of stripping bark from living trees.[46]

However, it is evident from the examples cited that the attempt to verbalize "formerly tacit knowledge" in the case of peasants' agricultural expertise faced numerous difficulties, just as in any other field of eighteenth-century technical knowledge.[47] In browsing through the thousands of pages of eighteenth-century publications devoted to agricultural innovation, it seems that no systematic attempt was made to translate peasants' knowledge, even if this was one of the intentions behind the prize contests. The most promising activities in that direction were attempts to use clergymen's contacts with peasants in their parishes, both to distribute proposals for innovation and to acquire their knowledge of methods of increasing crop yields.[48] Otto von Münchhausen, in his journal *Hausvater,* invited farmers to report their trials on new or local crops or weeds and asked them, in such cases, to attach a plant specimen to their letters.[49] In the end, however, the barriers of learned discourse must have been pretty much insurmountable in most cases. This resulted in a kind of discourse that reached primarily those practitioners who were learned landowners themselves.

## PRACTICES OF VALIDATING KNOWLEDGE

One of the most pressing tasks of the economic societies was to establish formal standards according to which the countless proposals for agricultural improvement could be judged.[50] Cultivation trials for new agricultural products were time-consuming. To some extent, trials and experiments, known to contemporaries as *Experimental-Ökonomie* (experimental economy) simply followed established academic practices, but once again, the emphasis on immediate local implementation made it necessary to include additional considerations.[51] There is as yet no systematic study of such field trials for the period before the nineteenth century, though trials of cultivation had been

46. *Göttingische Anzeigen* 1756, no. 141, 1275–76. For a list of prize contests held by the Leipzig economic society between 1764 and 1789, see Eichler 1978, 381–86.

47. Mokyr 2005, 298.

48. For a case study, see Konersmann 2005.

49. Reported in *Göttingische Anzeigen* 1765, no. 32, 260.

50. For related procedures, see the chapters by Klein and Nieto-Galan in this volume.

51. See, for example, Eckhart 1754.

undertaken as early as the seventeenth century. In 1653, for example, Walter Blith in England monitored crop yields on lands to which different kind of natural fertilizers had been applied.[52] Following the tradition of such trials on private lands, the economic societies set up experimental gardens on the lines of the botanical gardens of academies. One of the first measures of the founders of the Palatinate society in around 1770 was to set up a model apiary, a linen manufactory, and a model farm in villages near Kaiserslautern. The results of trials performed on the farm were to be used to instruct local peasants. While these operations ceased after just a few years because of organizational difficulties, such field trials became ever more extended during the late eighteenth century. For example, the forest administrator Friedrich von Burgsdorf began the long-term experiment, beginning in the 1780s and lasting for about twenty years, of planting a grove containing some four hundred species of domestic and foreign trees in order to understand their growth differences under similar conditions. In an attempt to achieve the standardized collection of data, he also devised a detailed table for potential emulators, so as to control for weather conditions and a number of other factors influencing tree growth. Burgsdorf himself claimed to be "researching the economy of plants," something that he judged "more important and more useful than sticking to names, inventories, and artificial systems," a critique of academic versions of natural history.[53] Friedrich Casimir Medicus's interest somewhat later in *Robinia*, discussed above, can be viewed in this light and so too can the measure taken by the Altötting economic society, who by their 1791 statutes obliged every member to "produce one experiment, communication, or observation annually."[54]

It is easy to imagine that generating a standard interpretation of the results achieved in this way proved to be quite difficult, especially with regard to transfers of plants between different locations and soils. In such cases, eyewitness testimony was of crucial importance. The difference between such endeavors and similar experiments in the arts and crafts concerned not only the investment of time and the issue of space, but also the problem of communicating results. At one end of the spectrum were Burgsdorf's studies of tree growth, while at the other was the author of a treatise on "new trials and samples for using the plant kingdom economically for papermaking," who simply added paper samples made from different plants to the publication in

52. Lowry 2004, 92; for other examples, Fussell 1976; Böhm 1990; Schaffer 1997a; Denis 2001.

53. Burgsdorf 1785, 236; see also Lowood 1991, 254.

54. Lowood 1991, 48.

which he described his activities in detail.[55] This special problem of accreditation where agricultural innovations were concerned was often also stressed when it came to making them known, for example through teaching. As a reviewer of Johann Beckmann's groundbreaking agricultural treatise *Grundsätze der teutschen Landwirthschaft* remarked in 1769, lectures in *Naturlehre,* in this case concerning the visible characteristics of plants, could be made comprehensible by demonstrations using specimens; in agricultural practice, by contrast, one often still depended on the oral or written testimony of some authority.[56]

The "technical," material-related character of all these efforts is made clear by comparison with one of the most spectacular field trials, carried out after 1770 under the auspices of the Margrave of Baden and widely commented upon in learned discourse. Inspired by French physiocratic writings, the Margrave had converted three villages into experimental sites for carrying out structural reforms, including the realization of advanced agricultural techniques, but also, and more especially, far-reaching tax reform and a redistribution of land that still left traditional power structures intact. However, these efforts were abandoned within a few years, both for lack of motivation and because it proved impossible to isolate the communities concerned completely from their traditional economic contexts. Because this failure was widely publicized, it might have reinforced efforts to concentrate exclusively on the technical measures described here.[57]

## PRACTICES OF DISSEMINATING "USEFUL KNOWLEDGE"

It is evident from many of the examples cited so far that the relationship between the members of the economic societies and agricultural materials was mediated by several layers of intervention. This is most evident in the way that successful agronomic projects were communicated. In theory, economic societies sought to address farmers directly and to persuade them to accept the societies' superior knowledge. Though such conversations certainly took place, contacts with peasants were usually limited to the distribution of seed from useful plants, like clover, either for free or for a small payment. These

55. Reported in *Göttingische Anzeigen* 1766, no. 16, 122–26.

56. The author expressed the hope that the economic garden about to be installed at Göttingen University would at least turn lectures somewhat more toward practical considerations (*Göttingische Anzeigen* 1769, no. 52, 475).

57. For a study emphasizing the process, rather than the outcome, of these reforms, see Metzler 2001.

exchanges were often accompanied by written instructions describing the best conditions for sowing and growing the plants in question.[58] Monetary recompense was an additional incitement to involvement in such agricultural actions as the successful planting of a particular plant.

Besides these direct measures for spreading advanced knowledge, economic societies produced a prolific literature of textbooks, pamphlets, contest announcements, and agricultural calendars containing "useful information" about all manner of agricultural issues.[59] Indeed, agricultural treatises had offered landowners better soil exploitation methods since antiquity. Since the sixteenth century, the German husbandry literature (*Hausväterliteratur*) and similar treatises on agricultural improvement published elsewhere in Europe had proposed ways of increasing productivity in agriculture and stock breeding.[60] However, in the context of the Enlightenment's "instructive zeal" (*Belehrungswut*), attempts to increase agricultural production were made known to broader strata of the population and certainly also reached decision makers in administration, agriculture, and the crafts and trades. In subscription lists, such as that for *Bemerkungen der kuhrpfälzischen physikalisch-ökonomischen Gesellschaft* (1774), the main categories of subscribers, apart from booksellers acting as middlemen, were government officials and pastors.[61]

Around 1750, ways of achieving a qualitative and quantitative rise in agricultural productivity became a common theme in the so-called *Intelligenzblätter* and other journals supplying regional markets with general information.[62] One year after its foundation in 1768, the editor of the *Lausitzisches Magazin*, "humbly following his readers' wishes," promised to provide more information about practical matters. From then on, advice on increasing agrarian production featured regularly. The journal named merchants who imported seed of different kinds of clover from France, Spain, and England—clover so productive it could be cut six to seven times a year. It reported on a certain kind of grass from Virginia, already being cultivated in England and Russia, that was especially suitable for fodder—the article closed with the explicit wish that cultivation of this plant might propagate in Lusatia as well. Methods of destroying "harmful" animals, like moles or field mice, were discussed,

58. Lowood 1991, 101.

59. See Bödeker 1999.

60. See Fleischer 1981; Sieglerschmidt 1999.

61. See Hess 1987, 61.

62. For the general context, see Fischer, Haefs, and Mix 1999; Doering-Manteuffel, Mancal, and Wüst 2001. For a case study, see Gerber 1995, 108–10; the list of articles on such issues is at Gerber 1995, 363–419.

and a brochure distributed free by local governors in the 1770s in order to encourage the cultivation of tobacco in Lower Lusatia was cited at length.

Besides the incorporation of economic issues in general journals, the closing decades of the eighteenth century also saw the appearance of a new genre of journals addressing issues of commercial and artisanal practice. Mostly published by the economic societies themselves, they offered a forum exclusively devoted to economic issues.[63] One reviewer spelled out the central aim of Leipzig's *Ökonomische Nachrichten* in 1753: "to employ each of Nature's products in the most favorable way possible."[64] The economic societies' "publicity campaign" included details of their own activities, such as lists of books and objects received during the year.[65] Other than this, such published materials usually consisted of empirical, ad hoc procedural accounts of successful or unsuccessful achievements in their state, or simply of suggestions regarding the cultivation of some foreign plant or other under local conditions. A typical example is an enthusiastic account written by a member of the Palatinate economic society who had visited an estate in a village near Heidelberg where an enlightened administrator, installed by the owners, had implemented many up-to-date measures, turning the estate into a profitable earthly paradise.[66] In addition to journal contributions, a large body of independent works, again often written by the members of patriotic and economic societies, covered such topics in more detail.

All such publications sought to encourage readers to put the proposals in their pages into practice. They appealed to the enlightened patriot to contribute to the common weal and also underlined the possibilities of individual financial gain. For example, the renowned "Upper Lusatian Physical-Economic Bee Society" saw its central aim, the promotion of large-scale bee husbandry among the peasantry, as a useful supplementary economic activity at a time when honey was still the main locally produced sweetener, because the sugar beet had yet to be introduced. The tradition of keeping quantities of beehives in hollow tree trunks in the forest was increasingly viewed as harmful to the forest, in the context of the authorities' efforts to administer forestry more systematically. Teaching peasants to keep bees in the vicinity of their homes instead was thus representative of efforts to do away with uncontrolled forest use by peasants, while at the same time directing them toward

63. For case studies, see Hammermayer 1995 and Stuber, in press.

64. *Göttingische Anzeigen* 1759, no. 28, 256; my translation.

65. Lowood 1991, 68.

66. Jung 1787.

new sources of revenue.[67] Within the common framework of late eighteenth-century attempts to enlighten the people (*Volksaufklärung*), many other texts directly advertised new potential sources of revenue, be it for the individual landowner or for the nation as a whole. In 1775, for example, the reviewer of a short treatise entitled "The Utility of Importing Foreign Animals, Trees and Plants as Foods and for Manufactures" reported on the advantages of growing foreign plants: "A soil yielding one pound of rye is able to yield eighteen pounds of potatoes. One acre of 180 square rods (*Ruten*), which yields rye worth three *Reichsthaler*, yields up to 10 [*Reichsthaler*] of tobacco, and thereby improves the soil. With potatoes, the profit rises to eighteen [*Reichsthaler*], and with mulberry trees, to thirty [*Reichsthaler*]."[68] Praise for an English treatise on agriculture in 1770 was short and to the point: "This book is practical indeed. Its sole aim is to extract as much from a given estate as possible."[69] On another occasion, the use of dyer's weed as a surrogate for indigo was highlighted: "Perfecting this invention would earn Germany millions a year, which, for the most part, it presently has to pay in customs duties to France."[70]

## Conclusion: Limitations to the Mediatory Approach of the Economic Enlightenment

In sharp contrast to the nationwide aims of the economic enlightenment, the efforts to publicize economic goals and procedures outlined in the preceding paragraphs were challenged by seemingly simple problems of verbalization and communication inherent in the process of creating publicly accessible knowledge concerning agricultural materials. On a basic level, for example, there remained the problem of how to identify and obtain seeds or young plants of species successfully propagated in print. Authors often pleaded for consistency in the naming of plants and pointed out wrong denominations, but in the end it must have been tacitly presupposed that anybody interested in obtaining specimens of the plant in question would contact the editors. Only rarely was an address supplied for a merchant who, as in the case above, sold clover seed of a particular variety in great quantities. The immense prob-

67. From its 1768 foundation onward, the *Lausitzisches Magazin* regularly reported on the work of the local economic society.

68. *Göttingische Anzeigen* 1776, no. 12, 95; my translation.

69. *Göttingische Anzeigen* 1771, no. 31, 257; my translation.

70. *Göttingische Anzeigen* 1752, no. 40, 407; my translation.

lems of transmitting knowledge on agricultural matters impersonally, via the economic societies' publications, can also be deduced from instructions for the prospective authors of economic texts. Journals repeatedly stressed the need to develop a new style of writing characterized by simplicity and comprehensibility, so that transfers into practice would not be impeded merely for lack of proper communication. The Leipzig *Ökonomische Nachrichten* provided detailed instructions on a "good style of economic writing" in the preface to its first volume in 1750. Contest announcements also emphasized that competitors should write their essays in the vernacular and that they should not seek to use the essays to demonstrate their erudition or scholarship, but rather concern themselves with practical utility and concise description.[71] At the same time, there were attempts to teach readers alternative standards for a "scientific" approach to agricultural knowledge. Individual pride was to be done away with. The decision not to publish failed experiments could obstruct progress by causing others to waste their time repeating the same mistakes. Discussing trials of plowing at different depths and at different times, the *Göttingische Gelehrte Anzeigen* remarked: "Mr. Möller conducted various trials, the account of which is perhaps the more useful because they did not succeed."[72] In other respects, too, the readership was reminded of the standards prevailing in learned culture: the reviewer of a 1771 treatise on silkworm rearing severely criticized the author for having supplied the insects with human faces in his illustrations.[73]

Although seemingly banal, such problems make it clear that economic enlighteners were faced with a need to determine the properties of certain agricultural materials that was not limited to knowledge of plant physiology or soil composition. The broad approach they adopted to the collection, validation, and communication of knowledge according to the standards of contemporary learned or scientific practice generated a number of additional challenges. Thus, as has already been indicated, the economic enlightenment figured as a time of transition and development within the longer term epistemic history of agriculture. By around 1800, the institutional framework of the economic societies was losing its attraction. This change was not least a result of the insight that increasing technological or scientific expertise in the multiplication of agricultural resources did not lead of its own accord to an increase in agricultural production and income. Instead, it was necessary to adopt systematic approaches within administration and education,

71. *Göttingische Anzeigen* 1756, no. 141, 1274.

72. *Göttingische Anzeigen* 1753, no. 35, 322; my translation.

73. *Göttingische Anzeigen* 1771, no. 38, 326.

rather than merely appealing to the moral qualities of committed patriots. This trend might be understood as a learning process taking place among a generation of leading figures of the economic enlightenment. As Friedrich Casimir Medicus reflected in 1787, the institutional framework of economic societies itself, developed to bring together enlightened individuals from different professions and of different social standing, had confronted severe difficulties. The individual expertise of the beekeeper, estate manager, or horse breeder did not guarantee that the most advanced knowledge each of them possessed would be implemented; this had been a central experience of many improvers in the preceding decades. What was more, even if individuals willingly took up proposals for reform, public felicity was not guaranteed—if every estate were to take up tobacco cultivation in order to increase income, the price of tobacco would fall and the effort would be in vain. Thus, the crucial issue was to find ways to implement the technical knowledge that had been collected within state administrations, so that it could be evenly distributed across the country—and this made advanced professional education necessary.[74]

Thus, in the nineteenth century, innovative agricultural knowledge was generated in different kinds of settings: institutions for research and higher education, which increasingly aimed to turn out expert administrators rather than enlightened farmers. Among the first of these was the *Kameral-Hohe-Schule* (Cameral College), founded as an initiative of the Palatinate's Physical-Economic Society at Kaiserslautern in 1774.[75] This process of differentiation produced a range of nineteenth-century "specialized institutions—forestry academies, polytechnical schools, agricultural institutes, and, eventually, *Technische Hochschulen* (Technical High Schools)," which in the long run also undertook advanced research on increasing agricultural productivity.[76] On the other hand, new institutions for popularizing agricultural knowledge among the peasantry were also founded.[77] In the wake of the Napoleonic wars, there was increased political pressure toward the redistribution of land within the German territories, and from that time on the interest of peasant landowners to acquire additional revenues and the generation of advanced technical knowledge were to converge on the same goal of increasing agrarian production.

Despite their failures, for any long-term study of agricultural knowledge

74. Medicus 1787.

75. See Tribe 1988b.

76. Lowood 1991, 11; see also Harwood 2005.

77. See Pelzer 2004.

as directed toward providing a given society with a wide range of plant and animal resources, the economic enlightenment still appears as a decisive step. In England, the leading nation in the early modern "agricultural revolution," as elsewhere, innovations in the cultivation of agricultural products were usually generated spontaneously under particular local conditions and depended on local knowledge and personal transmission. In the second half of the eighteenth century, the economic enlightenment for the first time sought to channel, encourage, and accelerate such innovation processes by creating a reliable body of publicly accessible knowledge. Even if the immediate outcome of the activities of the economic enlightenment was less transformatory than its protagonists had hoped, they set in motion a process of knowledge production that was familiar to every administrative official in the late eighteenth century, to most members of the Republic of Letters, and even to the reading public of general newspapers. In the long term, the knowledge base for agricultural materials that was accumulated in the course of the eighteenth century without a doubt formed an indispensable foundation for what is traditionally known as nineteenth-century "scientific agriculture."[78] Renowned agronomists like Albrecht Thaer in Germany or Arthur Young in England fruitfully combined proposals for innovation, derived from many sources, with extensive field trials and educational measures.[79]

This chapter has sought to trace general traits of the "making of materials" with regard to agricultural products in the eighteenth-century German lands. I have argued that the attempts of economic improvers to generate an advanced knowledge base concerning agricultural materials far exceeded the scope of early modern sciences of material substances, which aimed at disclosing only the natural qualities of agricultural resources. Because such studies had only limited applications in the late eighteenth century, the economic improvers viewed extensive data collection and field trials as the most promising strategies for accomplishing their goal of directly increasing a state's wealth. At the same time, their efforts to compile publicly accessible knowledge concerning promising agricultural innovations faced numerous difficulties, starting with the attempt to represent agricultural experience in words. In the end, the economic enlightenment probably adopted more innovations from contemporary agricultural practice than it returned in terms of measurable economic benefit. Despite its shortcomings as regards direct application, the

78. See Klemm 1992; Wilmot 1990.

79. On Thaer, see Panne 2002; on Young, see Gazley 1973.

publicly accessible knowledge created in the economic enlightenment was of a completely different order than that which had previously generated—and would continue to generate—spontaneous agricultural innovation. The longer term consequences of the economic enlightenment may be sought in nineteenth- and twentieth-century agricultural science, which further developed the economic enlightenment's knowledge-based approach to agricultural materials according to the standards of learned culture. A principal legacy of that movement remains the establishment of an approach and an institutional setting in which agricultural resources were investigated as materials exclusively in order to facilitate their economic exploitation.

## Primary Sources

Secondary sources can be found in the cumulative bibliography at the end of the book.

Beckmann, Johann. 1769. *Grundsätze der teutschen Landwirthschaft.* Göttingen: Dieterich.

*Bemerkungen der physikalisch-ökonomischen und Bienengesellschaft zu Lautern* (various volumes).

Burgsdorf, F. v. 1785. Aufmunterung zu sorgfältiger Miterforschung der Verhältnisse, welche bey ihrer Vegetation die Gewächsarten gegen einander beobachten. *Schriften der Berlinischen Gesellschaft Naturforschender Freunde* 6:236–46.

Eckhart, Gottlieb von. 1754. *Vollständige Experimental-Oeconomie über das vegetabilische, animalische und mineralische Reich. . . .* Jena: Hartung.

*Göttingische Anzeigen von gelehrten Sachen* (various volumes).

Jung, D. J. H. 1787. Vom hohen Werthe eines rechtschaffenen staatswirthschaftlichen Landbeamtens, hergeleitet aus der landwirthschaftlichen Geschichte des freiherrlich Uxküllschen Guts zu Münchszell. *Vorlesungen der Kuhrpfälzischen physikalisch-ökonomischen Gesellschaft* 2:3–40.

*Lausitzisches Magazin* (various volumes).

Medicus, Friedrich Casimir. 1787. Über die Ursachen, warum ökonomische Gesellschaften nicht den Nutzen gestiftet haben, den man von ihnen erwartete. *Vorlesungen der Kuhrpfälzischen physikalisch-ökonomischen Gesellschaft* 2:285–328.

11

# The Crisis of English Gunpowder in the Eighteenth Century

SEYMOUR H. MAUSKOPF

In September 1883 the director of artillery wrote a memorandum on what he perceived to be a crisis in British munitions.

> The important question of the supply of powder is in a very unsatisfactory state. Rejections frequently take place in consequence of powder not passing the regulation proof, and the makers protest against War Office decisions, and urge that the results obtained in the proof guns are not trustworthy. The Superintendent Royal Gunpowder Factory is unable to give a satisfactory explanation, and it would seem absolutely essential both in the interest of the Service, and in fairness to Contractors who have hitherto made satisfactory deliveries, that the case should be thoroughly investigated without loss of time, and a solution arrived at which will remove difficulties in the future.[1]

The powder in question was prismatic powder, recently introduced from Germany and generally considered the most successful and up-to-date powder for large ordnance.

Two and a half years earlier, at the time prismatic powder was being introduced, the aforementioned superintendent of the Royal Gunpowder Factory (Waltham Abbey), Colonel C. B. Brackenbury, expressed serious concern to Lieutenant Colonel E. Maitland, superintendent of the Royal Gun Factory, over discrepancies in the determination of moisture and density of powder manufactured at the Royal Gunpowder Factory in two different (but nearby)

1. Quoted in Mauskopf 2006, 326. The director of artillery recommended that the Ordnance Committee call on the superintendents, R.G.P.F., and R.G.F., as well as the War Department chemist. He also urged that facilities be extended to the powder contractors to attend and offer suggestions.

English venues: "At present I do not know whether this is due to differences in the powder, or in the instruments, or in the methods of operation. Similar discrepancies have occurred frequently on other occasions and I am about to propose to Chemist W. that we should endeavour to ascertain the cause of the difference."[2]

By November 1884, Brackenbury was still plagued—more than ever, indeed—with problems of variability in prismatic powder proofing, although he now could see a personal light at the end of the tunnel:

> But please don't ask *me* to say whether "it is the gun or the projectile or what?" That is just the question which I have been charging the committee to investigate these many months, and all I get is the rather monotonous reply "It is the blending." This is a trifle unsatisfactory being that some of the most irregular results have come from *unblended* lots. Fortunately I am a bit of a philosopher and have the happy knowledge that on the 1st of July next some other happy person will enjoy the felicities which are at present almost too delightful for. Ever very sincerely, Brackenbury.[3]

I share this sequence on late nineteenth-century English powder manufacture to illustrate a historical theme of my paper: Right through to the very end of its long history, gunpowder (by this I mean black powder) was, in the useful phrase of Ken Alder, a "thick thing."[4]

This was already apparent in the eighteenth century, when European governments tried to bring scientific expertise to bear on the amelioration of gunpowder. This was done partly in response to perceived crises in both the supply and quality of gunpowder available for the protracted wars of the second half of the century and partly as a component of the Enlightenment program of rationalizing craft production.

In this chapter, I shall examine the salient features of eighteenth-century munitions production and improvement. I shall concentrate on gunpowder and focus on its production in England. Until quite recently, the history of "modern" (i.e., post-seventeenth-century) gunpowder had not received much attention. This was due, I think, partly to the relatively little interest in the study of material cultures as a part of the history of science and technol-

2. Letter from Col. C. B. Brackenbury to Superintendent, Royal Gun Factory (Lieut. Col. E. Maitland), 6 May 1881; quoted in Mauskopf 2006, 325. The "Chemist W." no doubt referred to the War Department chemist. The two locations were Waltham Abbey and Woolwich.

3. Letter of 17 November 1884 from Brackenbury to H. May; quoted in Mauskopf 2006, 326–27. It is one of a series between Brackenbury and Maitland over the variability of ballistic performance of prismatic powder.

4. Alder 1997, 13.

ogy, and of the history of chemistry in particular. It was also due to the fact that the development of firearms and gunpowder between their ascendancy during the late Middle Ages and Renaissance, and the late nineteenth century has been deemed to have been static and hence not very interesting.[5]

Although I would not dispute the absence of a major revolutionary development in the nature of gunpowder during the eighteenth century, I would not thereby characterize activities connected with gunpowder production and improvement as without interest. Like every century before and since, the eighteenth century witnessed almost constant warfare. The great midcentury Seven Years' War (1756–1763), often characterized as the "First World War" because of the far-flung sites of conflict, placed unusually severe pressures on the provisioning of gunpowder. The ensuing criticisms of the quantity and quality of the gunpowder supplied in that war and in the subsequent American Revolutionary War led to attempts both to reform the system of procuring gunpowder and to ameliorate gunpowder quality through systematic experimental study. Thus, in common with other chapters that focus on specific materials (or classes of materials) in the eighteenth century, mine focuses on the attempts by persons in authority to comprehend, rationalize, and, above all, standardize the materials. In the case of my material, gunpowder, the latter goal had a history long antedating the eighteenth century. Especially for military purposes, it was crucial to have some idea that gunpowder would perform reliably and effectively on the battlefield.

More than fifteen years ago, such a study as I propose—even at a superficial level—would have been quite difficult to carry out. Before that time, there was little research on any aspects of eighteenth-century munitions.[6] Moreover, in the eighteenth century, the English (unlike the French) published very little on munitions production: there were just two essays by active figures in the field,[7] as well as essays on various aspects of gunpowder by

5. Van Creveld 1989, 96–97; see also B. Hall 1997, 215. Hall characterizes certain eighteenth-century developments (e.g., the adoption of a single grain size for all artillery gunpowder by the British) as "absolutely regressive" (B. Hall 1997, 215–16).

6. A major exception is *The Rise and Progress of the British Explosives Industry* (1909). But, although comprehensive, the individual entries are often quite superficial.

7. Napier 1788; R. Coleman 1801. George Napier (1751–1804) was comptroller of the Royal Laboratory (Woolwich) from 31 May 1782 to 8 April 1783. A soldier who had served in the American Revolutionary Wars, Napier obtained the position through his brother-in-law, the Duke of Richmond, who was master general of the ordnance. (Stearn, n.d., "Napier"). R[obert] Coleman (no dates available) had various administrative roles with the Board of Ordnance, the most significant of which was his appointment as clerk of the cheque at the Waltham Abbey Royal Gunpowder Works (acquired by the government in 1787) in 1793. Although neither a military officer nor apparently trained in powder making, he was in charge of securing good quality

the cleric and chemist Richard Watson.[8] The first full-fledged English book on gunpowder did not appear, as best I can ascertain, until 1832, and it was first printed for British artillerymen in India.[9] Thanks to the recent work of Jenny West, Wayne Cocroft, and Brenda Buchanan, however, we are much better informed than previously about virtually all of the aspects of organization, production, and attempts to improve gunpowder in the eighteenth century.[10] I am heavily indebted to their research.

This chapter will be organized around a number of topics in eighteenth-century English gunpowder production: national organization of gunpowder production, procurement, and proving; manufacture of gunpowder; proof of gunpowder to ascertain whether it met threshold standards; government reforms in gunpowder manufacture, testing, and improvement; and systematic and experimental investigation of gunpowder. I have placed the principal focus of this chapter on the challenges faced by military authorities in gaining mastery over gunpowder through analysis and measurement. As has already been noted, gunpowder is a complex material whose behavior is affected by a bewildering number of factors, including the very instruments with which it is being tested and the weapons in which it is subsequently employed. To maintain my focus, I give less attention to some issues taken up by the other chapters on specific materials included in this volume. For instance, I do not treat in any detail the producers per se of gunpowder or, for that matter, its military consumers. This is, it should be said, partly the result of the comparative dearth of detailed contemporary historical studies of individual eighteenth-century powder producers.[11] Nor do I deal with what I shall term the "culture of the material," by which I mean the configuration of material substances and objects with cultural and social meanings and significance. This would make an interesting sequel to this study, especially with a widened

---

wood and charcoal for the Waltham Abbey Works in the 1790s (personal communications from Leslie Tucker and Brenda Buchanan, 13 June and 15 June 2006, respectively, for which I thank them). I thank Les Tucker for providing Coleman's Christian name.

8. Watson 1800. Watson (1737–1816), bishop of Llandaff, Regius Professor of Divinity, and erstwhile professor of chemistry at Cambridge, developed the method of charring wood by distillation.

9. Braddock 1832. Braddock, son of the master refiner of saltpeter, Royal Powder Mills, Waltham Abbey, of the same name, was trained there but spent his career in India. Braddock wrote: "there is not a single work extant in the English language that discusses the manipulation of Gunpowder, and the best and most accurate methods of ascertaining its strength and quality" (Braddock 1832, iii).

10. West 1991; Cocroft 1996; Buchanan 1995–1996. See also Buchanan 1996a; 2006a.

11. An exception is the work of Brenda Buchanan (1995–1996; 1996b; 2005) on eighteenth-century Bristol powder producers.

functional purview of gunpowder to include its other uses, for example, in blasting, mining, and hunting. But it is beyond the scope of this chapter.

## National Organization of Gunpowder Production

Military powder was produced in England for most of the eighteenth-century by privately owned gunpowder mills under contract with the Ordnance Office.[12] The organization of this office was ratified by a Warrant of 1683, and it persisted until the reforms of the Crimean War. The Ordnance Office was headed by a Board of Ordnance consisting of five principal officers,[13] the chief function of which was to oversee the provisioning and maintenance of military stock for the armed forces. The entire Ordnance Office staff consisted of about forty persons, including aides to the board members, and storekeepers at the garrisons and outposts. At the Tower of London were about fifteen staff members, including the storekeeper of the saltpeter, and there were two proof masters of powder at Greenwich.

London and Greenwich were the principal centers of activity for the Ordnance Office. Its offices were located at the Tower and the Palace Yard, Westminster; and London was the principal point of entry for saltpeter (from India) and sulfur (mainly from Italy). Greenwich was the principal magazine for the reception, proving, and storage of gunpowder for most of the century.[14] As a result, the mills supplying gunpowder to the military were all located in the southeast of England, near London. This was for convenience in securing some of the imported raw materials, minimizing the dangers of explosion involved in transporting gunpowder, and the utility of proximity in business negotiations with the Ordnance Board.

These privately owned powder mills were economically precarious operations, depending to a large extent on the vicissitudes of foreign policy and state of war as well as the hazards of destruction of mill buildings and equipment by inadvertent explosions in the course of production.[15] During wartime, powder mills had all the government orders they could want (and more), but when peace came, government contracts ceased abruptly. More-

12. West 1991, 13. See A. Forbes 1929, vol. 1, chap. 5, "The Board of Ordnance"; and Buchanan 2005. Most of the administrative details are taken from this latter work.

13. Master general as chairman (sometimes but by no means always prominent military commanders such as Marlborough and Ligonier), lieutenant general, surveyor general (responsible for the proving of powder), clerk of the ordnance, storekeeper, and clerk of the deliveries.

14. Later in the century, proving was usually done further downriver at the powder magazines constructed in the early 1760s at Purfleet (Cocroft 1996, 2).

15. As a result of which, they were also difficult to insure.

over, at the start of a given war, neither the Ordnance Office nor the powder entrepreneurs could have any notion as to how long or extensive that war might be. As a result, powder makers were usually men "either connected with other trades, of prominent social standing, in positions of influence in the City commerce, or a combination of these."[16]

There was a means of compensating for the fluctuations of government powder orders; this was the commercial sale of gunpowder on the private market, particularly for use as barter in the slave trade.[17] Indeed, private sale tended to be more lucrative for the powder makers because the quality of the gunpowder supplied to private merchants needn't be of the quality of that demanded by the Ordnance Office, and its price on the private market was higher than what the government was willing to pay.

Through the eighteenth century (down to the reforms of the early 1780s), there were usually about ten powder mills to whom the government offered contracts to produce gunpowder.[18] The government would contract with each powder maker for a certain quantity of powder of a specified manufacture and of proof quality. West cites an example from just before the start of the Seven Years' War of the government specifying details of powder manufacture to Ewell mills: "The Board reported that each barrel [of 300 contracted for] should be prepared from 80 ¼ lb double refined saltpetre, 15 lb charcoal, and 12 ¾ lb refined brimstone; the total composition was to be divided into three separate parcels of 26 ¾ lb saltpetre, 5 lb charcoal and 4 ¼ lb brimstone, each of which was to be worked for five hours to reduce the entire quantity from 108 lb to 100 lb. The makers were to abide by the Board's decision at the testing or proof of the powder and to collect that which failed."[19]

16. West 1991, 22. In appendix 1, West provides brief histories and descriptions of the mills supplying the Ordnance Office in the eighteenth century and brief accounts of the mill owners.

17. Ibid., 36.

18. The following mills had contracts with the Ordnance Office during the Seven Years' War: Bedfont mills, Middlesex; Chilworth mills, Surrey (added during the war); Dartford mills, Kent; Ewell mills, Surrey; Faversham mills, Kent (became state owned, 1759); Hounslow mills, Middlesex (added during the war); Molesey mills, Surrey; Oare mills, Kent; Waltham Abbey mills, Essex; Worcester Park mills, Surrey (West 1991, appendix 1, 197–211). There were other gunpowder mills in Britain that did not supply the government with military powder, for example, in the Bristol area, where powder was made for commercial use and for trade (e.g., barter in the slave trade). Bristol powder makers tried but failed to get Ordnance Board contracts during the Seven Years' War (Buchanan 1996b).

19. West 1991, 28, 34. During the Seven Years' War, contracts were for either 242 ½ or 485 100-lb. barrels (per month), according to the choice of the maker. Buchanan (1995–1996, 150) gives the 1753 Ordnance Board's requirement as: 74.3% saltpeter, 11.8% sulfur, 13.9% charcoal.

Although this arrangement seems to have worked tolerably well during the wars of the first half of the century, it functioned much less well during the Seven Years' War: "The Ordnance Office, faced with an unprecedented demand for gunpowder, did not receive the quantity expected. Throughout the war it made various unsuccessful attempts to obtain more cooperation from the powder makers, who themselves had considerable problems in achieving a product of a satisfactory standard and in the necessary quantities, especially when trying to balance the demands of government and private trade."[20]

One expedient resorted to early in the war and continued until 1759 was to purchase gunpowder from the Netherlands.[21] The government also attempted various strategies to ameliorate the problem of domestic gunpowder supply. One was to curtail all mills under contract with it from selling powder on the private market (attempted in 1760 with little success). Another was to offer financial incentives to powder makers who met their contracted quota reasonably well with powder of acceptable quality.[22] Finally, in 1759, the government purchased one of the contracted powder mills, the Faversham mills in Kent, so that powder production at one source, at least, would be directly under the control of the Ordnance Office. Hopes that this would result in better production were not fulfilled,[23] but this did set a precedent that was to be expanded in the 1780s with the government purchase of the Waltham Abbey mills.

---

William Congreve gave an equivalent pre-1785 recipe:74 lb., 8 oz. saltpeter; 10 lb. sulfur; 15 lb., 7 oz., 4 4/11 dr. charcoal. In 1785 the proportions were fixed at 75 saltpeter, 15 charcoal, 10 sulfur (Congreve 1788, 17–18; my pagination). A good deal of the powder supplied was reprocessed old stock supplied to the mills by the Ordnance Board.

20. West 1991, 77–78.

21. Ibid., 50–52. At one point (January 1758) there were 14,470 barrels of Dutch powder in store compared to 2,840 barrels of English powder! Diplomatic shifts in favor of France toward the end of the decade and concern over the safety of transport (in secret) across the sea led to cessation of gunpowder import. At the same time that it sold gunpowder to the English, the Netherlands was exporting saltpeter, obtained from India, to the French at twice the domestic price (Gillispie 1980, 56).

22. For example, from January 1757 until November 1762, the Ordnance Board offered to pay £0 2s 6d over contract price (£1/barrel instead of £0 17s 6d) for every 80 out of 100 barrels of powder that passed proof. The results were disappointing (West 1991, 61–63). Appendix 2 in West has extremely valuable data on gunpowder submitted to the Board of Ordnance during the Seven Years' War.

23. See West 1991, chap. 9.

## Manufacture of Gunpowder

### NATURE AND PREPARATION OF INGREDIENTS

Gunpowder is a mixture of three ingredients: saltpeter ["niter," ($KNO_3$)], charcoal, and sulfur. Two of the ingredients (saltpeter and charcoal) originated from organic materials; the processes of their production and preparation were, consequently, more complex than that of sulfur. The proportions of these ingredients can have a latitude of values, but in England, the Ordnance Office set them at 75, 15, and 10 per 100 parts of gunpowder.[24] By mid-century, the saltpeter and sulfur were being imported respectively from India and Italy. Indeed, the chemist Richard Watson knew of no saltpeter generation or collection taking place in England. He worried about the elimination of domestic production in terms that resonate strongly today regarding Middle Eastern oil: "The French very wisely keep up their establishment for the making of saltpetre; the revolutions which have formerly taken place in India, render it not improbable, that similar ones may take place again; and England would feel the distress which would attend the non-importation of saltpetre from the East Indies, more sensibly than any other state in Europe."[25]

Soil and climatic conditions in India were nearly ideal for the production of copious quantities of high-grade saltpeter. Refined once at the collection site, it was then exported under the name of "grough petre." To establish the purchase price for the imperfectly refined saltpeter, the Ordnance Office subjected it to "refraction" to gain a first approximation of the amount of impurities present.[26] The saltpeter was issued to the mills according to require-

24. Congreve 1788. This is corroborated by Coleman 1801, 357. However, there must have been some fluctuation in the official proportions, because Congreve stated that the proportions prior to 1785 were 74 lb., 8 oz., 11 7/11 dr. saltpeter; 15 lb., 7 oz., 4 4/11 dr. charcoal; 10 lb., 0 oz., 0 dr. sulfur. Since 1785, they were 75 saltpeter, 15 charcoal, 10 sulfur (Congreve 1788, 17–18).

25. Watson 1800, "Of the manner of making Saltpetre in the East Indies," 1:325–26. In another essay, Watson noted that there had been schemes to revive this industry, for example, by the Society for the Encouragement of the Arts and Manufactures during the Seven Years' War. But, although one saltpeter works was set up, it was soon abandoned; "the proprietors having been experimentally convinced, that they could not afford to sell their saltpetre for less than four times the price of that imported from India." Watson attributed the high cost of making saltpeter in England to the costliness of wood ashes (also used in soap manufacture) and of labor to collect and make the saltpeter (Watson 1800, 1:292–93).

26. "Refraction" involved breaking a cake of saltpeter which had been melted and noting the appearance of the crystalline fracture, "which varies in appearance with the amount of impurity present, a property which at one time was used to roughly indicate the value of any sample of the

ments stipulated in the contract to supply powder.[27] At the mill, the saltpeter was refined twice more.[28]

Imported sulfur was purchased directly by the powder makers. Generally, its production and purification was the most unproblematic of the three components. Napier recommended purification and "sublimation" by melting in an iron pot under gentle heat and then straining the melt through a double linen cloth. However, he did warn that prepared sulfur was often adulterated with wheat flour, leading to fermentation and decomposition, a state of affairs that, in his opinion, was the "principal cause of British gunpowder being less durable now than formerly."[29]

There was debate over the function of sulfur in the gunpowder reaction and therefore the necessity of including it,[30] especially because it was known to contribute to the fouling of guns. John Braddock succinctly summarized the view of the role of sulfur:

---

salt, and was called the 'refraction' of the nitre" (West 1991, 172–73). For a history of England's trade in Indian saltpeter, see Buchanan 2006b.

27. Each mill had to deposit £1,200 for each contract of 485 barrels of powder (West 1991, 173).

28. A contemporary account that is lively and concrete: "The Method of Refining Salt Petre in the Years 1784 and 1785. One Ton of Petre, either Grough or that extracted from damaged Powder, is dissolved in 280 Gallons of Water, in a large Boiler, and made to boil as soon as possible, one Man attending to take off the Scum as it rises, and frequently throwing cold Water in small Quantities into it, to make the filth rise freely, it continues boiling till entirely free of Scum, and judged to be of a proper Thickness, which is known by pulling the Scummer into it, and if the Petre adheres to it, so as to form a drop on the edge of about an Inch in Length it is thought to be of a proper Consistence for Chrystallisation; it is then pumped or ladled into a spout from which it runs into a trough with four Brass Cocks, through which it runs into the Filtering Bags, made of double Canvas, which have sand thrown into them, to prevent the Liquor from passing too freely; when one Bag is full and the Liquor runs through clear the remainder of the Bags are, [*sic*] filled, and the Liquor that runs from them before it is clear is returned into the Trough, and the clear Liquor put into a large Pan to Chrystallize, in which state it stand during the same time as the Whole Liquor from the Extracting Rooms, and when the Liquor is drawn off it, part or the whole if there is room is carried and mixed with the whole Liquor in the Extracting Rooms. The Petre is rinsed and set to drain for twenty four hours and then refined again, in the same manner as above, when this last Petre is drained it is carried to the Store House to dry, except when ordered to be melted. If there is no Extracting, the Liquor from the refined Petre is reduced, and the Petre produced from it, is considered as a single refined Petre" (Caruana, n.d., 47–48; Caruana argues against Congreve's authorship of these notes). In France, saltpeter was refined three times.

29. Napier 1788, 102–3. Napier spoke of this as having "purified and sublimed" the powder. Very impure sulfur was refined by sublimation (Coleman 1801, 357). Braddock wrote that sulfur was purified simply by melting and then essentially gave Napier's method (Braddock 1832, 27).

30. Watson 1800, "Of the Composition and Analysis of Gunpowder," 2:5–8.

> Gunpowder made without sulphur has, however, several bad qualities; it is not, on the whole, so powerful, nor so regular in its action; it is porous and friable; possesses neither firmness nor solidity; cannot bear the friction of carriage, and in transport crumbles into dust. The use of the sulphur, therefore, appears to be, not only to complete the mechanical combination of the other elements, but being a perfectly combustible substance, it increases the general effect, augments the propellant power, and is thought to render the powder less susceptible of injury from atmospheric influence.[31]

Although the Board of Ordnance stipulated no specific wood source, British powder makers subscribed to the general and traditional preference for soft, light wood, such as alder, willow, and black dogwood. Wood for charcoal was obtained from the Wealden areas of Surrey and Kent.[32] This component became the principal challenge for British powder makers because of the increasing scarcity of timber. Watson asserted that experiments showing that hard wood could make as effective charcoal for gunpowder as the preferred soft woods (like willow or alder) offered promise for English powder makers, "as it is not always an easy matter for them to procure a sufficient quantity of the coal of soft wood."[33]

The wood was traditionally charred by the open-pit method: "It consists in the wood [stripped of its bark] being cut into lengths of about three feet, and then piled on the ground in a circular form (three, four, or five cords of wood making what is called a pit), and covered with straw, fern, &c. kept on by earth or sand to keep in the fire, giving it air by vent-holes as may be found necessary."[34] But in 1786, Richard Watson proposed an improved way of making charcoal adapted from the physician and chemist George Fordyce: charring by distillation in closed iron cylinders. In a contemporary description: "The wood to be charred is first cut into lengths of about nine inches, and then put into the iron cylinder, which is placed horizontally. The front opening of the cylinder is then closely stopped: at the further end are pipes leading into casks. The fire being made under the cylinder, the pyro-ligneous acid, attended with a large portion of carbonated hydrogen gas, comes over. The gas escapes, and the acid liquor is collected in the casks. The fire is kept up till no more gas or liquor comes over, and the carbon remains in the

31. Braddock 1832, 2. Up to the end of the black powder era in the late nineteenth century, the role (and necessity) of sulfur continued to be debated. Just before the end of this era, in the 1880s, "cocoa powder," with very low sulfur content, came into widespread use.

32. West 1991, 14.

33. Watson 1800, 2:4. His authority was the French chemist Antoine Baumé.

34. Coleman 1801, 357–58.

cylinder."[35] I shall return to Watson's innovation when I discuss William Congreve's experimental research.

## PULVERIZATION AND INCORPORATION OF INGREDIENTS

The three components of gunpowder were separately ground finely, sieved (to remove foreign matter), mixed in the proper proportions by weight, and sent to the powder mill to be "incorporated," in other words, intimately mixed. Napier believed that the preliminary mixing ought to be carried out: "in clear dry weather; a lowering sky and humid atmosphere, being found inimical to that thorough blending of the materials which ought to precede their being worked in the mill."[36]

In the mill, the ingredients were much more comprehensively mixed together; this was indeed considered to be as critical as the purification of the constituents to the quality of the resultant powder.[37] There were two types of incorporating mills used in the eighteenth century: stamping and cylinder (or edge-runner) mills. The former (which continued to be used in France) was "a large mortar, in which a ponderous wooden pestle moved by men, by horses, or by water, performed the operation very perfectly, but with obvious danger to the workmen."[38] The perceived danger was explosion due to overheating, and in 1772 stamping mills were outlawed in Britain except in certain Sussex mills producing fine sporting powder.[39] By this time, most of the powder mills in the London area had already switched over to the alternative method of incorporation.[40] Performed in a "slight wooden building and boarded roof,"[41] this incorporation mill: "consists of two stones vertically placed, and running on a bed-stone. On this bed-stone, the composition [i.e., the preliminary mixed ingredients] is spread, and wetted (not with sal-ammoniac, urine, &c. as some authors state, but) with as small a quantity of water as will, together with the revolutions and weight of the runners, bring it into a proper body, but not into a *paste*. After the stone runners have made the proper number of revolutions over it, and it is in a fit state, it is taken

35. Ibid., 358.

36. Napier 1788, 105.

37. For example, see Watson 1800, 2:8–9.

38. Napier 1788, 106.

39. However, one French authority deemed cylinder mills "more dangerous" than stamping mills (Renaud 1811, 109).

40. Buchanan 1995–1996, 144.

41. Coleman 1801, 359.

off. . . . These mills are either worked by water or by horses." Only a limited amount of material (40–50 lb.) was worked at a time to obviate danger of explosion due to sparks between the runners and the bed-stone.[42] There was no specified threshold time for incorporation. Napier (who was critical of edge-runner incorporation) gave a common time of "seven or eight hours."[43]

During the latter part of the century, a further procedure was introduced. The incorporated powder was subjected to intense compression in a screw press. The powder was placed between copper plates and subjected to intense pressure much like the arrangement in a printing press.[44]

### CORNING

The incorporated (and compressed) powder was taken to the corning house to be "corned," in other words, reduced to grains of specified size. This was done by means of an apparatus consisting of a horizontal wheel to which were affixed perforated parchment sieves whose holes corresponded to the intended grain size. Lumps of the powder were placed in the sieve, along with a flat circular piece of *lignum vitae* (guiacum wood). The rotation of the wheel entrained the *lignum vitae* piece to move along the sieve, breaking up the powder lumps and forcing them through the sieve. The powder grains were then separated from smaller grains and powder dust by additional sieving.[45] Until the regime of William Congreve, military powder was of only one grain size. Congreve introduced production of two grain sizes: small-grained powder for small arms (musket powder) and for cannon priming, and a large-grained powder for use in cannon.[46]

42. Ibid., 359. Although water mills were common, horse mills continued to be built—for example, at the government-owned Waltham Abbey mills—because of lower cost and reluctance to put overmuch demand on the flow of water through the mills (*Royal Gunpowder Factory, Waltham Abbey, Essex* 1994, 34).

43. Napier 1788, 106.

44. The greater density engendered by the press may have been as significant a factor in increasing the ballistic force of gunpowder as the introduction of closed cylinder charcoal (see Howard 1996, 12–13).

45. The account is an amalgam of those given by Napier (1788, 109–10) and Coleman (1801, 359).

46. Congreve 1811, 22–23. Although grain size was demonstrated to be of great importance ballistically in the mid-nineteenth century, it was not so apparent early in the eighteenth century. Braddock, for example, wrote, "but neither large nor small sized grain appears to possess any advantage with reference to the regularity of practical effect: in any given number of firings, the ranges of the one vary as much as those of the other" (Braddock 1832, 51).

## GLAZING AND DRYING

The process of "reeling,"[47] or glazing, was designed to maintain the durability of powder by removing or attenuating the rough edges on the powder grains, hardening the grains, and giving them a gloss. The attenuation was the result of frictional action of the powder grains on each other. It was thought that glazing (along with pressing) also reduced the hygroscopic characteristic of the powder and thus preserved it.[48] Glazing was done by rapidly tumbling the powder in casks or drums attached to the axis of a water wheel.[49] The practice of glazing with graphite was not introduced until the nineteenth century. Although Napier recommended glazing for the durability it imparted, he also asserted that it reduced the ballistic strength of the powder.[50]

In the eighteenth century, powder was generally dried in a "gloom stove": "This species of stove consists of a large cast-iron vessel projecting into one side of a room, and heated from the outside till it absolutely glows."[51] The powder itself was spread on cases placed around the room. A thermometer was placed in the door of the stove to regulate the temperature.[52] A sheet of copper, covered in turn by a canvas screen, was placed over the vessel when powder was introduced or removed from the room.[53]

47. The term is used by Coleman (1801, 360) and refers to the turning apparatus, or "reel," used for the glazing.

48. For example, Braddock 1832, 56–57.

49. Napier 1788, 113.

50. On the basis of nearly "*six hundred* experiments" (author's italics), Napier found that good powder was reduced by one-fifth and inferior quality powder by nearly one-fourth (1788, 113). Braddock reiterated and confirmed this assertion (1832, 58). However, he wrote: "Time, however, reverses this action; the porousness of the light powder makes it more susceptible of injury; it imbibes more humidity than the dense powder, and what it gains in immediate effect, it loses by age and long keeping" (60).

For a recent positive evaluation of the benefits of glazing, see Howard 1996, 18: "Glazing is more important physically than the innovators or even some of the current makers realize. Several things happen. First, the rough edges of the cut (corned) powder are rounded. Second, the long run glaze produces a powder that is harder and more uniform. Third, a hard glaze will be more moisture resistant (graphite not helping). Fourth, there seems to be a migration in the glaze, whereby the nitrate works to the surface of the grain, aiding combustion."

51. Coleman (1801, 360, footnote of the editor), who went on to recommend heating by steam "passing through steam-tight tubes" to prevent accidents caused by powder coming in contact with the gloom.

52. Ibid., 360; Braddock (1832, 60) recommended a temperature of 140–150 degrees Fahrenheit.

53. W. Simmons 1963, 18. The covers for early glooms were 10 feet by 12 feet.

The Royal Commission on the Historical Monuments of England (RCHME) archaeological survey of the Waltham Abbey powder mills gives a good overview of the physical plant for gunpowder manufacture near the close of the Napoleonic era:

> A list of works and repairs proposed to be carried out at Waltham Abbey in 1814 provides a useful snapshot of the full war-time establishment of the factory. To ensure purity of the ingredients the Board refined its own saltpetre within its own refinery on site. Surprisingly, no mention is made of a sulphur refinery at this date, although sulphur mills are shown on near-contemporary maps and additionally Coleman in his description of the manufacture of gunpowder in 1801 states that the sulphur was refined at the Waltham Abbey factory. The most logical explanation is that the refining of sulphur took place within another building that was identified under a different name. Cylinder charcoal was at this date manufactured in Sussex, although only Fisher Street was mentioned by name as a works where expenditure was required. The gunpowder was mixed in one of five composition mills and incorporated in one of eleven water-powered mills or nine horse-powered mills that were in operation. Additionally another two water-powered mills of unspecified function are described as lying on Millhead. Further processing took place in the four Corning Houses, before the powder was dusted and glazed and finally dried either in one of the two Gloom Stoves or in the more advanced Steam Stove. Four receiving magazines, six charge magazines and the Grand Magazine were provided for the temporary storage of powder. A figure is also quoted for the erection of four water-driven mills for grinding charcoal and an associated reservoir. The gunpowder was moved between the process buildings by a flotilla of boats, comprising five barges, nine powder boats, two ballast barges and six punts. Also attached to the factory were a number of ancillary buildings made up of store houses and lodgings.[54]

## Proof of Gunpowder

Ballistic testing of gunpowder was carried out under the supervision of two proof masters of the Ordnance Office,[55] first at the powder magazine at Green-

54. *The Royal Gunpowder Factory, Waltham Abbey, Essex* 1994, 34–35.

55. Mockridge cites a document from 1620, which mentions a "Master Gunner out of the Ordinary of the Office, by Title of Proof Master, for proving of ordnance, powder, saltpeter and match" (Mockridge 1950, 83). Although the proof masters obviously should have been technically adept, they appear to have been a heterogeneous lot in the eighteenth century. Some were military officers, some referred to as "gentlemen." With the reform of the early 1780s, the proof masters were artillery officers (Mockridge 1950, 82–83, 89).

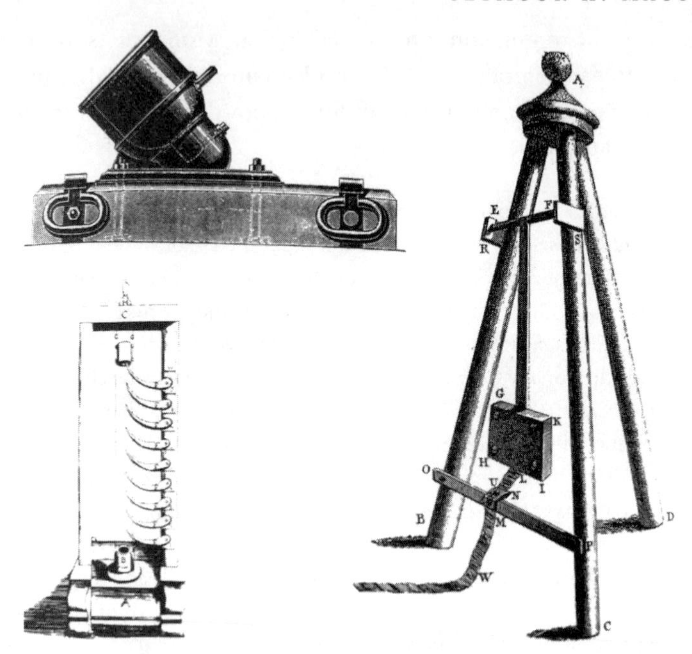


FIGURE 11.1. Gunpowder testers: John Babington's weighted trier of 1635 (bottom left); mortar, its axis at an angle of 45 degrees (top left); Robin's ballistic pendulum (right) (Guttmann 1906; 1895, vol. 2; Robins 1742).

wich and, from the 1760s, further east at Purfleet. There were a variety of test instruments and procedures, no one of which seems to have had precedence (see figure 11.1). One of the most common instruments employed was the vertical eprouvette, which tested how high a standard weight (20 lb., 7 oz.) could be raised by a test weight of powder (2 dr.). But the report that stipulated these conditions for the vertical eprouvette (from 1740, on comparative proof of English and Spanish naval gunpowder) also noted that this was one of "three divers ways" to prove powder. The other two involved: "Firing a 12 pdr. shot out of a 5 2/3 inch mortar with a quarter ounce of powder. Firing a half pound shot out of a swivel gun with two drams of powder." Although all the results favored the English powder over the Spanish, the report pointedly noted that "many people who had been to Portsmouth affirm that the shot of the Princessa [the Spanish warship] was much sharper thrown than those of our Men of War."[56]

56. Cited in Mockridge 1950, 86. In the vertical eprouvette test, English powder raised the weight "from 4 inches to 6 inches and some tenths"; Spanish powder "no higher than 1 inch and some tenths." In the swivel gun trial, the Spanish powder "had not enough strength to

Moreover, the different methods of measuring ballistic force did not always agree, as they apparently did in this case. A report of a reformist Parliamentary commission in 1782, noted a sharp discrepancy between the results of the vertical eprouvette and mortars: "A powder that with two drachms would raise the Vertical Eprouvette 4.5" would, with three pounds, range a shell from a 13" mortar 1108 yards only; while another powder, that with the same quantities would raise the Eprouvette only 1.9" would range the shell 1112 yards."[57]

The mortar superseded the vertical eprouvette in the last decades of the century, especially because it enjoyed the vigorous approbation of William Congreve. However, it did not achieve complete hegemony. As late as 1832, John Braddock noted that the vertical eprouvette, the mortar, and several other test instruments and tests were still all in use.[58] Braddock used mordant language in his overall evaluation of methods of proof: "Gunpowder is invested with difficulties, from the earliest stage of its manufacture to the very last result obtained by trial for ascertaining its explosive powers. And the many methods that have been invented, and that are still in use for proving it, constitute a sufficient testimony of the difficulty attending this indispensable, final check upon its manufacture."[59]

## Scholium

It should be apparent that gunpowder production and testing was a complex process, with virtually all of the elements contested by powder makers and writers. Moreover, there were geopolitical considerations in this process, particularly regarding the importation of saltpeter from the Orient.[60] The nature of gunpowder as an intractable, unpredictable "thick thing" stems from a number of factors. One is the material itself, a mechanical mixture of three components, two of which originate from complex organic matrices. As a result, gunpowder is an *inherently* variable material, and the challenges to

throw it out of the gun; and the tryal by the mortar turned out equally in favour of the English powder."

57. Cited in Mockridge 1950, 87.

58. Braddock 1832, 89–90. These latter included the ballistic pendulum invented by Benjamin Roberts and developed by Charles Hutton. This came into more general use in the nineteenth century (see Mauskopf 1998).

59. Braddock 1832, 73. This is the introductory paragraph to his section on "proof of gunpowder."

60. I have not come upon any evidence of problems in importing sulfur from Italy in the eighteenth century.

standardize it remained acute up to the end of the black-powder era, as my quotation from Colonel Brackenbury illustrates. For the eighteenth century, these challenges were all the more acute: "Each stage of the manufacturing process—the refining, weighing, and grinding of each of the three raw materials, then mixing together, incorporation, corning, and drying—presented opportunities for variability in the standard of the gunpowder produced. . . . Even the slightest inaccuracy in refining, drying, or weighing could affect the powder produced and overall it is likely that these were the main culprits and intentionally short measure less so."[61]

West relates the case of one mill owner's frustration because powder he had submitted had been rejected after testing "although it was the same composition and worked the same number of hours as that which had previously passed."[62] This highlights another factor contributing to unpredictability of results: environmental conditions under which the powder was produced, such as temperature, pressure, and moisture content of the atmosphere. They were noted by Napier and were still playing their mischievous role plaguing Brackenbury a century later.[63]

A third factor that came under serious scrutiny and study in the nineteenth century was the way the powder performs in the gun. This, after all, is the desideratum of gunpowder making and is at least as complex as the other two factors. It is also closely related to the challenge of proving gunpowder. The gun and ballistic test instrument are not simply passive recipients of gunpowder. Rather, they function as an interactive and complex system, whose parameters were still by no means well understood at the end of the black-powder era.[64] As some of the above quotations indicate, it was also not at all clear which ballistic test instrument (if any) really indicated how the powder would function in actual use on the battlefield.

## Reform

The unsatisfactory state of gunpowder during the Seven Years' War had apparently not changed during the American Revolution. The quality of English powder was notorious; its ballistic force was deficient, and it lacked durability.[65] The details of what then ensued were provided by one of the principal pro-

61. West 1991, 175–76.

62. Ibid., 176. This took place in 1762.

63. One of the German suppliers of prismatic powder excused its substandard quality by saying that it was made during very cold weather (Mauskopf 2006, 325).

64. Mauskopf 2006.

65. Cocroft 1996, 3.

tagonists, William Congreve. Following a serious explosion at the Faversham mills in 1781, the government of Pitt considered selling them off, "it having been represented to him that the powder merchants could make better gunpowder, and much cheaper, than the King's servants." However, the government was persuaded, in fact, to do just the reverse;[66] in 1787, it acquired another set of powder mills, those at Waltham Abbey in Essex. Moreover, in reforms of 1783, the supervision of gunpowder manufacture at the Faversham mills (and the Waltham Abbey mills, after their acquisition by the government) was placed in the hands of the comptroller of the Royal Laboratory, Woolwich.[67] Powder contracted with private powder makers now came under the standards and supervision the comptroller.[68] The motive force behind these changes appears to have been Deputy Comptroller, Captain William Congreve (1743–1814),[69] who in effect acted as comptroller during much of the 1780s and was awarded the office in 1789 (as Major Congreve).

During these same years, the procedure for proving gunpowder was also clarified. In the above-mentioned Parliamentary commission report of 1782, it was noted that, hitherto, the only established method of proof was by vertical

66. "Fortunately, however, for the country, His Grace the late Duke of Richmond, then Master General of the Ordnance, attended to the representations which Gen. Congreve, at that time Comptroller of the [Royal] Laboratory [Woolwich] [in fact, Deputy Comptroller at the time], thought it his duty to make; by which it was proved, that there existed a profit on the powder manufactured at the King's Mills; and that if this profit were for a few years properly expended in improving the works, the Ordnance would be enabled to make both stronger and more durable Gunpowder at the Royal Powder Mills, than had ever been previously made. This statement was confirmed by a course of experiments, proposed and carried on by the Comptroller; and in consequence, the idea of disposing of the Royal Powder Mills was not only abandoned, but the improvements suggested were carried into effect" (Congreve 1811, 9–10). This work appears to have been a reworking of Congreve 1788. One of the main points of this publication was to demonstrate that, by 1811, a large savings had been realized through government-supervised manufacture of military powder. It is clear in both writings that Congreve was fighting a running battle with the private powder makers for control of production of military gunpowder.

67. Fourteen or fifteen workers were retained, and workmen were brought in from the Faversham mills (now the King's Powder-Mill). forty-six workers were employed in October 1787 (seventy-nine, including an attendant clerk, surgeon, carpenter, cooper, and five apprentices are cited in a Board of Ordnance document of 16 February 1789 as necessary to run the mills; Hogg 1963, 2:1066). Water power was not substituted entirely for horse power until 1814; in 1791, double horse mills were still in use. Powder was sent regularly from Waltham Abbey to Purfleet (Thames) for proving (Nathan 1909, 318–20).

68. This included such matters as providing powder makers with fully refined saltpeter (Congreve 1811, 21–22).

69. 1783–1789. For his role, see West 1991, 177–78; Cocroft 1996; Hogg 1963, 1:458–92. See also Congreve 1811, 15, where he says that the Waltham Abbey mills were acquired "in consequence of his repeated remonstrances."

eprouvette but that recent experiments with mortars had seemed to indicate that testing with a mortar afforded "a certain proof of the strength and goodness of the powder."[70] The procedure of testing (all carried out at Purfleet) was laid out in detail. Most notable was the entrusting of the supervision of gunpowder proof to the comptroller of the King's Laboratory at Woolwich: "After every proof, a report of the state of the powder proved, signed by the Comptroller, the two Firemasters and the Storekeeper is made to the Master General and the Board; who, in consequence of that report, direct what powder shall be received as serviceable into the King's Magazines."[71]

The Royal Laboratory was founded in the seventeenth century, primarily as a venue for making and testing weapons.[72] In 1746, with the intent of establishing the laboratory "on a proper footing," the office of comptroller was established, with Charles Frederick installed as the first incumbent, and a staff of technically trained workmen built up.[73] With a background and vocational interest in classical studies, Frederick would seem to have been an inauspicious choice for this position. Nevertheless, in his long tenure as comptroller (1746–1782) and surveyor general of the ordnance (1750–1782), he presided over the challenging efforts to maintain sufficiency and quality of munitions during the Seven Years' War and the American Revolutionary War. Although the success in these efforts has hardly been viewed positively (as Congreve's evaluation testified), Frederick himself has recently come in for rehabilitation by Brenda Buchanan. Examining correspondence between Frederick and provincial powder makers (Bristol), Buchanan sees Frederick as a knowledgeable ordnance official and, in particular, as fostering "a consciously scientific approach to gunpowder making" by encouraging the powder makers to "experiment" (his word) on different parameters of powder making, such as composition, time of incorporation, and so forth.[74]

In 1741 the Royal Military Academy was also established at Woolwich to train cadets in artillery and engineering. In the early 1770s, reform and educational upgrading took place. In 1773 Charles Hutton was appointed professor of mathematics and soon instituted "a course of experiments on fired gun-

70. Mockridge 1950, 87.

71. Ibid., 87.

72. The only (and brief) account of its history is Barlow 1909, 307–15. He mentions that the first reference to a "Royall Laboratory" is in 1688 and to it at Woolwich is in 1694 (308).

73. Consisting of a chief firemaster, a firemaster's mate, a clerk, and workmen to produce fireworks and gunpowder-based weapons (Buchanan 2005, 259).

74. Ibid., 262–65.

powder and cannon balls."[75] In 1777 Isaac Landmann was appointed professor of fortification and artillery.[76]

In the late 1770s, the facilities of the Royal Laboratory were expanded and a "Repository for Military Machines" was established to give theoretical instruction to artillerymen. By the early 1780s, first under the comptroller Colonel George Napier (1782) and then, more comprehensively, under Congreve, actual experimentation was carried out at the Royal Laboratory, with the intention of ameliorating the manufacture of military powder produced at what were now royal powder mills. Although Frederick's own orientation toward experimentation in his advice to private powder makers may have set the stage for it, nothing like this function seems to have been carried out at the Royal Laboratory under his tenure as comptroller.

## Systematic and Experimental Investigations

It should be evident from what has been said that an important component of the reform activities in the last quarter of the eighteenth century consisted in the application of "science" to gunpowder improvement. I have put the term in quotes here because of the obvious—and perennial—definitional problem of what this multifaceted term means in connection with industry. This problem is compounded for the eighteenth century by the relative lack of clear-cut professional boundaries between scientists and others and by the use of scientific terms for rhetorical ends. I shall be fairly straightforward in what I mean: the attempts to bring to bear on the improvement of gunpowder production (and product) systematic testing, experimentation, and, if possible, theoretical explanation. Part of the reason for these attempts was that so many of the procedures for making and evaluating gunpowder were contested.

Since the seventeenth century, it was recognized that the explosive force of gunpowder arose from the sudden release of a great volume of "air" from the solid powder. An experimental tradition to measure the volumetric ratio of "air" to gunpowder and to determine the exact nature of this "air" spanned much of the eighteenth century and, in England, included such notables as Stephen Hales and Benjamin Robins. With the discovery of new

75. The course was begun in 1775 and was focused on exterior ballistics based on the work of Benjamin Robins ("the determination of the actual velocities with which balls are impelled from given pieces of cannon, when fired with given charges of powder"; Hutton 1812, 2:306).

76. Hogg 1963, vol. 1, chap. 10, esp. 377ff. This is a topic badly in need of further study.

gases, such as "fixed air" and "inflammable air,"[77] and the subsequent development of pneumatic chemistry, refined understanding about the chemical nature of the explosive reaction of gunpowder and of the gases produced became possible.[78]

An interesting example of an intermediate understanding of the chemistry of gunpowder explosion is given by research carried out by the Dutch-born scientist Jan Ingen-Housz (1730–1799) during a brief residency in England and published in the *Philosophical Transactions of the Royal Society of London* in 1779. Joseph Priestley had discovered "dephlogisticated air" (oxygen) five years earlier, and it was soon demonstrated that this gas was a component of nitric acid and, therefore, of saltpeter. Charcoal was thought to be a compound substance, one of whose components was "inflammable air" (hydrogen), often equated in England with the inflammable principle, phlogiston. The explosive force of gunpowder explosion was explained by Ingen-Housz thusly: "Nitre [saltpeter] yields by heat a surprising quantity of pure dephlogisticated air. Charcoal yields by heat a considerable quantity of inflammable air. The fire employed to inflame the gunpowder extricates these two airs, and sets fire to them at the same instant of their extrication."[79]

The explosive reaction of gunpowder was particularly violent because it involved the release (and reaction) of these two gases from their compressed state in the solid powder. But Ingen-Housz gave no account of exactly what was the resultant gaseous product of this burning of dephlogisticated air and inflammable air. Of course, the product of that reaction is water, as was determined in 1781 by Henry Cavendish. But water (or water vapor) is not the gaseous product of gunpowder explosion. The actual products seem first to have been ascertained two years later by Tiberius Cavallo (1749–1809), another foreign scientist residing in London. In a rather crude but effective experiment, Cavallo found two gases produced by gunpowder explosion: fixed air ($CO_2$) and phlogisticated air ($N_2$).[80]

77. Respectively, $CO_2$, discovered by Joseph Black (1756), and $H_2$, discovered by Henry Cavendish (1766).

78. Mauskopf 1988.

79. Ingen-Housz 1779, 398.

80. He exploded musket powder, collected the gases in a large bladder, and measured a volumetric ratio of phlogisticated air (nitrogen) to fixed air (carbon dioxide) of slightly less than 3:1. Being a phlogistonist, Cavallo readily explained the production of nitrogen (phlogisticated air) as the union of the dephlogisticated air ($O_2$) in the saltpeter with the phlogiston from the highly flammable charcoal. Less explicable was the production of fixed air by this explosion (Cavallo 1781, 810–15).

Ingen-Housz's and Cavallo's experiments appear to have had a purely scientific context and intent. Neither seems to have had any connection with the Ordnance Board, the Royal Laboratory, or the production of gunpowder in England.[81] Most importantly, the focus of their research and publications was not on military matters. But in the eighteenth century there were men with interest and facility in experimental natural philosophy who *did* direct their research primarily to military matters, especially artillery and ballistics. Some, like the Englishmen Benjamin Robins and Charles Hutton, were not military officers but became associated with the military through their publications (e.g., Robins's pioneering book on artillery, *New Principles of Gunnery,* 1742) and through their appointments (Hutton was professor of mathematics at the Royal Military Academy in Woolwich). Hutton developed a scaled-up form of the ballistic pendulum that Robins had invented, one in which a cannon served as the pendulum. He also claimed to have instituted the first "course of experiments on fired gunpowder and cannon balls" at Woolwich in 1775.[82] Others, like Count Giuseppe Angelo di Menusiglio in the Kingdom of Sardinia (Turin), Patrice d'Arcy in France, and the colonial expatriate, Benjamin Thompson (Count Rumford) in England and Bavaria, were military officers, held positions of authority in military establishments, and carried out research and publications in artillery and ballistics.

## WILLIAM CONGREVE (1743–1814)

The experimenters at the Royal Laboratory in Woolwich, George Napier, comptroller in 1782, and especially William Congreve, fall into this latter category. Until West's book and the recent preliminary study of Congreve's role at the Royal Laboratory by Wayne Cocroft, Congreve has reposed forgotten and certainly unstudied. He is clearly emerging as the major figure in the reform of munitions in England in the last decades of the eighteenth century.

Coming from a military family, Congreve himself entered the military (Royal Artillery) in 1757, when he was fourteen. He saw service in the Seven Years' War and had risen to the rank of captain by 1772.[83] In 1773 he

81. But see below relating to Congreve's adoption of two powder grain sizes.

82. Hutton 1812; Mauskopf 1996, 278–81; Steele 1994; Steele 2006, 296–99; West (1991, 177–84) gives a brief but interesting treatment of the work of Robins, Hutton, and, above all, Thompson.

83. He was promoted to major in 1785 and eventually reached the rank of lieutenant-general (1808) and was created a baronet in 1812. The only detailed account of a part of his life (down to 1779) is found in a chapter of a study of his son by Major J. P. Kaestlin (n.d.). The chapter

became a founding member of a "Military Society" devoted to "improving military, mathematical and philosophical knowledge" at Woolwich, where he had been posted. In connection with an improvement in gun carriage design he made in the early 1770s, he was commended by his commanding officer to the Ordnance Board for his scientific prowess: "His knowledge of the mechanical powers, so very necessary in our service, has always been his delight and study; in which my humble opinion is that he stands foremost among us."[84]

Although Congreve did not actually become comptroller of the Royal Laboratory until 1789 (as Major Congreve), he had been appointed deputy comptroller in 1783 and had probably functioned as the leader since the comptroller was in India.[85] There is considerable evidence, some of it suggested above, that he was the principal instigator of many of the changes taking place from the late 1770s, including the turn of attention to gunpowder improvement at the Royal Laboratory in the 1780s, carried out in tandem with the new association between the laboratory and the government-owned powder mills. However, powder making remained a mixed private-government enterprise in Britain, and nothing like the French administrative system of the *Régie des Poudres* nor its associated education structure developed there.

In the interest of ameliorating the quality of the military powder supplied to British armed forces, Congreve brought to bear his scientific bent. During the 1780s, he carried out "experiments" (his word) on virtually every aspect of gunpowder production and testing at the order of the master general of ordnance (but one suspects that initiative may have come from Congreve).[86] It is exceedingly fortunate that a contemporaneous account of these experiments, signed and dated by Congreve himself, survives, as do others at the Public Record Office used by Cocroft.[87]

What Congreve instituted were series of comparative tests and parameter-variation tests. These tests extended to the measuring instruments them-

devoted to our William Congreve is titled, "Remote Control or The Start of the Royal Military Repository."

84. Kaestlin, n.d., 7.

85. Napier was comptroller for just ten months, and his successor, Colonel James, was in India (see Cocroft 1996, 2).

86. Congreve was instructed "to make the king's gunpowder as strong as possible" (Cocroft 1996, 2).

87. Congreve 1788. Pro Supply 5/866 (Quality of gunpowder miscellaneous letters, 1781–1809), Supply 5/65 (Faversham Letter Book), Supply 5/870, and MP HH 597 (Faversham plan and section of subliming furnace, 21 July 1790) are the Public Record Office documents Cocroft cites. I have not examined them.

selves, namely, those used to determine the ballistic capability of the powder. I shall conclude this chapter with an examination of Congreve's research as detailed in the manuscript dated 17 January 1788 and signed by Congreve. Nineteen pages long, and written in a very legible hand,[88] it recounts the experiments in the third person. The document (and hence, the experiments) had a polemical objective: to justify the production of powder at the government mills. The final thought makes this clear: "Yet the Gun Powder which is made by the Ordnance will cost Government considerably less than it can be purchased for, from the Contractors."[89] Part of Congreve's strategy was to show the "rottenness" (i.e., poor workmanship and staying quality) of the gunpowder supplied by private contract. But his experiments went beyond mere polemics.

The experiments are arranged chronologically. The first set of 1782–1783, ascribed to Napier, highlighted the polemic; they "proved the rottenness of the Grain of the Gun Powder, and its tendency to imbibe Moisture."[90] All of the others were Congreve's. Some of these continued the polemic against the private contractors,[91] but most explored contested issues regarding gunpowder manufacture and proving. One of the first sets (1783–1784) concerned the traditional employment of the vertical eprouvette to prove the powder. Congreve asserted that his experiments: "Clearly ascertained the Imperfection of the Mode of proving Gun Powder by the Vertical Eprouvette, as by it durable good Powder was often condemned, and very tender grained bad Powder, received into the Royal Magazines as serviceable."[92]

The most ambitious tests were the comparative ones—of powders from different countries as well as of powders made from different proportions of ingredients, or (in the case of charcoal) different woods and methods of charring the wood. In 1783 Congreve carried out comparative tests on powders of

88. Congreve 1788. The handwriting of the manuscript is different from Congreve's signature. In my quotations, I have retained the spelling and capitalization of the manuscript.

89. Ibid., 19–20. The context of the acquisition of the Waltham Abbey mill in 1787 needs to be explored.

90. Ibid., 1.

91. For example: "Major Congreve had two Experiments tried at the Royal Laboratory, one in 1783 and the other in 1785, to ascertain the Quantity of SaltPetre which was contained in 100 lbs of each Merchants GunPowder, from which Experiments it appeared, that none of the Merchants Powder contained more than 74 lbs 13./4 ozs of pure SaltPetre in each 100 lbs of Powder" (ibid., 3).

92. Ibid., 2. It is clear from subsequent passages that Congreve favored a 13-inch mortar as the proof instrument (ibid., 5). By 1788, he had also come to favor its use in combination with Hutton's eprouvette (Cocroft 1996, 2).

the compositions used in France, Italy, Sweden, Poland, and Russia.[93] Similar comparative tests were carried out in 1785 between English, Hanoverian, French, and Swedish compositions. The 1785 comparative tests were particularly interesting because of their complex and comprehensive nature. Congreve was ordered (or elected) to:

> Make some gunpowder at Faversham, with different proportions of Ingredients, to ascertain whether Wood charred in the common Way, or after the Method practiced at Battel in Sussex, would make the best Gun Powder; whether black or white Dogwood, white willow, small perished, or large sound Alder, Hazle, Lime Tree, or Horse Chestnut, would make the best Charcoal for Gun Powder—. Also whether Saltpetre which is extracted from damaged Gun Powder, would be as fit for use as that which was refined from the grough Saltpetre brought from the East Indies—Whether by melting Saltpetre to form it into cakes, its Strength would be injured—Whether glazed Gun Powder is as strong as unglazed Powder, and which of the following Proportions of Ingredients is the best, Viz.
>
> The English.
> Hanoverian.
> Swedish.
> French, with several other new Proportions.[94]

As can be seen, these captured most of the contested issues of gunpowder manufacture that have been explored in this paper.

In late 1786, a new series of experiments was carried out exploring yet other issues.[95] The most novel of these concerned the method of charring powder: "To determine whether the same Wood charred in a common Pit, in iron Pots, as practiced at Battel in Sussex, or in a Furnace invented by Doctor George Fordyce, and recommended by the Bishop of Landaff [*sic*] [Watson], would make the best Gun Powder."[96]

Finally, perhaps the most complicated series were carried out in the fall

93. Congreve 1788, 3.

94. Ibid., 3–4. The reference to "Battel" (subsequently spelled both this way and "Battle" in the manuscript) is to the Battle mill in Sussex, where the wood was charred in iron pots.

95. "Whether the Mill Cake of first Powder unpressed, or that which is pressed and corned; or Powder made from Dust will make stronger Powder, than that manufactured from green Composition pressed. To ascertain whether GunPowder formed into spherical, or angular Grains, would be the strongest, and if glazed GunPowder would not be more durable, and nearly as strong, as unglazed Powder" (ibid., 4–5).

96. Ibid., 4. Cocroft has shown that these tests were actually begun early in 1785 at the government mill at Faversham (1996, 3–4).

of 1787. In these, powders with charcoal charred in different manners and in different proportions[97] were made and tested in a "13 inch Mortar, loaded with 3 lbs of each Sort of Gun Powder, and iron Shells."[98]

Although no ballistic results were provided, what follows is a series of eleven "Essentials towards the Improvement of the manufacture of Gun Powder for His Majesty's Service." Here are some of the conclusions. Regarding the comparatively best powder, Congreve asserted that "the English had the advantage of the others"[99]—this at just the time that Lavoisier claimed French powder to be "the best in Europe"![100] As for the wood source, black dogwood yielded the strongest powder, followed by white willow. However, for charcoal made by a new method proposed by Richard Watson, alder wood of eleven years' growth and small white willow of three, four, and six years' growth yielded powder of nearly equal ballistic strength. In 1791 black dogwood, white willow, and alder were specified as the wood sources to be used for gunpowder charcoal.[101]

Melted saltpeter produced powder as strong as that produced with ordinary saltpeter (dried in the air), and melted saltpeter extracted from damaged powder was "rather stronger" than refined grough saltpeter. Angular grained powder was found to be stronger than spherical grained; powder "moderately glazed and the Angles not broke off their Grains" seemed to be as strong as unglazed powder.[102]

There were a few "essentials" that are particularly significant because they anticipate changes introduced into English gunpowder manufacture. One concerns the employment of the "closed-cylinder" or "furnace" method of charcoal proposed by Richard Watson. Congreve commented laconically that

97. "Between the 24th of September and the 17th of November, Major Congreve was ordered by the Duke of Richmond [master general of the ordnance] to make some Gun Powder with the following Sorts of Charcoal, and with the English Proportion of Ingredients, Viz. the old Charcoal in Store at Faversham, Coal charred at Blythe in September 1787, and with Coal made in a Furnace; also with a new Proportion of Ingredients Viz. 76th of SaltPetre, 15th of Charcoal and 9th of Brimstone" (Congreve 1788, 5).

98. Ibid., 5.

99. Ibid., 7. Tests conducted in 1787 found gunpowder made in the proportion 76% saltpeter, 15% charcoal, 9% sulfur to be superior to either the pre-1785 composition or the one adopted in 1785 (75% saltpeter, 15% charcoal, 10% sulfur) (Congreve 1788, 18).

100. A.-L Lavoisier 1789, "Mémoire sur l'établissement, les produits et la situation de la Régie des poudres et salpêtres," quoted in Mauskopf 2005, 306. The ordained French proportions were 75% saltpeter, 12 1/2% charcoal, 12 1/2% sulfur.

101. Cocroft 1996, 6. Cocroft notes that Congreve took a personal interest in this set of experiments and went down to Faversham to supervise the charring of the wood (1996, 3–4).

102. Congreve 1788, 6.

this method "makes much stronger Gun Powder than that which is charred in a common Pit, or after the Method practiced at Battle."[103] Comparative ballistic tests of traditionally charred wood and charcoal made by Watson's method, in fact, were decisively in favor of the latter. Congreve communicated to Watson that ballistic data from tests in both the vertical eprouvette and the mortar yielded a ratio of ballistic force for powder made with closed cylinder charcoal and pit charcoal of 100:60 or 5:3 "in round number."[104]

The closed-cylinder method of making gunpowder charcoal was adopted by the government powder mills in the 1790s; by 1800 it was noted that the increased force of cylinder-charcoal gunpowder had led to a reduction in ordnance charges of one-third (and an annual saving to the government of £100,000).[105] Indeed, Congreve himself noted years later that, even with reduced charges, the powder was so powerful that it had resulted in bursting guns and even loss of life.[106]

Another conclusion, also put very laconically, was to have as great significance as the change in method of charring wood: "When large Quantities of Gun Powder are used, large grain'd Powder is stronger than small grain'd."[107] Although different sized powder grains (for different caliber guns) had been employed in earlier times, by the eighteenth century, only one type of military powder was employed for all types of weapons. Under the aegis of Congreve, different types of gunpowder were introduced: a smaller grained powder for muskets and a larger grained one for cannon. In 1811 Congreve elaborated somewhat on his earlier assertion, stating that he had "ascertained by experiment, that although the small-grained powder is stronger in *small* quantities, and therefore fitter for musquetry, the large-grained powder is better for the charges of cannon."[108] Cocroft recounts experiments carried out under Congreve's direction at the new government powder mill at Waltham Abbey in 1789 that tested both different sizes of powder grains and different proving instruments.[109]

103. Ibid., 6. Congreve took a close personal interest in the tests of Watson's method (Cocroft 1996, 4).

104. "Obituary of Bishop Watson" 1816, 274.

105. Cocroft 1996, 4.

106. Congreve 1811, 38 and footnote, where it was noted that bursting mortars had occurred in Alexandria, at the Dardanelles, in Basques Roads, and at Cadiz, where twenty men were killed and wounded.

107. Congreve 1788, 6.

108. Congreve 1811, 22–23.

109. Cocroft 1996, 2, where he also asserts that Congreve had introduced three sizes of powder grains. West has suggested that a source of the idea of gunpowder of different grain sizes for

In the course of the nineteenth century, with the growth in size of both guns and powder charges, and the consequent increasing problem of bursting guns, the issue of the size, shape, and physical characteristics of powder grains and cartridges as a means of controlling the rate of powder burn was to become ever more important. Prismatic powder, with which I began this chapter, represented the final effort of black powder researchers and makers to suit the powder to the very large guns of the 1880s.[110]

There is one last, general area where Congreve apparently effected important changes. This was in the domain of the purification of the ingredients. Cocroft has studied and written on the archeology of the structures built under Congreve's direction at Faversham specifically for purifying sulfur and saltpeter.[111] This concern for purity had its parallel in the munitions enterprise in France presided over by Lavoisier, where the watchword was "exactitude." In the 1788 manuscript, Congreve noted that the master general of the ordnance had proposed to the private mill owners that: "They should admit some Officers of Ordnance to visit their Mills, to take Samples of the several Ingredients, to be analized if any Doubts should be entertain'd of the Purity of the Sorts each Merchant had provided to compleat their Contract with the Ordnance." According to Congreve, his proposal was apparently turned down by the mill owners.[112]

## Concluding Thoughts

Congreve justified the experiments recounted in the 1788 manuscript as "having clearly pointed out the slovenly mode hitherto made use of in manufacturing Gun Powder for His Majesty's Service."[113] Like his French colleague, Lavoisier, Congreve viewed his mission in terms of the ideals of Enlightenment public officials: government administrators deploying exper-

different guns might have lain in Ingen-Housz's paper of 1779 (Ingen-Housz 1779, 406), where he postulated that quickness of inflammation depended upon the size of the interstices between powder grains; "Hence the size of the grains of gunpowder must be proportionate to the size of the fire arms to which it is destined, the greatest fire arms requiring in general grains of the largest size" (West 1991, 180). However, it is, I believe, going too far to say that "the scientific experiments of Ingen-Housz confirmed the increased effect possible with graded sizes for different purposes" (West 1991, 171).

110. Mauskopf 2006.

111. Cocroft 1996, 5.

112. Congreve 1788, 8. The master general of the ordnance at the time was the Duke of Richmond. This would then have a *terminus post quem* of 1784 and would most likely have been at the instigation of Congreve.

113. Congreve 1788, 7.

iment-based "useful knowledge" to reform and improve industry. Indeed, in both cases, their mission seems to have been more to supplant private craft production with more rational government enterprise than just to improve it.

Thus, the context for Congreve's texts of 1788 and 1811 was his concern with proving the economic value of supporting the government gunpowder factories at Faversham and Waltham Abbey against the private powder makers. Referring specifically to the value of the Faversham powder makers: "It would have been highly impolitic to have discharged such a set of experienced Powder Makers, who are now making such kind of Gun Powder as the Ordnance have never been able to prevail upon the Powder Merchants to make for the King's service, although they have had such a prodigious Profit allowed them for encouragement: but the Merchants being indolent to go into the dirty work of their Mills, have generally left the superintendance thereof, to artful but ignorant Foremen, who probably made a very considerable Profit by their Masters innattention [*sic*]."[114]

There were two desiderata for gunpowder in the eighteenth century: strength (ballistic force) and durability. These were the improvements that Congreve claimed to have effected in government-made gunpowder. He also reintroduced the hitherto obsolete desideratum of suiting the powder to guns of different calibers by making it in different grain sizes.[115] Congreve attributed the greater ballistic force to the introduction of Watson's closed-cylinder method of making charcoal. Aside from Congreve's own experimental support,[116] the claim for ballistically stronger gunpowder is credible because of the incontrovertible trade-off that this "improvement" entailed: bursting cannon. Congreve had already noted this effect; when the French adopted English manufacturing procedures after the Restoration, this became an acute problem for them, and later for the Americans as well.

But gunpowder was denominated at the beginning of this chapter as a "thick thing," and this comes out in assessing even the very cause of change. For example, Congreve's and his successors' explanation for the cause of this greater brisance has been called into question recently. Namely, it has been argued that the greater compression of powder resulting from the introduc-

114. Ibid., 14.

115. Congreve 1811, 19–20, 22–23, 24–25.

116. Context is also important here, since one of Congreve's arguments was that greater ballistic force meant a financial savings because less powder was needed for the same ballistic effect (see above, note 105).

tion of the screw press may have been as important as the method of making charcoal in augmenting ballistic force.[117]

The second claim of durability was based on "a variety of circumstances," by which Congreve specifically meant attention to the purity of the ingredients used and to the procedures used in producing the powder.[118] I certainly believe (or *want* to believe) that this made a positive difference, at least in the uniformity of the product. But, as the quotation from Colonel Brackenbury with which I began the chapter illustrates, ensuring uniformity of gunpowder remained a challenge that was, at best, only *partially* solved until the end of the black-powder era (and beyond).

The final claim of suitability for guns of different calibers was based on his reintroduction of different powder grain sizes. Viewed retrospectively, Congreve was correct but probably not for the reason he gave. Namely, he argued that bigger powder grains were suitable for larger caliber guns because: "When large Quantities of Gun Powder are used, large grain'd Powder is stronger than small grain'd."[119] By the late nineteenth century, in the wake of the bursting cannon crisis and the investigations of T. J. Rodman, grain size was enlarged for large guns to *control* the rate of burn and the build-up of pressure in the bore of the gun, not to increase the power.[120]

Some of Congreve's other experimental claims are more problematic. One is his claim that the proportions of components in English powder produced the "best powder." I have noted that this was matched at almost exactly the same time by Lavoisier's equally positive claim for French powder. These claims are difficult to evaluate: there is an issue of justification of position and role involved in each case. Moreover, the whole question of which proportion of gunpowder components yielded the greatest ballistic force was a vexed question; the most elaborate contemporary experimental study of the issue, by the French chemist J.-L. Proust, appeared to confirm the traditional French proportions of 75 percent saltpeter, 12.5 percent charcoal, and 12.5 percent sulfur.[121] The definition of "best" proportion entails the issues of ballistic force and durability, both of which, as we have seen, depend upon a complex

117. Howard 1996, 12–13.

118. "The great care now taken to render the ingredients most perfectly pure, the increased attention paid to working the powder under the runners, to the pressing of it, and to the improved modes of dusting and glazing it" (Congreve 1811, 24).

119. Congreve 1788, 6.

120. Mauskopf 1996, 285–89.

121. Mauskopf 1990, 398–426.

array of factors. In particular, the measurement of ballistic force remained a highly contested issue for the greater part of the nineteenth century.[122]

Despite Congreve's condemnation of the "Powder Merchants," the British gunpowder industry remained a mixed one of both government and private production. Although Congreve complained about the defiance of the private mill owners to government supervision, nevertheless, it appears that his changes in powder making instituted at the government mills were also soon adopted by the private mills; this, of course, was due to the fact that all gunpowder had to pass government ballistic testing to be accepted for purchase.[123]

But the situation of gunpowder production in the nineteenth century is in need of much further research. It is almost a century now since *The Rise and Progress of the British Explosives Industry* was published. This was the last (and first, I might add) synoptic study of this industry, and it is desperately in need of an update. In particular, I would especially like to know how the larger private companies (such as Curtis's and Harvey) evolved and maintained their competitive edge from the eighteenth through the nineteenth century.

I would mention one final area that certainly needs study: the continuation (or not) of the munitions research tradition that Congreve instituted in Britain in the nineteenth century. There is evidence that it was not maintained. He was succeeded as comptroller of the Royal Laboratory by his son and namesake. The historical fame of Congreve *fils* has been more robust than that of his father by virtue of the rocket system that he developed. Although the author of the article on the younger Congreve in the *Oxford Dictionary of National Biography* credits him with inventing "an improved process of gunpowder manufacture" (nature unspecified),[124] he has not hitherto figured at all in the history of English powder or powder making.

In the aftermath of the Napoleonic Wars, the Royal Laboratory was severely downsized.[125] Although many of the changes introduced into gunpowder production by Congreve *père* were adopted in other countries, notably

122. Mauskopf 1998, 234–36.

123. Cocroft 1996, 6. He evaluates Congreve's success as follows: "Through his systematic practical research into the manufacture of gunpowder and his ability to enact change, Congreve had transformed British powder from one of a notorious quality to a world standard."

124. Stearn, n.d., "Congreve."

125. Barlow 1909, 309–10, 314, table 2, "Numbers Employed in the Royal Laboratory at Various Periods." Between 1813 and 1817, employment dropped from 1,451 to 461. By 1835, it was down to 126.

France and the United States, in subsequent decades, the tradition of research instituted by him in Britain in the 1780s seems to have largely ceased, at least in the decades before the Crimean War.[126] Although research on munitions revived after 1860, gunpowder itself remained a remarkably intractable "thick thing" until the very end of its military use, as my opening vignette demonstrates.[127]

### Primary Sources

Secondary sources can be found in the cumulative bibliography at the end of the book.

Braddock, John. 1832. *A Memoir on Gunpowder; in which are discussed, the Principles both of its Manufacture and Proof.* London: Reprinted by Permission of the Hon. Court of Directors, Printed at Madras at the Expense of the Indian Government for Use of the Artillery.

Caruana, Adrian B. n.d. Laboratory Papers 1790. Compiled from a Manuscript in the possession of the Royal Artillery Library, Woolwich.

Cavallo, Tiberius. 1781. *A Treatise on the Nature and Properties of Air and other Permanently Elastic Fluids.* London: for the author.

Coleman, R. 1801. On the Manufacture and constituent Parts of Gunpowder [Read before the Askesian Society, May 1801]. *Philosophical Magazine* 9:355–65.

Congreve, William. 1788. A State of Facts relative to the Grounds on which the late and present Master General have had so much Reason to doubt the goodness and durability of the Gun Powder which was delivered into the Royal Magazine for the King's Service. Royal Artillery Library, Woolwich.

———. 1811. *A Statement of Facts, Relative to the Savings which have Arisen from Manufacturing Gunpowder at the Royal Powder Mills; and of the Improvements which have been made in its Strength and Durability since the Year 1783.* London: Printed by James Whiting.

Guttmann, Oscar. 1895. *The Manufacture of Explosives.* London: Whittaker & Co.

———. 1906. *Monumenta Pulveris Pyrii.* London: Artists Press.

Hutton, Charles. 1812. *Tracts on Mathematical and Philosophical Subjects, Comprising, Among Numerous Important Articles, The Theory of Bridges, . . . Also . . . The Force of Gunpowder, With Applications to the Modern Practice of Artillery.* 3 vols. London: Printed for F. C. and J. Rivington.

Ingen-Housz, John. 1779. Account of a new Kind of Inflammable Air or Gass, Which Can be Made in a Moment without Apparatus, and is as Fit for Explosion as Other Inflammable Gasses in Use for That Purpose; Together with a New Theory of Gunpowder. *Philosophical Transactions* 69:376–418.

Kaestlin, Major J. P. n.d. Firemaster. MD 213/14, no. 6. Royal Artillery Library, Royal Artillery Institution, Woolwich.

126. An anonymous writer asserted in 1872 that the development of gunpowder had essentially ceased in Britain between Congreve's work and the introduction of rifled gunnery around 1860 (cited in Mauskopf 2006, 307). This was in contrast to France and the United States, where ordnance research became institutionalized in the first part of the nineteenth century.

127. Mauskopf 2006.

Napier, George. 1788. Observations on Gunpowder. *Transactions of the Royal Irish Academy* 2:97–117.

Obituary of Bishop Watson. 1816. *Gentlemen's Magazine*, September 1816.

Renaud, L. 1811. *Instruction sur la fabrication de la poudre.* Paris: Magimel.

*The Rise and Progress of the British Explosives Industry. Published under the auspices of the VIIth International Congress of Applied Chemistry by its Explosives Section.* 1909. London: Whittaker.

Robins, Benjamin. 1742. *New Principles of Gunnery.* London: J. Nourse.

Watson, Richard. 1800. *Chemical Essays (1781–87).* 7th ed., 4 vols. London: Printed for J. Johnson et al.

12

# Between Craft Routines and Academic Rules: Natural Dyestuffs and the "Art" of Dyeing in the Eighteenth Century

AGUSTÍ NIETO-GALAN

Over the eighteenth century, the materials used in the art of dyeing, both the chemical substances and the coloring matters, were numerous and complex.[1] Extracted mainly from animal and vegetable sources, natural dyestuffs were applied to a wide variety of surfaces. Indigenous sources of red (madder), blue (woad), and yellow (weld) competed with exotic products such as cochineal for reds, indigo for blues, quercitron for yellows, and a long list of "American woods." Dyeing processes also required gums, astringents, bleaching liquors, metallic salts (mordants), acids, and alkalis. The great variety of substances in the workshops seriously complicated the organization of the art. Controversies often arose over the sources of colors, between indigenous dyes and exotic plants from the colonies.

Natural dyestuffs and their chemicals graduated from small-scale workshops to *manufactures* and factories. In the case of calico printing, mechanization had a great impact on the methods of preparation and application of colors. In a European network, experts, samples, manuscripts, and printed texts circulated widely and played a crucial role in the construction of the body of knowledge of the art of dyeing. Some of these practices constituted learned forms of knowledge, which were often debated in terms of the acquisition and demonstration of expertise over production. But who was the real expert? This was the crucial question that remained unresolved throughout the century, even in the final decades, after Lavoisier's new chemistry. Craft traditions were powerful and firmly rooted in the old guild systems; since the

1. This chapter updates some of the main points published in Fox and Nieto-Galan 1999; Nieto-Galan 2001. Original quotations in French have been translated into English.

Renaissance, tacit knowledge and oral transmission of recipes had been fundamental elements of the world of the trades. Members of enlightened scientific societies from all over Europe fiercely competed for expertise and authority with prestigious dyers and printers in manufactories and factories. In the process, recipes, tests, colors, and dyed samples all contributed to the establishment of complex mechanisms of communication and consensus, which demand further exploration.

Dyers and academic experts perceived and applied learned knowledge in different ways, and the ways ideas and practices were circulated between the two groups did not fit into simple patterns of diffusion and passive reception.[2] Dyers in the workshops often criticized academic works out of hand, while many skilled workers resisted the rhetoric of the new academic chemistry at the end of the century.[3] Popular books on dyeing, aimed at simple dyers, went largely unconsulted. There is evidence from this period that academic chemists were aware of the difficulty of properly explaining and organizing this complex set of skills; for their part, the dyers often felt uncomfortable with academic language and style.

Hopefully, this chapter will contribute to producing a more symmetrical history of the relationships between learned concepts and the tacit knowledge of the artisanal tradition of dyeing. Together, the rhetoric of printed texts and public addresses and the informal culture of the dyers probably constituted the real essence of an art that cannot be easily qualified either as academic or as workshop culture. As I have argued in *Colouring Textiles*, dyeing practices in the eighteenth century were no simple, mechanical repetitions of old craft routines; nor did they fit neatly within the rules of the academy.[4] For the case of dyeing, at least, it seems necessary to contrast various ways of knowing and doing that did not coincide with the traditional categories of theory and practice, academy and workshop.[5] Is it possible to talk about the "art" of natural dyestuffs in the eighteenth century in a broad sense? Perhaps a reasonable answer will progressively emerge if we are able to describe in detail some of these ways of knowing and doing in indeterminate areas.[6] Following Klein's definition, eighteenth-century "technoscience *avant la lettre*"—of which the art of dyeing was a significant part—must have been

2. Secord 2004.
3. For similar concerns about authority and expertise, see Spary's chapter in this volume.
4. Nieto-Galan 2001.
5. Klein 1996.
6. Pickstone 2000.

a complex combination of experimental history, arts and crafts, and philosophical knowledge that involved skilled techniques, academic judgments, and systems of organization.[7]

## The Workshop, the Manufacture royale, and the Factory

Our interest in the material aspects of the art of dyeing is closely linked to the sites in which the craft was practiced. Here, the arts and crafts tradition coexisted with different aspects of learned culture. In preindustrial times, natural dyestuffs were used in small workshops, supervised by a guild organization, where tacit knowledge played a very important role.[8] As Sidney Edelstein described some years ago, a visitor to an eighteenth-century dyeing house would find the following actors and objects:

> The dyer, dressed in black homespun, and wearing a heavy leather apron, ushers us into the main dye house room and proceeds to point out the apparatus. In the center of the dirt floor are two caldrons, set into a brick furnace. Both are about 70 gallons of capacity, and one is in copper, and the other in iron. . . . Beside the iron kettle a boy is stirring a dark bubbling liquor. . . . At another kettle we see an older man busily winding some woolen cloth over a winch and through a red liquor. . . . The dyer asks us to come with him to the drug room, where we can see the different dyes and chemicals he uses.[9]

Dyeing procedures required a highly skilled workforce able to handle natural dyestuffs and their regional variations, coloring matters from the colonies, and ancillary products (mordants, bleaching liquors, gums, and astringents). A list of ingredients for dyeing and printing would include lead acetate, alum, copper acetate, nitric acid, ammonia and ammonium chloride, potassium carbonate, potassium tartrate, gallic acid, astringent resins, gums, bleaching lyes, nitric acid, hydrochloric acid, sulfuric acid, and arsenic salts. Calico printers in particular used alum, carbonates, sulfates, and acetates.

The small-scale workshop culture progressed gradually to larger sites of production such as the *manufactures royales*, manifestations of the French

7. Klein 2005c.

8. Donnelly 1994, 125. In Anthony Travis's view, "the role of the individual in the creation and transfer of new knowledge remained paramount until the third quarter of the 18th century" (Travis 1994, 98–99).

9. Edelstein 1974, 19–20. See Nieto-Galan 2001, 44.

protectionist model established by the minister Jean-Baptiste Colbert (1619–1683) in the second half of the seventeenth century. Under the official protection of the monarchy, skills from different guilds were concentrated and reorganized in bigger spaces. Samples of dyed fabrics were systematically compared in terms of their different shades and colors; operations such as heating, degreasing, mordanting, dyeing, washing, drying, and wringing were tested first in the workshop and later extended to large-scale production in the *manufacture* as a whole. The workshop also contained a laboratory for small-scale tests, chemical analysis, quality control, and the classification of dyestuffs and side substances.[10] The École de teinture (dyeing school) publicized the activities of the *manufacture* and attracted new audiences interested in the art.[11] The amphitheater held public lectures and dyeing courses for young students wanting to learn the secrets of the craft, later to be applied in workshops and *manufactures* in France and abroad. The *manufactures* also played an important role in the education of provincial dyers, and general inspectors acted as mediators between the academy and the workshop.[12]

Although the guild workshop tradition was long lived, the emergence of larger, more centralized spaces of production such as the *manufactures,* as well as the increasing influence of new kinds of experts, made the art of dyeing a far more complicated procedure. The protagonists, materials, and objects all underwent significant changes. The spectacular growth of *indiennes,* printed cotton goods, led to the emergence of block printers, copperplate printers, pencilers, bleachers, and engravers, who worked together with the expert dyers in a "factory system."[13] In the second half of the eighteenth century, calicoes went through a long series of operations:

10. Gastinel-Coural 1997, 67–80.

11. In 1819, the Comte de la Boulaye Marillac, *Directeur de Teintures* at the Gobelins, and director of the École royale de teinture, made a spectacular demonstration in front of all his former students of the use of new machines for dyeing pieces of cloth in different colors (Archives Nationales [A.N.], Commerce et Industrie, F/12/2260; A.N., Commerce et Industrie, F/12/2437).

12. We know that in Spain, for instance, the general inspectors were commissioned with supervising workshops and *manufacturas.* The *Ordenanzas de Tintes* set out their responsibilities: (1) visits to the *manufacturas,* (2) technical education of dyers, (3) a compilation of all learned and craftsman knowledge, (4) a set of chemical tests, (5) promotion of indigenous dyeplants (Nieto-Galan 2002).

13. Parkes 1815, 2:110–36. A manuscript written by the Swiss calico printer Samuel Rhyner is clear proof of the dynamism of the technology in the middle years of the eighteenth century (Simon 1994, 117).

> *Bleached* cotton had to be *dressed* by singeing off its nap, then *steeped*, soaking the pieces for 24 hours in a vessel of weak alkaline lye at 100°, *ashed*, treating them with potash to remove any impurity, *cleansed*, by washing them to remove all the alkali, *soured*, with diluted sulphuric acid, *washed* with water and *dried*, then *calendered* (ironed) through a set of rollers at the final stage of the preparation phase, to provide calicoes ready for printing. *Blocks* were *cut* to apply the mordant on the surface of the cotton in a *tearing* and *sightening* process using various sorts of gums. After 24 hours in the *stove*, *animal dung* would later remove the portion of mordant, which was not combined with the cloth. After a second *washing*, the calicoes were ready to receive different sorts of dyestuffs, and numerous finishing operations were applied before the final cotton fabric was ready to be sold.[14]

As described by the English chemical manufacturer Sheridan Muspratt: "The art of the calico-printer . . . not only comprehends that of the dyer . . . but also that of the artists, for the designing of tasteful and elegant patterns, that of the engraver, for transferring those patterns to the metal used to impress them on the cloth, and that of the mechanician, for the various mechanical processes of engraving and printing. Taste, chemistry and mechanics have been called the three legs of calico printing."[15]

The imitation of printed patterns from the East, with the rise of new mechanical, aesthetic, and chemical skills, placed natural dyestuffs at the point of conjuncture between the old tacit knowledge of the workshops and *manufactures* on the one hand, and the mechanical culture of the "factory system" on the other.[16] Materials for dyeing were now complemented by new materials for printing. After singeing, bleaching, and calendering, the wood block received the color from a sieve and was then placed on the calico and struck with a wooden mallet in order to transfer the color. Copperplate printing involved coating the plate with the color, then removing the color from the surface, leaving it in troughs in the plate from which it was transferred to the cloth. Copper plates were used to print large surfaces, something which could not be done with the wooden blocks because they did not produce a continuous pattern. Cylinder printing was introduced progressively, begin-

14. Parkes 1815, 2:63–182 ("On calico printing," 109–31).

15. Muspratt 1853–1861, 2:680.

16. " . . . with the onset of the industrial revolution and the advent of mechanical spinning and weaving devices, an additional stimulus was given to the need for suitable chemical preparations for printing, dyeing and finishing of the vast quantity of cloth made available by the new machines" (Melvin Kranzberg, introduction to Ron 1991, 8).

ning in the final decades of the eighteenth century. It is worth reproducing Sheridan Muspratt's details of the block-printing process:

> The block printing is performed in an oblong apartment, termed the printing shop, lighted with numerous windows at each side, and having a solid table opposite to each window. . . . The table is formed by a strong plank of hardwood, mahogany, or marble. . . . At one end of the table the legs carry brackets for supporting the journals of the roller, from which the calico to be printed is unwound as the work proceeds, and passing along the table is carried successively over the hanging rollers. . . . The workman charges the block with colour by pressing it upon a surface of woollen cloth stretched tightly over a wooden drum, which is called the sieve . . . the sieve is covered with the colouring matter by a child, called the tearer . . . who takes up with the brush a small quantity of colour from the pot, and spreads it uniformly over the surface of the sieve. . . . The printer unfolds a length of the piece upon this table. . . . He presents the block towards the end, to determine the width of the impression, and this width he marks along the piece on the table by drawing a line with his tracing point. The spreader or tearer now besmears the sieve with the colour; the printer seizes the block with his right hand, and daubs it twice in different directions upon the sieve cloth; then he transfers it to the calico within the line marked, and generally strikes it on the back with a wooden mallet, in order to transfer the impression fully. Having done so, he again charges the block from the sieve; and to enable him to make the next impression exactly join or fit in with the last, the block is furnished with small pins at the corners, two of which he inserts in the marks that were made by the other two during the previous impression. Thus, by repeated applications of the block, a pattern is produced in one colour. . . . If the pattern contains five or more colours; there must in general be as many blocks as there are colours, all of equal size, the raised portions of one, which take out colour, corresponding with the depressed portions in the others, which do not.[17]

The evolution of the art of dyeing and printing through the eighteenth century was strongly influenced by large-scale production, mechanization, and the growing division of labor and expertise. It was a complex process, which went much further than a simple application of academic knowledge into the workshop. The new sites of production acted as an ideal environment for the fruitful interaction of a great variety of skills.[18]

17. Muspratt 1853–1861, 2:683–84.

18. As the Catalan calico printer Carles Ardit stated in an early nineteenth-century treatise: "If large scale operations are to be considered as much more advantageous, due to the fact that they can be subdivided and every expert can devote himself to a single object, thus acquiring

## "Good" and "Bad" Dyestuffs: Quality Control, Tests, and Other Experiments

The "art" of natural dyestuffs in the eighteenth century is probably closer to the Baconian tradition of experimental natural history than to the mainstream account of the physical sciences as experimental natural philosophy. As Klein has recently pointed out, "experimental natural history was a socially acknowledged style of experimentation in the eighteenth century academic institutions that evolved around mundane, most useful objects of inquiry. . . . Whereas in natural history facts relied on the observations of things 'given by nature,' experimental history studied all kinds of human arts and crafts and their outcome."[19] In fact, much historical evidence shows how the formation and development of the art of dyeing depended on strategic appropriations of academic knowledge, which had a great influence on the making of the commodity under investigation. The work done by the French inspectors of dyeing and the central role of the Manufacture royale des Gobelins provide an excellent starting-point for a detailed analysis of practices that were associated with the attempts to control the multiple processes of making and applying natural colors (see figures 12.1 a, b, and c).

In 1669 Colbert's *Réglement des teintures* divided natural colors into *bon teint* and *petit teint. Bon teint* dyes, such as indigo and cochineal, and the dyeing procedures associated with them, were characterized by high quality and price, and resistance to sunlight, atmospheric air, and aging. *Petit teint* dyes were the fugitive colors, of lower quality and poor solidity, for instance, Brazil wood, archil, and curcuma.[20] Samples were colored with all the available dyestuffs, and after twelve days of exposure, the quality of each swatch was checked in relation to a standard. Though the *petit teint* colors might be considered uninteresting because of their low quality, they were easy for the workers to handle, they were cheaper, and they often provided very light

---

skill, aptitude and perfection, and saving time, this principle should be applied without doubt to dyestuffs. The colourist should define a plan of operations which are closely related to each other, in order to optimise the use of the raw materials: time, combustibles and labour" (Ardit 1819, 1:87–88; cited in Nieto-Galan 2001, 43).

19. Klein 2005c, 233.

20. Hellot 1750. Along the same lines, Susan Fairlie (1964–1965, 490) divided reds, blues, and yellows into two categories: fast blues (woad and indigo), fugitive blues (logwood in alkaline solution); fast reds (cochineal, madder, kermes), fugitive reds (orchil, Brazilwood, safflower, logwood in acid solution); fast yellows (weld, fustic, quercitron), fugitive yellows (yellow berries and turmeric).

FIGURE 12.1. The art of dyeing at the Manufacture royale des Gobelins. (*a*) The dyeing workshop at the Manufacture royale des Gobelins (*Teinture des Gobelins*, planche 1). (*b*) Details of the dyeing process at the Gobelins (*Teinture des Gobelins*, planche 8). (*c*) A workshop for dyeing silk (*Teinturier de Rivière*, planche 1). (Diderot and D'Alembert 1772, vol. 10.)

shades, whereas *bon teint* colors were closer to the tradition of luxury goods of the French *manufactures*.[21]

However, in the early decades of the eighteenth century, Colbert's regulations became too rigid, and protests began to be heard. Dyers were forced to specialize in a single color, fiber, or procedure, and only a few retained a comprehensive view of the art. The rigid regulations threatened the mercantilist policies of the monarchy and in fact made it difficult to control potential fraud by foreign suppliers of dyes.[22] In 1737 the Académie des Sciences asked Charles-François Du Fay (1698–1739) to review Colbert's old regulations.[23]

21. "Commonly one accepts as good dyestuffs all those which, when exposed to the sun and dew for twelve entire days, do not change at all, or which take on a slightly darker color without losing their principal shade. Any color which, on the contrary, fades and alters during the same period, is considered a poor dye. But since this test, which is the only truly valid one, and which ought to be sufficient on its own, cannot be implemented in cases where one needs to judge on the spot whether a material on sale at a fair or elsewhere is a good dye, if its price demands this, it has been necessary to find ways of making it lose all the colour it would lose when left in the sun for twelve to fifteen days in a few minutes" (Hellot 1741, 38).

22. Hellot 1740, 126–27.

23. "However, experience having shown that these methods were insufficient for certain colours, that M. du Fay has worked for many years in order to find some which were either

In order to speed up the process of quality control, especially when dyed and printed cloth had to be sold in the markets quickly, dyed samples were now treated with different chemicals to classify them in various degrees of solidity, and the experimental conditions of each test were extremely detailed compared to Colbert's procedures, always emphasizing the changes that had been introduced.[24] This was the origin of the *débouillis*, a quality test for dyestuffs, which played a fundamental role in the rationalization of the art at that time.

Perceived as seekers of immediate profit, artisans were often reluctant to change their routine procedures in the vats, and some manufacturers strongly opposed Du Fay's new regulations of 1737.[25] This was the starting point of a complex process of negotiation.[26] Du Fay spent long periods in the *manufactures* trying to understand and describe all the details of the recipes of dyeing. This was considered a necessary step on the way toward an agreement

---

more general or more reliable; and it is on the basis of a great many experiments which he has conducted with great care, that the new instructions for these sorts of tests, to which the name of *débouillis* has been given, were written. The instructions are printed after the *Réglement* of 1737, on Dyestuffs. . . . Some of these *débouillis* are carried out with soap, others with alum, others with red tartar, and others with a mixture of red tartar and alum. But . . . it is clear that these rules, though reputed to be excessively general, are also too narrow in several cases where pale colours would require salts with less activity than saturated colours, which can take up a considerable quantity of the particles that dye them, without very remarkable changes becoming evident. It would therefore have been necessary to prescribe a *débouilli* for almost every shade, something which was impossible, given their variety. So air and sunlight are the true test, and any colour which resists them for a certain period, or which acquires what the dyers call *depth*, should be regarded as a good dye" (Hellot 1740, 40).

24. "It is known that the Art of Dyeing is in the hands of different classes of workers, and that in the principal Cities of the Kingdom there are dyers who are only allowed to dye in *petit teint*. . . . These precautions were judged necessary by M. Colbert, to whom the State owes the establishment of its principal *Manufactures*. The *Réglement* bearing his name, published in 1669, was observed for a considerable time, and throughout this period, trade in our Cloth in other countries lost none of its profitability. But finally so many abuses were gradually introduced, either in production or in the dyes of the cloth which was produced, that our neighbors seized. this favourable opportunity to establish a profitable trade in Italy, in the Levant, and elsewhere. . . . It was necessary to order that the old *Réglements* be rigorously executed, and to remedy the ills which had not been foreseen with new *Ordonnances*" (Hellot 1740, 126).

25. "For the artisans . . . anything that does not tend towards their own particular utility is seen as a waste of time" (Du Fay 1737a, 253).

26. "The artisan, who ordinarily has nothing but his hands and his tacit knowledge, was not suitable for the Ministry's views: a natural philosopher who could operate and reflect was required, it was thought that he could only be found within this Company, whose object is the perfection of the arts, as well as discoveries in the sciences, and M. Du Fay was chosen" (Hellot 1740, 127).

between tacit knowledge in workshops and the new official regulations.[27] Du Fay's death in 1739 interrupted the project, and the comprehensive treatise on the art of dyeing, which was part of the original plan of the Académie, remained unpublished. In 1740 Jean Hellot (1685–1766) was appointed to update the regulations of 1737. From his privileged position at the Académie, Hellot supervised samples of dyed fabrics sent by numerous French dyers and reviewed and completed Du Fay's *débouillis*.[28] He described his method in detail in his *L'art de la teinture des laines* (1750), a valuable source for the historical reconstruction of these practices. His "Instruction sur les débouillis" responded to the need to improve dyeing regulations, but criticized contemporary books on dyeing, such as *Le teinturier parfait*, which he regarded as a list of inexact descriptions of the art.[29] Following Hellot's guidelines, swatches were exposed to sunlight and atmospheric air to classify them in terms of their solidity, and they were divided into three different categories. The first had to be dipped in a boiling solution of alum, the second in white soap solution, and the third in red tartar, and the time of the test and the quantity of reagent to be used were recorded. Hellot also provided a general description of the behavior of each natural dye with each *débouilli*.[30]

In the entry on *teinture* (dyeing) in Diderot and D'Alembert's *Encyclopédie*, Hellot's *débouillis* were considered insufficient.[31] Recipes were hard to

27. "Dyeing is one of those arts which can provide the most singular experiments for those who wish to study it as Natural Philosophers. Up to now, it has been the sole province of the artisans who mastered it, and it even appears that there was not one among them who knew all its aspects. . . . Dyeing has been separated into different classes and those who are skilled in one part are banned from knowing the work of others" (Du Fay 1737a, 253).

28. In 1749, for example, the Abbé Menon, the secretary of the Angers academy and a corresponding member of the Parisian Académie Royale des Sciences, sent Hellot a sample of Prussian Blue. Turkey Red samples were sent in by a handkerchief workshop in Montpellier, and Saxon green ("Vert de Saxe"), imitated by Koederer in Strasbourg, was also covered in Hellot's study (A.N., Commerce et Industrie, F/12/2259).

29. De Lormois 1800. As Hellot put it: "Those who have no idea of this subject may believe that they will find some enlightenment in the books which have covered it, but it is only too certain that nothing is to be learnt from these. *Le teinturier parfait*, which went through several editions, . . . is nothing more than a monstrous collection of receipts, imperfect, wrong, or described in an unintelligible fashion" (Hellot 1750, viii).

30. "After seeing the effect of air on each good or bad colour, he tested different kinds of *débouillis* on the same cloth, stopping when he found one which produced the same effect on this colour as air: and, recording the weights of the drugs, the quantity of water, and the duration of the test" (Hellot 1750, 34).

31. "As it has been acknowledged that the old method prescribed for the *débouilli* of dyes is insufficient for judging the goodness or badness of several colours accurately; that this method might sometimes cause mistakes, and give rise to disputes, several experiments were performed

reproduce, local water conditions affected the preparation of dyeing procedures, and the transportation of plants, seeds, and swatches also made any standardization difficult. In spite of the rational image of dyestuffs and operations in the *Encyclopédie,* including an impressive collection of engravings that reproduced every step of the dyeing process in the workshop, the article "Teinture" described the difficulties in finding common criteria for a general understanding of the art.[32] After the presentation of the dyes ("Description de leur origine, culture, nature, qualité, espèce . . . "), a sort of dictionary of colors, textile fibers, dyeing objects, and practices was added. The *Recueil des Planches* contained fourteen reproductions of dyeing practices at the Gobelins, showing details of the dyeing workshop, water pumps, heating vats, dyeing vats, and drying areas. The entry described the technological itinerary of the art of dyeing in detail, as well as giving a precise description of all the colors and their shades.[33]

Detailed descriptions of Hellot's *débouillis* acquired a status independent of more academic discussions on the nature of the union between the dye and the fiber.[34] Hellot advocated a mechanical explanation in which the particles of the color were introduced into the holes or pores of the fiber, but the quality test of his *débouillis* was probably closer to the workshop culture, as Homassel would acknowledge some years later.[35] In fact, Du Fay, the "father" of the *débouillis,* had often resisted any theoretical speculation about the na-

by order of His Majesty on the wools destined for the production of tapestries in order to determine the degree of goodness of each colour, and the *débouillis* best suited to each" ("Teinture," in Diderot and D'Alembert 1751–1780, 16:27).

32. Gillispie 1959.

33. "One should add in finishing this article on dyeing that every day people are to be found who possess some secret in this art, now so widespread and delicate. . . . The Dutch produce violets on silk for which we can only produce imitations . . . for silks, black shades from Genoa and others from Italy are more beautiful than those from France . . . since their vats depended on the towns where the dyeing is done, they did not undergo deterioration, and were better maintained and run than if they had belonged to private citizens. Water, in addition, makes not a little contribution to the perfection of this art; dyestuffs transported by sea may diminish in quality, or not produce the same effect in a different climate. . . . It is up to the natural philosophers to instruct us about all these purported phenomena: no-one has yet undertaken to address this subject in France: perhaps someone might be found to be skilful enough to give an explanation for them and by that means put our dyers on the same level as these foreigners" (Diderot and D'Alembert 1751–1780, 16:31).

34. Bensaude-Vincent and Nieto-Galan 1999. See also Beer 1960.

35. "All the invisible mechanism of dyeing consists in dilating the pores of the material to be dyed, in depositing the particles of a foreign matter there, and in holding them in place by a sort of coating which neither water nor rain nor sunlight are able to alter; in choosing coloring particles of such tenuity that they can be retained, sufficiently bedded down in the pores of the

ture of the union between the dye and the textile fiber, preferring to focus his work on more empirical terms.[36]

Experimental natural history was not incompatible with a growing interest in philosophical knowledge about the union between dyes and fibers. Controversy arose between the inspectors of dyeing and academic experts, who often tried to justify their philosophical position with the "materiality" of some experiments with dyes. However, mechanical explanations never totally displaced chemical explanations. Pierre-Joseph Macquer (1718–1784) adopted a position intermediate between the two in the first edition of his *Dictionnaire de chimie* (1766) and an eclectic position in the second edition (1778).[37] In his *Éléments de l'art de la teinture* (1791), Claude-Louis Berthollet (1748–1822) presented the chemical affinity between the fiber, the dye, and the solvent as the key explanation for the different degrees of fixation of dyes on fibers.[38]

In a well-known book on the art of dyeing silk, published in 1763, Macquer reported that a dyestuff sample treated with hot water yielded, in an "imperfect" process of separation, a soluble part (*matière extractive*) composed of mucilage, gums, salts, and oily matter, while the insoluble part contained other oily parts, resinous substances, and earthy materials.[39] In Macquer's view, these "imperfect" separations provided useful information for identifying a particular dyestuff and for estimating its purity and any possible adulteration and became a useful practice of quality control. He also attempted to divide natural dyestuffs into three classes based on their chemical composition: (1) those containing extractive, resinous earthy matter, (2) those without earthy matter, and (3) those soluble in water by fermentation.[40] Although Macquer expressed frequent doubts about the validity of Hellot's mechanical explanations of dyeing and tended to introduce chemical terms, he acted as a referee in prize competitions at the Académie and played an active part in the École de teinture at the Gobelins, a pole of attraction for young men from families of dyers or calico printers in France and abroad. He was well aware of the difficulties of "translating" practical, tacit knowledge of dyeing proce-

subject, opened up by the heat of boiling water, then constricted by the cold . . . so that the colouring atom is retained there rather like a diamond in the setting of a ring" (Hellot 1750, 42).

36. Du Fay 1737a, 254.

37. Macquer 1766, 1:546–47; 1778, 4:26–27.

38. "Berthollet's *Elements* of this art must undoubtedly be regarded as constituting the first truly philosophical treatise on the subject" (Berthollet 1824, xxiv).

39. Macquer 1763, iv.

40. This classification was later borrowed by the chemist Thomas Henry in Manchester (Henry 1790, 364–66).

dures into academic works.[41] Macquer would ask his pupils at the Gobelins to design a "Plan" to organize and classify dyeing knowledge and in this way bring some order to the world of chemicals, colorants, gums, and soaps. In such teaching practices, learned knowledge was linked to the history of dyeing from the ancient times, to the Newtonian explanation of the primitive colors, as well as to the description of natural dyestuffs within the framework of the three kingdoms of nature, with only a few comments on new chemical tests (*débouillis*).

As we find some years later in Berthollet's *Eléments* (1791) (see figure 12.2), alongside this general framework, the teaching of dyeing is said to involve long, detailed descriptions—the recipe tradition—of making different colors and applying them to wool, silk, cotton, and linen. In spite of the strong influence of the recipe tradition, Macquer was unwilling to produce simple collections of practical procedures; he rejected the secrecy and the underground dealings of the anonymous *compilateurs,* who, in his view, made profits selling adulterated formulas.

Because they played a fundamental role as the substrates of natural colorants, textile fibers—silk, wool, cotton, and linen—also provided useful information concerning the process of dyeing. Thus, dyes were often classified in relation to their solidity on the different fibers, a practice that was closely related to the classification criteria of *grand teint—petit teint.*[42] It is no coincidence that Hellot's and Macquer's books were devoted to the art of dyeing wool and dyeing silk respectively. Le Pileur d'Apligny's *L'art de la teinture des fils et étoffes de coton* appeared later, in 1776, but the focus on a single fiber was common throughout the eighteenth century. In fact, Berthollet's *Éléments* was an ambitious attempt to overcome these divisions, with the specific aim of extrapolating chemical affinity to the problem of the color-fiber affinity in solution in dyeing vats. But, as stated above, he was obliged to write a second volume of the *Éléments* devoted exclusively to detailed descriptions of dyeing procedures.

In 1794 the English chemist Edward Bancroft (1744–1821) published his *Experimental Researches Concerning the Philosophy of Permanent Colours,* a

41. "Most dyers have no knowledge of chemistry . . . and are not in the habit of reading. . . . A book on dyeing that brought together theory and practice would be hardly of any use to them. . . . In general, one must give artists very simple rules" ("Plan d'un Traité de l'Art de la Teinture," sent to Macquer by a student of the "Manufacture des Gobelins," [ca. 1760], A.N., Commerce et Industrie, F/12/2259, Paris).

42. Fairlie 1964–1965, 491. In the mid-eighteenth century, for instance, after distilling wool, silk, and cotton, the German chemist Gaspar Neumann obtained characteristically different fractions for each fiber (Neumann 1759, 428–30).

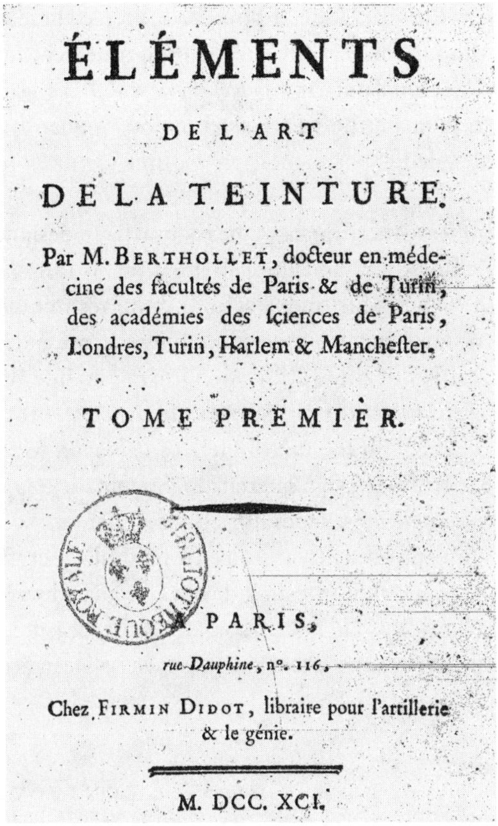

ÉLÉMENTS
DE L'ART
DE LA TEINTURE.
Par M. BERTHOLLET, docteur en médecine des facultés de Paris & de Turin, des académies des ſciences de Paris, Londres, Turin, Harlem & Mancheſter.
TOME PREMIER.

A PARIS,
rue Dauphine, n°. 116,
Chez FIRMIN DIDOT, libraire pour l'artillerie & le génie.
M. DCC. XCI.

FIGURE 12.2. Berthollet's *Eléments de l'art de la teinture* (1791a).

treatise on dyeing with goals similar to Berthollet's *Éléments.*[43] In 1775 Bancroft had introduced quercitron to England, a new yellow dyestuff from a black oak, over which he had a monopoly until 1799. Curiously, in the book, Bancroft described frequently repeated experiments testing the quality of quercitron. A cotton cloth was divided into four equal pieces impregnated with four different chemicals. The same amount of quercitron solution was applied to (1) a piece with iron liquor and gallnuts, (2) a piece with iron liquor alone, (3) a piece of cloth with iron liquor and acetite of alumina, and (4) a piece with *acetite* of alumina alone. Because a single natural dyestuff can produce several colors, depending on the chemicals used in the process, the four samples yielded different results. The same criteria were applied to

43. Bancroft 1794.

test the solidity of Prussian Blue, a popular source of blue and an alternative to indigo. Although Bancroft used (without much enthusiasm) the new French chemical nomenclature in his *Experimental Researches,* the book was less theory-laden than Berthollet's *Éléments,* and frequently performed experiments and detailed experimental procedures can be found in many of the chapters.

To what extent were these quality control practices, designed by academic chemists, accepted by dyers and implemented in workshops and *manufactures*? This is still a debatable question, and more testimonies from the artisanal side are needed, but from what we know at present, it appears that leading representatives of the workshop culture—mainly the expert dyers, who should be distinguished from simple *ouvriers* or workers—seemed to feel quite comfortable with the use of Hellot's *déboullis,* which were not far removed from Bancroft's tests on quercitron. Systematic quality control tests in workshops, *manufactures,* laboratories, and markets were intrinsically linked to the eighteenth-century culture of natural dyestuffs, in the same way that areometers and liquor testing devices circulated widely in the making of wines and brandies. Therefore, some of these "learned practices" survived in the subtle but crucial intermediate space between academia and the workshop.

## THE BERTHOLLET-HOMASSEL CONTROVERSY

The development of such intermediate spaces for dyeing practice and knowledge did not prevent misunderstandings and controversies between different constituencies. Influenced by Antoine-Laurent Lavoisier's *Traité élémentaire de chimie* (1789), which had appeared just two years before, Berthollet's *Éléments* marked the beginning of a new genre in the treatises on the art of dyeing.[44] In his introduction, Berthollet proclaimed that the new chemistry would play a crucial role in "enlightening" the art. However, after a series of chapters devoted to chemical affinity, oxygen, and combustion, the second volume of the *Éléments* focused entirely on detailed descriptions of dyeing procedures. Although Berthollet was determined to use the new chemistry in his description of the art of dyeing, he saw the tacit knowledge of the craftspeople in the *manufactures* as a vital source of information for his work. While writing the book, he made many attempts to visit dyeing workshops and to learn and to describe different procedures side by side with the *ouvriers*—as the majority of eighteenth-century French texts called the "real

44. Lavoisier 1789; Berthollet 1791a, 1804; Keyser 1990.

practitioners" of the art. However, his attempts were not always successful; in fact, he often encountered resistance.[45] After a disappointing visit to Lyons, he sent out a questionnaire to dyers and manufacturers all over France, but as far as we know only a small group responded. Although there are records of his visits to leading calico-printing factories, he was never fully accepted into workshop culture.[46]

In 1784 Berthollet was appointed general inspector of dyeing. He followed in the footsteps of Du Fay, Hellot, and Macquer, who had all contributed, with varying degrees of success, to the introduction of academic knowledge into the artisan practices of the art of dyeing. Berthollet was asked to write a general "Traité théorique et pratique de teinture," a text in which the new chemistry would be the basis for a new way of understanding and diffusing dyeing expertise. In the period 1784–1791, Berthollet was especially keen to acquire the necessary knowledge and experience to publish this new text. To do so he needed to establish close contacts with dyers and printers all over France. This turned out to be a difficult enterprise, and his experiences show us how little academic and workshop cultures intersected, even in the late eighteenth century.

Some of the strongest resistance to the intrusion of the academic chemists came from Homassel, the *chef d'atelier* ("workshop head") at the Manufacture royale des Gobelins in Paris between 1778 and 1787.[47] In the preface to his *Cours théorique et pratique sur l'art de la teinture* (1798), Homassel published details of the controversy between the academic chemistry issuing from Lavoisier's circle and the workshop culture of expert dyers. For Homassel, the desire of academic chemists to visit *manufactures* and to standardize dyeing procedures in the Gobelins was merely a vehicle for their personal promotion; in his view, they had no sincere interest in the culture of dyeing in the workshop.[48] Homassel must have seen Berthollet as an author of aca-

45. "I wish I were able to follow all dyeing procedures in the workshops, but the secretiveness of the dyers prevented me from profiting from the observation I tried to make. . . . A journey to Lyons was practically useless" (A.N., Commerce et industrie, F/12/1329).

46. Nieto-Galan 2000, 171.

47. Homassel 1798; 1807.

48. Homassel 1798: "At that time, there was only a general complaint about the poor practices used in dyeing, in disregard of Colbert's regulations; and although the inspection of this *manufacture* had long been entrusted to the most famous chemists of Paris, who used up the state's finances more in order to satisfy their cupidity and their chemical ambition by multiplying their experiments without any goal, rather than in improving or correcting the procedures used in dyeing" (vii). "The revolution, which has fomented intrigue, has placed the dregs of Parnassus in the most eminent of positions, that is to say, those greedy, immoral second-rate experts who, mediocre and snivelling, have . . . burnt their odourless incense at the feet of the

demic texts supposedly based on workshop experience, but inspired by the new chemistry that he considered of no value to dyers.[49] Though very critical of the new chemical nomenclature of Lavoisier and his circle, Homassel's contempt for the new chemistry did not prevent him from paying tribute to the old seventeenth-century Colbertian tradition of classifying dyes and dyeing procedures—*grand* and *petit teint*—and to the later amendments to the system made by Du Fay and Hellot.[50] In fact, in spite of Homassel's criticisms of the attempts by Parisian supporters of the new chemistry to control dyeing procedures in the workshops, he freely expressed his indebtedness to Colbert, Du Fay, and Hellot.[51]

Homassel's account is one of the rare printed works to present the history of eighteenth-century dyeing from the expert dyers' perspective. It reveals how they shared a different chemical culture far from the representatives of academia.[52] Some years ago, Frederic L. Holmes used the phrase "cultures of chemistry" to describe the sites and values shared by groups such as students of medicine and surgery, apothecaries, glass and pottery craftsmen, drysalters, dyers, distillers, and members of the scientific academies for the promotion of agriculture and industry. They all shared some chemical knowledge, but their tacit practices and values were not homogeneous, and each "culture," such as dyeing, preserved its own identity. In Holmes's terms: "Those who studied, practised, or used chemistry can be divided into categories such as those who taught the subject in medical schools, those who taught in other contexts,

---

Vosges Apollo" (xii). "Let us leave the experts of this calibre to grow rich by leading the artists into error, by advising them prematurely to introduce changes in their procedures. . . . The career of the arts should receive consideration proportionate to its utility and the knowledge it requires. But when will this fine day dawn in France? When will the reign of the schemers and rogues come to an end?" (xvi).

49. Keyser 1990.

50. "I have not used the new chemical nomenclature in this work, firstly because at sixty one does not easily renounce the language of one's fathers, secondly because I was always more eager to learn things than words: thirdly, because supporting a large family obliges me to prefer the useful to the agreeable" (Homassel 1798, xv; see also Homassel 1798, vii, x).

51. "Some scheming chemists, after twisting the words of the famous Hellot to appropriate his ideas for themselves, incessantly appoint and dismiss their lowly and inept protégés at the Gobelins *manufacture*" (Homassel 1798, x).

52. Nieto-Galan 2000. In 1786, for instance, a provincial French dyer expressed his concern to the experts at the Manufacture royale des Gobelins in the following terms: "It is well known that one salt dissolves or fixes, another salt transmits a greater or lesser degree of intensity or brightness, while another . . . would destroy the particles . . . but the use of all these drugs varies enormously due to the different modifications to which they are susceptible, due to the immense quantity of hands that use them, due to the lack of study and often the misconceptions of the manufacturer" (A.N., Commerce et Industrie, F/12/1330).

those who learned it as medical students or as future bureaucrats, curators of museums who provided public demonstrations, those who applied it to agricultural or industrial problems, or apothecaries for whom chemistry was an adjunct to the preparation of drugs."[53]

## CHLORINE BLEACHING

On the market, clean, white surfaces of textiles were important to attract new customers and stimulate new sales. For this purpose, in the early eighteenth century, raw cloths were steeped or bucked in alkaline lyes, ashed, and treated with some diluted acids—sulfuric, hydrochloric, or lactic—before undergoing prolonged exposure to air and light in large, open fields next to workshops.[54] However, after pneumatic chemistry gave rise to the discovery and industrial application of a new air, these long-drawn-out processes gradually came to an end. Late in the century, another new material, chlorine, was discovered through academic experimentation and soon transferred to artisan practice. In 1785 Berthollet made public his discovery of the bleaching power of "oxygenated muriatic acid" (chlorine). Bleaching was particularly crucial for the calico-printing industry, in which multicolored patterns required white surfaces. The new gas provided high-quality bleached cloth far more quickly than the old methods of lye treatment and prolonged exposure to air and light. Berthollet presented the results before the Académie des Sciences and in a first paper in the *Journal de Physique* in 1785, and the procedure was later explained in 1788 in the *Annales de Chimie* and in the second edition of Berthollet's *Éléments* (1804). In 1788 the manufactory of Javel, near Paris, developed a new commercial product called *Eau de Javel* by treating chlorine with an alkaline solution. French dyers and calico printers tested the new bleaching liquor in their workshops, and Berthollet himself ran large-scale tests.[55] That same year, the English chemist Thomas Henry (1734–1816) remarked, "The new mode of bleaching, by means of the dephlogisticated marine acid, . . . promises to be of great utility . . . not only by shortening the time required for the process, which has been generally extended from one to two months, may now be reduced to a few hours, but

53. Holmes 1995, 276.

54. Parkes 1815; Clow and Clow 1952; Musson and Robinson 1969, 251–52; Gittins 1979. A typical bleaching procedure consisted in steeping the cloth in water with bran or a lye, then heating and bucking it in an alkaline solution, later exposing it to sunlight and atmospheric air, washing, souring in buttermilk, and scrubbing with soap (see Nieto-Galan 2001, 65–71).

55. Raymond-Latour 1836, 2:33.

by sending up the goods in a state much better adapted to the subsequent processes."[56]

In 1789 the industrial chemist Charles Tennant (1768–1838) patented a new material at Saint Rollox Works at Glasgow. It was a more manageable solid bleaching lye, a new powder of chlorine, absorbed in slaked lime.[57] This new solid lye had important advantages for the transport, price, and stability of the product and avoided the danger to workers' health posed by aqueous solutions of chlorine. Tennant's solid lye produced a reduction in labor, time, and waste, a greater certainty and security in executing the process, and a comparatively small interest on investment capital.[58]

However, large-scale chorine bleaching did not obviate the need for the cloth to undergo lengthy treatment processes. In preparation for dyeing, fabrics often required a long bleaching process, which included stamping, singeing, washing, bucking with lime, souring, washing, immersing in bleaching liquor, souring, steeping, bucking with caustic soda, souring, washing, mangling, starching, and calendering.[59] In fact, old and new bleaching techniques tended to be presented as a single package in books devoted to the art of dyeing and printing. Chlorine was a strong oxidizing agent that jeopardized the stability of the textile fibers, often damaging their structure.[60] Calico printing required a mild bleaching agent that could produce an evenly white cloth, and dyers required a balanced combination of rapid bleaching and fiber protection.[61] In fact, bleaching was a bottleneck in the process of coloring textiles.[62]

The bleaching process also provided new criteria for the quality control of dyes. This was the case, for instance, for the Rouen chemist François-

56. Henry 1790, 361. In 1788 Charles Taylor, the secretary of the Royal Society of Arts, learned of the power of the new gas by reading reports of Scheele's early experiments. Under the supervision of Richard Kirwan, Taylor seems to have quickly set up a large-scale bleaching process at Rikes, near Bolton (Rees 1819, 1:177–90).

57. "Tennant uses 5 1/2 parts of black oxide of manganese, 7 1/2 parts of common salt, and 12 1/4 parts of sulphuric acid . . . diluted with equal quantity of water, to make the chlorine gas, with which he impregnates a layer of slacked lime, some inches thick in a stone chamber" (Baines 1966, 249).

58. Parkes 1815, 4:87.

59. Muspratt 1853–1861, 1:303.

60. "On visiting several manufacturing towns, it was not without surprise and pain that I found . . . processes of bleaching and dyeing whose use was dangerous for the lifespan of the cloth, even allowing that it was entrusted to skilled hands" (Gréau 1834, 2).

61. Parkes 1815, 4:111, 127.

62. A substantial part of this story from the eighteenth century to the early years of the nineteenth century has been covered by J. Smith 1979, 113–91. For an interesting approach to the history of bleaching, see O'Neill 1877, 4: 3–13, 5: 63–74, 6: 223–36.

Antoine-Henri Descroizilles (1751–1825), who developed a volumetric analytical test for dyestuffs. He used a standard dye solution (0.1 per cent indigo) to be decolorized by a bleaching liquor in an instrument called the "Bertholimeter." Apart from dyestuffs, a chlorine solution could remove resinous, fatty, saline, earthy, or ferruginous matters from the fibers, but the "real" chemical composition of bleaching powder and its decomposition in solution were controversial issues. Again, practical tests from the everyday practice of the art took priority over "learned" discussions.[63] Chlorine circulated from small flasks in the laboratory to industrial settings. It returned to the laboratory as a reagent for quality control of dyed samples. In this journey, it acted as a dangerous oxidizing gas, a liquid lye, and a solid powder. It was a raw material for large-scale bleaching, but also a specific product for chemical analysis. No other material attested so well to the complexities of the art of dyeing.

## Organizing the Workshop: "Natural Histories" of the Art of Dyeing

Historiographical discussions have clearly moved on from the old debate of the 1960s and 1970s on the role of science in the Industrial Revolution. Nevertheless, some ideas from this old analytical framework may still be useful in the case of dyes.[64] I would like to refer especially to Gillispie's canonical paper "The Natural History of Industry."[65] In his view, though it is hard to establish a clear connection between scientific knowledge and technological changes during the eighteenth century, taxonomical traditions for classifying, naming, and ordering natural species, objects, chemicals, or procedures were intrinsically linked to several industrial enterprises.

Although most protagonists of the art of dyeing agreed on the need to organize all the materials of everyday work in the workshop, there was much controversy over the best way of doing so. Many valid criteria could be applied.[66] Materials could be classified according to the three kingdoms of nature or to the colors obtained at the end of the dyeing processes. Coloring matters were often classified in terms of their composition and solidity.[67] Procedures for the application of dyestuffs to different surfaces could also be taken into account, or perhaps the chemical substances used, the colored textile fibers, and so on. In spite of the growing influence of chemical knowledge

63. J. Smith 1979.

64. A book stemming from that debate is Musson and Robinson 1969.

65. Gillispie 1957.

66. See Klein 2003b.

67. See Nieto-Galan 2001, chap. 1.

at the end of the eighteenth century, practical results in workshops seemed to reflect experimentation by trial and error, more than learned knowledge of affinity or the new nomenclature. In attempting to reconcile old artisanal practices with the emerging chemical influences, contemporary treatises on dyeing and printing with natural dyestuffs sought eclectic combinations of chemical nomenclature, botanical names, and taxonomies of the procedures of dyeing and printing.

Colors were also classified according to other criteria: following the practical use that dyers made of them, considering all the combinations of simple colors that created a definite sample, or even establishing numerical scales.[68] Macquer defined three categories of natural dyestuffs: (1) those that do not require prior preparation for dyeing, (2) those that require preparation before their application to the textile fiber, (3) those that do not themselves require preparation, but need a mordant to be applied to the cloth before the dyeing process.[69] Bancroft distinguished between "substantive" and "adjective" colors. The former became "permanent colors," fixed on the cloth without the use of a mordant (indigo, quercitron, purple, metallic salts); the latter needed intermediates to attain color solidity (madder, cochineal, Prussian Blue, American woods). Dyestuffs were classified here in terms of the logic of their making and applying, rather than according to their chemical composition or their botanical description.[70] If we compare Berthollet's *Éléments* (1791) and Bancroft's *Experimental Researches* (1794) (see table 12.1), in both cases, the question of who was the real expert was at the core of the intermediate space between academic chemists, on the one hand, and dyers, chemical manufacturers, and *compilateurs* on the other.[71] New chemical criteria were introduced to organize knowledge and objects in mineralogy, crystallography, medicine, geology, and botany, but their use in the art of dyeing and printing textiles was ambiguous. The textbooks on dyeing formed a specific and controversial genre, and their claims were not easily extrapolated to other books on theoretical and practical chemistry. The complex interplay between different substances of animal, vegetable, and mineral origin and the multiplicity of variables needing to be taken into account required a very particular balance.

68. Scheffer 1787, 17–19. Numerical scales of color were developed during the nineteenth century by the French chemist Michel-Eugène Chevreul in particular.

69. Macquer 1778, 4:30.

70. Bancroft 1794.

71. Nieto-Galan 2000.

TABLE 12.1. Organizing and classifying: the taxonomy of the art of dyeing in Berthollet's and Bancroft's texts

Claude-Louis Berthollet, *Eléments de l'art de la teinture* (1791)

Volume 1: *Des propriétés générales des substances colorantes*

A. Des opérations de la teinture en générale
   1. Des différences entre la laine, la soie, le coton et le lin, et les operations pour les teindre
   2. Des ateliers et des manupulations de l'art de la teinture
   3. Des combustibles utilisés dans l'atelier
   4. Des moyens per lesquels on constate la bont, d'une couleur

B. Des agents chimiques dont on fait usage en teinture

Volume 2: *Des procédés de l'art de la teinture*

1. Du noir
2. Du bleu
3. Du rouge
4. Du jaune
5. Du fauve
6. Des couleurs composés

Edward Bancroft, *Experimental Researches concerning the philosophy of permanent colours, and the best means of producing them, by dyeing, calico-printing* (1794)

1. Of the Permanent Colours of Natural Bodies
2. Of the Composition and Structure of the Fibres of Wool, Silk, Cotton and Linen
3. Of the Different Kinds of Properties of Colouring Matters employed in Dyeing and Calico Printing
4. Of the Substantive Animal Colours
5. Of Vegetable Substantive Colours
6. Of Mineral Substantive Colours
7. Of Adjective Colours Generally
8. Of Prussian Blue
9. Of Adjective Colours from European and Asiatic Insects
10. Of the Natural History of Cochineal
11. Of the Properties and Uses of Cochineal; with an Account of the new Observations and Experiments calculated to improve the Scarlet Dye
12. Of the Properties and Uses of Quercitron Bark
13. Of the Properties and Uses of Juglans Alba or American Hiccory; of the Weld Plant, Fustic and other Vegetables, affording Yellow Adjective Colouring Matters

Demand for efficient dyeing recipes was often yoked to the claims of academic chemists to be able to rationalize the art of dyeing and rescue it from the dark cave of secrecy, guild traditionalism, and mechanical repetition. It was this pursuit of systematization, which, for example, induced Macquer to train his pupils to organize that practical knowledge in a way that lent itself to reproduction, later including a standard nomenclature and new technical concepts. As a result, dyeing textbooks did not fully satisfy either dyers or chemists. The latter were aware of their inability to explain and organize the complex set of skills involved in dyeing in an adequate, scientific manner, and the former often felt uncomfortable with the language and style of elite academic authors. Bancroft's and Berthollet's works were an obvious consequence of the various cultures of chemistry linked to the art of dyeing and printing. Even *Experimental Researches,* which was closer to the empirical British tradition and whose author did not enjoy Berthollet's advantages of state protection and promotion, was also criticized as too theoretical by dyers and particularly by the authors of simple collections of dyeing recipes.[72]

As for the new chemical nomenclature of Lavoisier and his circle, its application to the art of dyeing was uneven and controversial. Here we should remember Homassel's radical opposition to the introduction of names that were alien to the traditions of the art, but also Bancroft's coolness toward the new language in his book. In the Spanish translation of Berthollet's *Éléments,* published in 1795, the new chemical nomenclature was used throughout as a symbol of modernity. Nevertheless, others adopted more eclectic positions, often using the new formulas for teaching or academic texts but keeping traditional names for informal discussions with dyers in the workshops.[73] In practice, only pigments (metallic oxides), mordants, and some other chemicals could be efficiently named using the new nomenclature in the final years of the eighteenth century. Dyestuffs of vegetable and animal origin belonged to an obscure realm of natural history in which the new chemistry had made few inroads: the plant and animal kingdoms.

On other fronts, the notebooks of dyers and printers often contained huge collections of swatches attached to detailed formulas in which slight changes in the composition of raw materials resulted in changes in color shades. Samples were elegantly arranged to display the best patterns for potential customers. In other cases, dyed and printed swatches of colored cottons appeared next to systematically modified formulas.[74] In the late eighteenth century, Charles O'Brien's *The British Manufacturer's Companion* provided a numbered classification of geometric patterns for new designs in calicoes, corresponding to wooden blocks also numbered to ensure the correct sequence of application in complex printing processes.[75] In this regard, notebooks and books of swatches represented an original experimental practice in their own right, but at the same time they also acted as taxonomical strategies: they are ideal objects for displaying the quantifying spirit of the eighteenth century and the range of different "cultures of chemistry."[76]

72. Nieto-Galan 2000.

73. Nieto-Galan 1995.

74. J. Koch 1984, 33.

75. O'Brien 1795.

76. Frängsmyr, Heilbron, and Rider 1990.

The art of dyeing was not restricted to binomial nomenclatures in botany or chemistry; nor could it be understood through simple descriptions of production methods or final colors obtained. Dyeing taxonomies included complex sets of classifying criteria that made the art particularly original. Thus, as I emphasized in my *Colouring Textiles,* the "language" of the art of dyeing embraced vegetable, animal, and mineral substances, operations, machines, textile fibers, finished cloth for the markets, and a heterogeneous group of skilled experts.[77]

## Circulating Knowledge and Materials of the Art of Dyeing

Because the Linnaean influence and the importance of natural history—which obviously includes dye plants—was highly relevant during the Enlightenment, we should not neglect botany when trying to analyze craft knowledge. Natural dyes mainly came from the vegetable kingdom, and agricultural treatises often covered dye plants.[78] Botanical gardens in a sense became laboratories of plants, including dye plants. They were centers for agronomic studies, which contributed to numerous projects for acclimatizing colonial dye plants to European conditions.[79] Colonial dyestuffs were studied in European gardens, where dye plants were acclimatized. Local dye plants were promoted, and agronomic treatises were consulted concerning their cultivation requirements.[80]

In the 1770s, several expert dyers reported a tension between local dyeing traditions and the colonial extension of the craft. Notwithstanding eighteenth-century mercantilist policies, indigo, cochineal, quercitron, and American woods occupied a crucial place in Western markets. Hellot, Macquer, Le Pileur d'Apligny, and Berthollet—the official experts of the French monarchy—praised overseas colors but warned of the possible dan-

77. See the introduction to Nieto-Galan 2001.

78. Thus, in the second volume of Henri-Louis Duhamel du Monceau's *Éléments d'agriculture* (1763), natural dyestuffs had a significant position. Their cultivation and harvesting, as well as methods of preparing the raw coloring materials, were described in detail. At this time, Louis-Alexandre Dambourney (1722–1795) was working on the extraction of the coloring matter from dye plants. His *Recueil de procédés* was the result of many years of study of indigenous and naturalized flowers, fruits, woods, plants, and roots that provided him with more than nine hundred solid color shades on wool. Equally, in 1785, Pierre-Joseph Buc'hoz published a *Traité de toutes les plantes qui servent à la teinture et à la peinture* (Duhamel du Monceau 1763, vol. 2, *Livre Onzième,* chap. 1, "De la Gaude"; chap. 2, "Du Pastel-Gueldes ou Vouede"; chap. 3, "Du Safran"; chap. 4, "De la Garance, Rubia"; Dambourney 1786, 2–3; Buc'hoz 1785).

79. Brockway 1979; Osborne 1994.

80. Nieto-Galan 2001, 12–13.

gers of dependency on colonial production.[81] The importation of indigo never stopped, and it mounted a serious challenge to woad, the indigenous source of blue dyes. In spite of woad's impurities, the extraction from its leaves was almost as good as indigo; it was known as *l'indigo du pastel,* to emphasize the equivalence of the two coloring materials. Cochineal was one of the finest reds, with madder as a rival, indigenous alternative.[82] Quercitron, another colonial dyestuff, threatened the position of European weld.[83] Acclimatization, colonial plantations, and the circulation of seeds from dye plants deeply transformed local workshop cultures. Local and universal knowledge circulated widely in a perpetual interaction in colonial and European networks. For Macquer, the art of coloring textiles required a fluid communication in each particular locality between experts and simple craftsmen, but also on a much higher geographical level.[84] The *manufactures royales* became useful vehicles for technology transfer, attracting foreign experts.[85] The French general inspectors of dyeing also had numerous personal contacts abroad. Com-

81. "The conquest of America has brought to our attention . . . cochineal and indigo. The love of novelty has allowed them to gain prominence over woad, madder, and kermes, which used to be produced in large quantities and which provide a greater profit than that derived from the indigo grown in our colonies, without mentioning the fact that we are dependent on the Spaniards who possess Mexico, from where cochineal comes, who also sell us Guatimalo indigo, superior in quality to that of our islands, at a high price" (Le Pileur d'Apligny 1770, 50).

82. Cochineal was extracted from the cochineal insect, found in the Spanish colonial territory of New Spain (mainly Mexico); in combination with a tin salt, it provided a spectacular crimson. The possibility of introducing madder as an alternative source of red was exploited extensively in Spain by Canals. He promoted the cultivation of madder in the area of Valladolid (Castile) but did not neglect efforts to improve the quality and availability of colonial cochineal. In Canals's view, the lack of an effective colonial policy allowed foreigners (mainly French and English) to obtain cochineal at a low price from the Spanish American colonies (Canals 1768).

83. In Bancroft's attempt to introduce quercitron in Western Europe, he said that weld was "the only colouring substance from which yellow could be obtained, but never would afford any such colour without the aluminous or some other basis very different of iron." Quercitron was extracted from a black oak in North America, and yielded a solid yellow color when treated with tin salts or alum. Bancroft had a monopoly over its introduction until 1799 and certain privileges as regards its use in France (Bancroft 1794, 207).

84. "No-one has complete knowledge of all the dyeing procedures: wool dyers only know the practices of dyers in silk thread and cotton very vaguely. . . . It is to be wished that the best artists in other branches of dyeing would share their own practices with others: it is the only way one can know the current state and requirements of this important art precisely" (Macquer 1763, 45).

85. For example, in the "peripheral" Spain of the Bourbon monarchy, Jean Rulière became director of the Royal Manufacture in Talavera in 1748 and recruited French artisans; John Berry and Thomas Milne came to Avila from Ireland in 1750 and appointed other Irishmen to work in the woolens mill at San Fernando. After 1789, skilled French craftsmen emigrated to Catalonia

pilations of practical recipes were already in wide circulation, while practical guides to the production of different colors, treatises by mid-eighteenth-century inspectors of dyeing, and treatises on calico printing became popular genres through the century.[86] Foreign texts were translated and adapted to local audiences.[87] In 1802 the German chemist Sigismund Friedrich Hermbstädt (1760–1833) published his *Grundriss der Färberkunst* using data not only from the German lands, but also from France, England, Sweden, Italy, Spain, and Denmark.[88] His book was based on the works of Hellot, Berthollet, and Bancroft, whom Hermbstädt considered the founding fathers of the new chemistry of dyeing.[89] Translated texts were also a useful vehicle for circulating innovations in the art of dyeing, often presented as the application of a science to the progress of the arts and manufactures but, as in Berthollet's case, retaining many aspects of the recipe tradition. Eighteenth-century periodicals such as *Annales de Chimie* and *Dinglers Polytechnisches Journal* also contributed to this circulation.[90]

Complex procedures for making certain colors that were particularly successful among customers challenged academic and artisan knowledge from different perspectives. In the eighteenth century, this was particularly true of Turkey Red, or *rouge d'Andrinople,* and Prussian Blue. The ingredients and procedures of the recipes were particularly difficult to standardize in

---

to escape the Revolution, and Spanish dyers traveled abroad in search of technical innovations, under the patronage of the monarchy (La Force 1965, 32–34).

86. H. T. Scheffer, *Essai sur l'art de la teinture;* Jean Hellot, *L'art de la teinture des laines;* Delormois, *Le teinturier parfait.* This last was originally published in 1716 but was reprinted as late as 1800, in Paris, as *Le nouveau teinturier parfait* (Delormois 1800).

87. Poerner 1791. It was a collection of dyeing procedures without any particular systematization. The first German edition was Poerner 1772–1773. A valuable source for the processes of multilingual translation of texts on the art of dyeing and printing textiles is Ron 1991. Berthollet's *Éléments* and Bancroft's *Experimental Researches* were cited in technical journals across Europe and translated into several languages. See, for instance, Berthollet 1791b, 1795b, 1824, 1841; Bancroft 1817–1818.

88. Hermbstädt 1802. In Germany: Bartholdi, Beckmann, Denso, Gmelin, Göttling, Gren, Hacquet, Hausmann, Hessler, Justi, Kortum, Kulenkamp, Neuenebahn, Pörner, Schreber, Seiffert, Succow, and Vogler; in France: Baumé, Bertaud, Berthollet, Chaptal, Dambourney, Dyonwall, Du Fay, Fourcroy, Duhamel, Hallancourt, Hellot, Macquer, and du Trône; in England: Forsyth, Fritsch, Hoile, and Kirwan; in Sweden: Gener, Hord, Kalm, Linees, and Westring; in Italy: Fabroni; in Spain: Vasco; in Denmark: Tyschen (Hermbstädt 1802, 60–61).

89. "Gern hätte ich mich beim Praktischen der Färbereien länger aufgehalten . . . aber das Buch würde denn noch stärker geworden sein, als es schon ist; und überdies findet man von Hellot, Pörner, Berthollet, Bancroft, u.a.m die Grundsätze dazu in Allgemeinen angegeben" (Hermbstädt 1802, xvi).

90. Nieto-Galan 2001, 132.

workshops and *manufactures,* and academic chemists were often unable to provide consistent explanations for the unexpected contingencies of the processes, which might affect color solidity, shade, or brightness.

The making of Turkey Red has its origins in the seventeenth century on the coast of Asia Minor. It was made in small workshops, using a local variety of madder and other local substances that produced a beautiful red shade that caught the attention of numerous Western manufacturers. In fact, from the early eighteenth century onward, craftsmen from Smyrna, Salonica, and Andrinople were appointed to French *manufactures,* and frequent journeys were undertaken in pursuit of the secret of this red.[91] Turkey Red recipes were particularly complex. Was it a question of using local raw materials (madder from Smyrna, alkalis and olive oil from Greece), or a problem of imitating local skills?[92] In France, the problem of the *rouge d'Andrinopole,* or Turkey Red, stimulated the scientific interest of well-known figures such as Macquer, who in 1773, in a report to the Académie des Sciences, attempted to simplify Eastern methods to design a new procedure that would preserve the quality of the color but could be easily applied on an industrial scale. In his own words: "After verifying the process which has been published by the Ministry for dyeing cotton in a red as fast and beautiful as the red made in India, several times, I acknowledge that this method only succeeds when all the original operations are executed exactly. But since this is a long, difficult and expensive process, I have decided to learn it in detail, as well as to identify the essential step of the whole process."[93]

The production of Turkey Red stimulated numerous technological journeys. French experts visited Andrinople and Smyrna to copy the making of a *rouge rosé* (a local variety of madder) and went on to Syria and Cyprus in search of new types of madder. The journey continued later to Sidon and Beirut, and through the Red Sea to India, bringing back to France a huge collection of seeds, roots, objects, and machines.[94] Indeed, it was only through these large international networks that a consensus could be reached on the best method for making Turkey Red.

Another puzzling color was Prussian Blue, a dye discovered in Berlin in 1704 and made with animal substances (mainly blood, as a source of prus-

91. Vitalis 1823; Kinini 1999.

92. Chateau 1876.

93. "Déposé au Secretariat de l'Académie des sciences de Paris, 11 août 1773" (Bibliothèque National de Paris, Manuscrits n.a. 2761).

94. A.N., Commerce et Industrie, F/12/1329.

siate compounds) and iron salts. In Edward Bancroft's words, the discovery occurred as follows:

> Diesbach, a chemist at Berlin, wishing to precipitate the colouring matter of cochineal from a solution or decoction, in which it was combined with a portion of green vitriol, or sulphate of iron, borrowed for that purpose from his neighbour Dippel, an alkali, upon and from which the latter had several times distilled an animal oil, which had thereby become impregnated with the animal colouring part of the Prussian Blue. The alkali, when mixed with the decoction of cochineal, or rather with the iron contained therein, immediately and most unexpectedly produced a very beautiful blue colour.[95]

Although in the mid-eighteenth century the new blue was extensively used in workshops and *manufactures,* learned explanations of its composition and behavior were unconvincing to contemporary experts. In 1725 Prussian Blue was considered to be an "assemblage bizarre des différentes matières," but at the same time it was praised for painting and as a substitute for colonial indigo.[96]

In 1756, in close contact with the artisans, Hellot tested several procedures for making Prussian Blue in the workshops. Red tartar, dried ox blood, potash, saltpeter, quicklime, alum, green vitriol, water, and spirit of salt were mixed in different proportions and under different experimental conditions to modify the yield and the shade of the resulting blue color.[97] For his part, Macquer treated Prussian Blue with mineral acids, alkalis, niter, and other reagents, judging it to be more beautiful than indigo and woad, for it penetrated deeper into the textile fibers, was a fast color on wool and silk, and was cheaper than colonial blues because of the availability of the raw materials, including animal blood.[98] Bancroft also emphasized the availability of

95. Bancroft 1794, 199. In Leiden in 1711, Johann Conrad Dippel (1672–1734) isolated an "animal oil" while distilling hart's blood. He had studied medicine in Amsterdam and graduated from Leiden (Partington 1961–1970, 2:378–79).

96. "They have kept this secret hidden for as long as they were able. But such a useful preparation, which passes through the hands of so many people who work with it, is unlikely to remain concealed for long" (É-F. Geoffroy 1725, 154).

97. Hellot used red tartar, dried ox blood, potash, saltpeter, quicklime, alum, green vitriol, water and spirit of salt. "This member of the Académie Hellot, having witnessed . . . the experiments of a private individual who knew how to make the most beautiful Prussian Blue that one could hope to use . . . has reported on the procedures before the Académie after having seen them in practice" (Table pour l'Histoire 1756, 57–59).

98. Macquer 1749, 1752.

animal raw materials for the making of the new color: "The cheapest way of preparing this animal blue is, by burning dried blood, horns, hooves, hides, tendons, and other animal substances, so as to reduce them to coal; which is afterwards to be calcined with three times its weight of pot-ash in an iron vessel."[99]

In the eighteenth century, well-known academic chemists did research on Prussian Blue but obtained equivocal results regarding its chemical composition. In spite of the optimism of Lavoisierian chemists, color was a challenging intellectual puzzle and a driving force in the development of new analytical techniques. In the academy, it gave rise to countless claims about intellectual impotence and experimenters' fatigue. Despite his battery of analytical tests, Macquer also noted: "Experiments performed to date by many skilled chemists . . . have already shed some light on the essential properties of Prussian Blue. However, I think there are still open questions and new research to be done."[100]

## Conclusion

Expertise in dyeing was the result of many years of close contact between workshops, chemical laboratories, and libraries of applied chemistry and the industrial arts. It also depended heavily on the cooperation of dyers, who supplied the samples and recipes that were compiled and organized. In the article "History of Dyeing" in Abraham Rees's *Cyclopaedia* (1819), the state of the art was described as follows: "Though we have to record no brilliant discoveries or improvements in the practice of dyeing, within these few years, yet the art has continued progressively to improve, the different processes have been simplified and amended; and what some years ago was considered a matter of chance and uncertainty, is now reduced to fixed principles."[101]

A certain suggestion of progress underlay this admission of an absence of brilliant discoveries, progress achieved through the application of fixed principles, and the simplification of processes in the intersection between learned knowledge and routine operation that took place in the workshop. This typified the art of dyeing in the eighteenth century: a complex combination of quality tests, frequently repeated experiments recorded in notebooks, analytical techniques, taxonomical criteria for classifying and ordering dyes,

99. Bancroft 1794, 199–200.
100. Macquer 1752, 67.
101. Rees 1819, 1:8.

chemicals, procedures, and colored patterns, attempts to import local and foreign tacit knowledge, and the circulation of books of swatches.

Beneath the rhetoric of academic debates about the use of scientific theories in the art of dyeing, all these practices produced a common space that could be shared by many chemists and dyers throughout the century and avoided polarized controversies such as the clash between Homassel and Berthollet. Although more primary sources still need to be explored in order to generate a more symmetrical historical account, I have summarized the main features of the art of natural dyestuffs in this chapter. The materials considered offer many excellent reasons for challenging traditional historical categories such as theory, technology, expert, and craftsman and dichotomies such as learned versus unlearned. In the final analysis, the effect of these materials on eighteenth-century European societies reflects the complexities of a colorful palette.

## Primary Sources

Secondary sources can be found in the cumulative bibliography at the end of the book.

Ardit, Carlos. 1819. *Tratado teórico práctico de la fabricación de pintados o indianas.* 2 vols. Barcelona: Viuda de Agustín Roca.

Baines, Edward. 1835. *History of Cotton Manufacture in Great Britain.* Facsimile edition, London: Frank Cass, 1966.

Bancroft, Edward. 1794. *Experimental Researches concerning the philosophy of permanent colours, and the best means of producing them, by dyeing, calico-printing.* London: Cadell T. and W. Davies.

———. 1817–1818. *Neues Englisches Färbebuch oder gründliche Untersuchungen über die Natur beständiger Farben.* Nürnberg: J.-L. Schrag.

Berthollet, Claude-Louis. 1791a. *Eléments de l'art de la teinture.* 2 vols. Paris: Firmin Didot.

———. 1791b. *Elements of the art of dyeing.* Trans. W. Hamilton. London: S. Couchman.

———. 1795a. *Description du blanchiment des toiles et des fils par l'acide muriatique oxigené et de quelques autres propiertés de cette liqueur relative aux arts.* Paris: Fuchs.

———. 1795b. *Elementos del arte de teñir.* Trans. Domingo García Fernández. Madrid: Imprenta Real.

———. 1803. *Essai de statique chimique.* 2 vols. Paris: Firmin Didot.

———. 1804. *Eléments de l'art de la teinture, avec une description du blanchiment par l'acide muriatique oxygéné.* 2 vols. 2nd ed. Paris: Firmin Didot.

———. 1824. *Elements of the Art of Dyeing.* Trans. A. Ure. London: Th. Tegg.

———. 1841. *Elements of the Art of Dyeing.* Trans. A. Ure. 2nd ed. London: Th. Tegg.

Berthollet, Claude-Louis, Antoine-François Fourcroy, Louis-Bernard Guyton de Morveau, and Antoine-Laurent Lavoisier. 1787. *Méthode de nomenclature chimique.* Paris: Cuchet.

Buc'hoz, Pierre-Joseph. 1785. *Traité de toutes les plantes qui servent à la teinture et à la Peinture.* Paris: Imprimérie de la Veuve Valade.

Canals, Juan Pablo. 1768. *Memorias que de orden de la Real Junta General de Comercio y Moneda se dan al público, sobre la grana kermes de España que es el coccum, o cochinilla de los antiguos.* Madrid: Viuda de Eliseo Sánchez.

Chaptal, Jean-Antoine. 1807. *L'art de la teinture du coton en rouge.* Paris: Detreville.

Chateau, Theodore. 1876. Critical and historical notes concerning the production of Andrinople or Turkey Red, and the theory of this colour. *The Textile Colourist* 1:172–78, 217–31, 276–82, 384–97; 2:27–33, 131–41, 191–200, 262–72.

Dambourney, Louis-Alexandre. 1786. *Recueil de procédés et d'expériences sur les teintures solides que nos végétaux indigènes communiquent aux laines et aux lainages.* Rouen: Pierres.

De Lormois. 1800. *Le nouveau teinturier parfait ou Traité de ce qu'il y a de plus essentiel dans la teinture, omis ou caché par l'auteur de l'ancien teinturier parfait: . . . avec un dictionnaire des principaux ingrédients, et des termes propres . . . Nouvelle édition corrigée* (1st ed., 1716). Paris: Ch. A. Jombert.

Diderot, Denis, and Jean D'Alembert, eds. 1751–1780. *Encyclopédie ou dictionnaire raisonné des Sciences, des arts et des métiers.* Paris: Briasson, David, Le Breton, Durand; Neuchatel: Samuel Faulche.

Du Fay, Charles François. 1737a. Observations physiques sur le mélange de quelques couleurs dans la Teinture. *Histoire de l'Académie des Sciences,* 253–68.

———. 1737b. Sur le mélange de quelques couleurs dans la teinture. *Histoire de l'Académie des Sciences,* 58–62.

Duhamel du Monceau, Henry-Louis. 1763. *Éléments d'agriculture.* 2 vols. Paris: Guérin, Delatour.

Fernández, Luis 1778. *Tratado instructivo y práctico sobre el arte de la tintura.* Madrid: Imprenta de Blas Román.

Geoffroy, Étienne-François. 1725. Observations sur la préparation du Bleu de Prusse, ou de Berlin. *Mémoires de l'Académie royale des Sciences,* 153–73.

Gréau, aîné. 1834. *De la destruction des tissus dans le blanchiment et la teinture et des moyens d'en prévenir les causes.* Paris: Bureau de la Société Polytechnique.

Hellot, Jean. 1740. Theorie chymique de la teinture des étoffes. *Mémoires de l'Académie royale des Sciences,* 126–48.

———. 1741. Theorie chymique de la teinture des étoffes. *Mémoires de l'Académie royale des Sciences,* 38–40.

———. 1750. *L'art de la teinture des laines et des étoffes de laine en grand et petit teint.* Paris: Pissot, Herissant.

Henry, Thomas. 1790. Considerations relative to the nature of wool, silk, and cotton, as objects of the art of dyeing, on the various preparations and mordants, requisite for the different substances; and on the nature and properties of colouring matter. together with some observations on the theory of dyeing in general, and particularly the Turkey Red. *Memoirs of the Literary and Philosophical Society of Manchester* 3:343–407.

Hermbstädt, Sigismund Friedrich. 1802. *Grundriss der Färberkunst oder allgemeine theoretische und praktische Anteilung zur rationellen Ausübung der Wollen, Seiden, Baumwollen und Leinenfärben.* Berlin: Friedrich Nicolai.

Homassel, M. 1798. *Cours théorique et pratique sur l'art de la teinture en laine, soie, fil, coton, fabrique d'indiennes en grand et petit teint, suivi de l'art du teinturier, dégraisseur et du blanchisseur, avec les expériences faites sur les végetaux colorants.* Paris: Courcier.

———. 1807. *Cours théorique et pratique sur l'art de la teinture.* 2nd ed. Paris: Courcier.

Lavoisier, Antoine-Laurent. 1789. *Traité élémentaire de chimie.* 2 vols. Paris: Cuchet.

Le Pileur d'Apligny. 1770. *Essai sur les moyens de perfectionner l'art de la teinture.* Paris: Laurent Prault.

Macquer, Pierre Joseph. 1749. Mémoire sur une nouvelle espèce de teinture bleue, dans laquelle il n'entre ni pastel ni Indigo. *Mémoires de l'Académie royale des Sciences,* 255–65.

———. 1752. Examen chimique du bleu de Prusse. *Mémoires de l'Académie royale des Sciences,* 60–67.

———. 1763. *L'art de la teinture en soie.* Paris: Didot.

———. 1766. *Dictionnaire de chimie.* 2 vols. Paris: Lacombe.

———. 1778. *Dictionnaire de chimie.* 4 vols. Paris: Didot jeune.

Muspratt, Sheridan. 1853–1861. *Chemistry, Theoretical, Practical and Analytical as applied and relating to the arts and manufactures.* 2 vols. London: W. Mckenzie.

Neumann, Gaspar. 1759. *The Chemical Works.* London: W. Johnston.

O'Brien, Charles. 1795. *The British Manufacturers Companion, and Calico Printers Assistant.* London: Hamilton & Co.

O'Neill, Charles, ed. 1877. Materials for a History of Textile Colouring (Bleaching). *Textile Colourist* 4:3–13, 5:63–74, 6: 223–36.

Parkes, Samuel. 1815. *Chemical Essays, principally related to the arts and manufactures of the British dominions.* 5 vols. London: Batwin, Cradock, Joy.

Poerner, Karl Wilhelm. 1772–1773. *Chymische Versuche und Bemerkungen zum Nutzen der Färbekunst.* 3 vols. Leipzig: M. G. Weidmanns Erben.

———. 1791. *Instruction sur l'art de la teinture et particulièrement sur la teinture des laines.* Paris: Chez Couchet.

Raymond-Latour, Jean-Michel. 1836. *Souvenirs d'un oisif.* 2 vols. Paris and Lyon: Isidore Person et Chez Ayné fils.

Rees, Abraham, ed. 1819. *The Cyclopaedia; or, Universal Dictionary of Arts, Sciences and Literature.* London: Longman, Hurst, Rees, Orme & Brown. Reprint, Newton Abbot: David & Charles, 1972.

Scheffer, Heinrik Teophilus. 1787. *Essai sur l'art de la teinture.* Paris: Chez Buisson.

Table pour l'Histoire. 1756. *Mémoires de l'Académie royale des Sciences,* 57–59.

Ure, Andrew. 1821. *A Dictionary of Chemistry.* London: T. and G. Anderwood.

Vitalis, Jean-Baptiste. 1823. *Cours élémentaire de teinture.* Paris: Galérie Bossagne Père.

# *Secondary Sources*


Adams, Frank Dawson. 1938. *The Birth and Development of the Geological Sciences.* New York: Dover.

Adams, Robert M. 1996. *Paths of Fire: An Anthropologist's Inquiry into Western Technology.* Princeton, NJ: Princeton University Press.

Adlung, Alfred, and Georg Urdang. 1935. *Grundriß der Geschichte der deutschen Pharmacie.* Berlin: Springer.

Albala, Ken. 2002. *Eating Right in the Renaissance.* Berkeley: University of California Press.

Alder, Ken. 1997. *Engineering the Revolution: Arms and Enlightenment in France, 1763–1815.* Princeton, NJ: Princeton University Press.

———. 2007. Introduction to Focus: Thick Things. *Isis* 98:80–83.

Allbutt, Thomas Clifford. 1913–14. Palissy, Bacon and the Revival of Natural Science. *Proceedings of the British Academy,* 234–47.

Allchin, Douglas. 1992. Phlogiston after Oxygen. *Ambix* 39:110–16.

Allen, Robert C. 2004. Agriculture during the Industrial Revolution, 1700–1850. In *The Cambridge Economic History of Modern Britain,* vol. 1, *Industrialization, 1700–1860,* ed. R. Floud and P. Johnson, 96–116. Cambridge: Cambridge University Press.

Amico, Leonard. 1996. *Bernard Palissy: In Search of Earthly Paradise.* New York: Flammarion.

Amouretti, Marie-Claire, and Georges Comet, eds. 1998. *Artisanat et matériaux: la place des matériaux dans l'histoire des techniques.* Cahier d'Histoire des Techniques 4. Aix-en-Provence: Publications de l'Université de Provence.

Anderson, Peter J., ed. 1889–1898. *Fasti Academiae Mariscallanae Aberdonensis: Selections from the Records of the Marischal College and University.* 3 vols. Aberdeen: New Spalding Club.

———. 1897. Aberdeen Grammar School Masters and Under Masters, 1602–1853. *Scottish Notes and Queries* 11:38–41.

Anderson, Robert G. W., James A. Bennett, and William F. Ryan, eds. 1993. *Making Instruments Count: Essays on Historical Scientific Instruments Presented to Gerard L'Estrange Turner.* Aldershot, UK: Variorum.

Anderson, Wilda C. 1984. *Between the Library and the Laboratory: The Language of Chemistry in Eighteenth-Century France.* Baltimore: Johns Hopkins University Press.

Appadurai, Arjun. 1986. Introduction: Commodities and the Politics of Value. In *The Social*

*Life of Things: Commodities in Cultural Perspective,* ed. A. Appadurai, 3–63. Cambridge: Cambridge University Press.

Ashworth, B. 2003. John Gregory and the Background to Medical Philosophy. *Journal of the Royal College of Physicians of Edinburgh* 33:67–69.

Ashworth, William B. 1996. Emblematic Natural History of the Renaissance. In *Cultures of Natural History,* ed. N. Jardine, J. A. Secord, and E. C. Spary, 17–37. Cambridge: Cambridge University Press.

Ashworth, William J. 2001. "Between the Trader and the Public": British Alcohol Standards and the Proof of Good Governance. *Technology and Culture* 42:27–50.

Bachelard, Gaston. 1958. Shells. In *The Poetics of Space.* Trans. Maria Jolas. Boston: Beacon.

———. 1987. *Die Bildung des wissenschaftlichen Geistes.* Frankfurt am Main: Suhrkamp. (Reprint of 1938 ed.)

———. 1991. *Le nouvel esprit scientifique.* 4th ed. Paris: PUF.

———. 1996. *La formation de l'esprit scientifique: contribution à une psychoanalyse de la connaisance.* Paris: Vrin. (Reprint of 1938 ed.)

Bachmann, Hans-Gert. 1993. Vom Erz zum Metall (Kupfer, Silber, Eisen)—Die chemischen Prozesse im Schaubild. In *Alter Bergbau in Deutschland,* ed. H. Steuer and U. Zimmermann, 35–40. Stuttgart: Theiss.

Balfour-Paul, Jenny. 1998. *Indigo.* London: British Museum Press.

Balland, Joseph Antoine Felix. 1902. *La chimie alimentaire dans l'oeuvre de Parmentier.* Paris: Baillière.

Ballot, M. J. 1924. *Documents de L'Arts/La Ceramique Francaise: Bernard Palissy et les Fabriques du XVI Siecle.* Paris: Editions Albert Morance.

Barlow, Hilaro W. W. 1909. The Royal Laboratory, Woolwich. In *The Rise and Progress of the British Explosives Industry: Published under the Auspices of the VIIth International Congress of Applied Chemistry by Its Explosives Section,* 307–16. London: Whittaker and Co..

Barnes, Barry, David Bloor, and John Henry. 1996. *Scientific Knowledge: A Sociological Analysis.* Chicago: University of Chicago Press.

Bartels, Christoph. 1988. *Das Erzbergwerk Rammelsberg.* Goslar: Preussag AG Metall.

———. 1989. The Development of the Turm Rosenhof Mine, 1540–1820, Clausthal, Upper Harz. In *History of Technology: The Role of Metals,* 46–64. Philadelphia: University Museum, University of Pennsylvania.

———. 1990. Der Bergbau vor der hochindustriellen Zeit—ein Überblick. In *Meisterwerke Bergbaulicher Kunst vom 13. bis 19. Jahrhundert,* ed. R. Slotta and C. Bartels, 14–32. Bochum: Deutsches Bergbau-Museum.

———. 1992a. *Das Erzbergwerk Grund.* Goslar: Preussag AG Metall.

———. 1992b. *Vom frühneuzeitlichen Montangewerbe zur Bergbauindustrie. Erzbergbau im Oberharz 1635–1866.* Bochum: Deutsches Bergbau-Museum.

———. 1994a. Lazarus Ercker und der Bergbau am Rammelsberg bei Goslar. In *Lazarus Ercker. Sein Leben und seine Zeit. Zur Geschichte des Montan- und Münzwesens im mittleren Europa.* Ed. Technische Universität Bergakademie, Freiberg, 26–31. Annaberg-Buchholz: Landesstelle für Erzgebirgische und Vogtländische Volkskultur.

———. 1994b. Soziale und religiöse Konflikte im Oberharzer Bergbau des 18. Jahrhunderts: Ursachen, Hintergründe, Zusammenhänge. In *Niedersächsisches Jahrbuch für Landesgeschichte,* 66:79–104. Hannover: Verlag Hahnesche Buchhandlung.

———. 1997a. Georg Winterschmidt's Water Pressure Engines in the Upper Harz Mining District, 1747–1763: Plans, Experiments, Problems, Results. *ICON: Journal of the International Committee for the History of Technology* 3:24–43.

———. 1997b. Strukturwandel in Montanbetrieben des Mittelalters und der frühen Neuzeit in Abhängigkeit von Lagerstättenstrukturen und Technologie. In *Struktur und Dimension. Festschrift für Karl Heinrich Kaufhold zum 65. Geburtstag,* ed. H.-J. Gerhard, 25–70. Stuttgart: Franz Steiner Verlag.

———. 2000. The Introduction of Gunpowder-Blasting. A Central Technical Change in Post-mediaeval European Mining: The Upper-Harz Example. In *Hombres, Técnica, Plata. Minería y Sociedad en Europa y América Siglos XVI–XIX,* ed. J. Sánchez-Gómez and G. Mira Delli-Zotti, 47–60. Sevilla: Aconcagua Libros.

———. 2002. Albertus Magnus und das Montanwesen des Mittelalters. In *Festschrift Rudolf Palme zum 60. Geburtstag,* ed. W. Ingenhaeff, R. Staudinger, and K. Ebert, 23–50. Innsbruck: Berenkamp Verlag.

———. 2004a. Der Bergbau im nordwestlichen Harz im 14. und 15. Jahrhundert. In *Der Tiroler Bergbau und die Depression der europäischen Montanwirtschaft im 14. und 15. Jahrhundert. Akten der internationalen bergbaugeschichtlichen Tagung Steinhaus,* ed. R. Tasser and E. Westermann, 19–44. Innsbruck: Studien Verlag.

———. 2004b. Die Ereignisse vor dem Riechenberger Vertrag und der herzogliche Bergbau im Oberharz. Entwicklungslinien im Vergleich. In *Der Riechenberger Vertrag,* ed. Rammelsberger Bergbaumuseum, Goslar, 65–90. Goslar: Verlag Goslarsche Zeitung Karl Krause.

———. 2004c. Die Stadt Goslar und der Bergbau im Nordwestharz. Von den Anfängen bis zum Riechenberger Vertrag von 1552. In *Stadt und Bergbau,* ed. K. H. Kaufhold and W. Reininghaus, 135–88. Cologne: Böhlau Verlag.

———. 2006. Entwicklung und Stand der Forschungen zum Montanwesens des Mittelalters und der frühen Neuzeit. In *Montan- und Industriegeschichte. Dokumentation und Forschung, Industriearchäologie und Museum. Festschrift für Rainer Slotta zum 60. Geburtstag,* ed. S. Brüggerhoff, M. Farrenkopf, and W. Geerlings, 171–210. Paderborn: Ferdinand Schöningh.

Bartels, Christoph, Andreas Bingener, and Rainer Slotta, eds. 2006. *Das Schwazer Bergbuch.* 3 vols. Bochum: Deutsches Bergbau-Museum.

Bartels, Christoph, Michael Fessner, Lothar Klappauf, and Friedrich Albert Linke. 2007. *Kupfer, Blei und Silber aus dem Goslarer Rammelsberg von den Anfängen bis 1620.* Bochum: Deutsches Bergbau-Museum.

Baxmann, Inge. 1989. *Die Feste der Französischen Revolution. Inszenierung von Gesellschaft als Natur.* Weinheim: Beltz.

Bayerl, Günter. 1994. Prolegomenon der „Großen Industrie". Der technisch-ökonomische Blick auf die Natur im 18. Jahrhundert. In *Umweltgeschichte. Umweltverträgliches Wirtschaften in historischer Perspektive,* ed. W. Abelshauser. Göttingen: Vandenhoek und Ruprecht.

———. 2001. Die Natur als Warenhaus. Der technisch-ökonomische Blick auf die Natur in der Frühen Neuzeit. In *Umwelt-Geschichte. Arbeitsfelder—Forschungsansätze—Perspektiven,* ed. R. Reith and S. Hahn, 33–52. Vienna: Oldenbourg.

Beer, John Joseph. 1960. Eighteenth-Century Theories on the Process of Dyeing. *Isis* 51:21–30.

Beisswanger, Gabriele. 1996. *Arzneimittelversorgung im 18. Jahrhundert: Die Stadt Braunschweig und die ländlichen Distrikte im Herzogtum Braunschweig-Wolfenbüttel.* Braunschweig: Deutscher Apothekerverlag.

Bennett, Jim. 1986. The Mechanics' Philosophy and the Mechanical Philosophy. *History of Science* 24:1–28.

———. 1987. *The Divided Circle: A History of Instruments for Astronomy, Navigation, and Surveying.* Oxford: Phaidon, Christie's Limited.

———. 1989. A Viol of Water or a Wedge of Glass. In *The Uses of Experiment: Studies in the Natural Sciences,* ed. D. Gooding, T. Pinch, and S. Schaffer, 105–14. Cambridge: Cambridge University Press.

———. 2002. Shopping for Instruments in Paris and London. In *Merchants and Marvels: Commerce and the Representation of Nature in Early Modern Europe,* ed. P. H. Smith and P. Findlen, 370–95. New York: Routledge.

———. 2003. Instruments and Instrument Makers. In *The Oxford Companion to the History of Modern Science,* ed. J. L. Heilbron, 406–10. Oxford: Oxford University Press.

Bennett, Jim, Michael Cooper, Michael Hunter, and Lisa Jardine. 2003. *London's Leonardo: The Life and Work of Robert Hooke.* Oxford: Oxford University Press.

Benninga, Harmke. 1990. *A History of Lactic Acid Making. A Chapter in the History of Biotechnology.* Dordrecht: Kluwer.

Bensaude-Vincent, Bernadette. 1992. The Balance: Between Chemistry and Politics. In *The Chemical Revolution—Context and Practices,* special issue, *Eighteenth-Century: Theory and Interpretation* 33:217–37.

———. 1993. *Lavoisier: Mémoires d'une révolution.* Pref. by Michel Serres. Paris: Flammarion.

———. 2000. "The Chemist's Balance for Fluids": Hydrometers and their Multiple Identities, 1770–1810. In *Instruments and Experimentation in the History of Chemistry,* ed. F. L. Holmes and T. H. Levere, 153–83. Cambridge, MA: MIT Press.

Bensaude-Vincent, Bernadette, and Ferdinando Abbri, eds. 1995. *Negotiating a New Language for Chemistry: Lavoisier in European Context.* Canton, MA: Science History Publications.

Bensaude-Vincent, Bernardette, and Anders Lundgren. 1999. *Communicating Chemistry: Chemical Textbooks in Europe, 1789–1900.* Canton, MA: Science History Publications.

Bensaude-Vincent, Bernadette, and Agustí Nieto-Galan. 1999. Theories of Dyeing: A View of a Long-standing Controversy through the Works of Jean-François Persoz. In *Natural Dyestuffs and Industrial Culture in Europe, 1750–1880,* ed. R. Fox and A. Nieto-Galan, 3–24. Canton, MA: Science History Publications.

Benzaquén, Adriana S. 2004. Childhood, Identity and Human Science in the Enlightenment. *History Workshop Journal* 57:35–57.

Beretta, Marco. 1993. *The Enlightenment of Matter. The Definition of Chemistry from Agricola to Lavoisier.* Canton, MA: Science History Publications.

Berg, Maxine, ed. 1991. *Markets and Manufacture in Early Industrial Europe.* London: Routledge.

———. 1994. *The Age of Manufactures, 1700–1820: Industry, Innovation and Work in Britain.* 2nd ed. London: Routledge.

Berg, Maxine, and Kristine Bruland, eds. 1998. *Technological Revolutions in Europe: Historical Perspectives.* Cheltenham, UK: E. Elgar.

Berg, Maxine, and Helen Clifford, eds. 1999. *Consumers and Luxury: Consumer Culture in Europe 1650–1850.* Manchester: Manchester University Press.

Berg, Maxine, and Elizabeth Eger. 2003a. Introduction to *Luxury in the Eighteenth Century: Debates, Desires and Delectable Goods,* ed. M. Berg and E. Eger. Basingstoke, UK: Palgrave Macmillan.

———. 2003b. *Luxury in the Eighteenth Century: Debates, Desires and Delectable Goods.* Basingstoke, UK: Palgrave Macmillan.

———. 2003c. The Rise and Fall of the Luxury Debates. In *Luxury in the Eighteenth Century: Debates, Desires and Delectable Goods,* ed. M. Berg and E. Eger, 7–27. Basingstoke, UK: Palgrave Macmillan.

Berkel, Klaas van. 1983. *Isaac Beeckman (1588–1637) en de Mechanisering van het Wereldbeeld.* Amsterdam: Rodopi.

Bertoloni Meli, Domenico. 2006. *Thinking with Objects: The Transformation of Mechanics in the Seventeenth Century.* Baltimore: Johns Hopkins University Press.

Bertie, David M. 2000. *Scottish Episcopal Clergy, 1689–2000.* Edinburgh: T & T Clark.

Bewer, Francesca G. 2001. The Sculpture of Adriaen de Vries: A Technical Study. In *Small Bronzes in the Renaissance,* ed. D. Pincus, 159–93. Washington, DC: Center for Advanced Study in the Visual Arts.

Beyerlein, Berthold. 1991. *Die Entwicklung der Pharmazie zur Hochschuldisziplin (1750–1875), ein Beitrag zur Universitäts- und Sozialgeschichte.* Stuttgart: Wissenschaftliche Verlagsgesellschaft.

Biagioli, Mario. 1993. *Galileo, Courtier: The Practice of Science in the Culture of Absolutism.* Chicago: University Chicago Press.

Bijker, Wiebe E. 1985. *Of Bicycles, Bakelites, and Bulbs: Toward a Theory of Sociotechnical Change.* Cambridge, MA: MIT Press.

Bijker, Wiebe E., Thomas P. Hughes, and Trevor J. Pinch, eds. 1987. *Social Construction of Technological Systems: New Directions in the Sociology and History of Technology.* Cambridge, MA: MIT Press.

Blanchard, Ian. 1994. *International Lead Production and Trade in the "Age of the Saigerprozess" 1460–1560. Zeitschrift für Unternehmensgeschichte, Beihefte,* vol. 85. Stuttgart: Franz Steiner Verlag.

Blocker, Jack S., Jr. 1994. Introduction to *Social History of Alcohol,* special issue, *Social History/Histoire Sociale* 27:225–39.

Bloy, Colin H. 1967. *A History of Printing Ink, Balls, and Rollers, 1440–1850.* London: Wynkyn de Worde Society.

Bödeker, Hans Erich. 1999. Medien der patriotischen Gesellschaften. In *Von Almanach bis Zeitung. Ein Handbuch der Medien in Deutschland 1700–1800,* ed. E. Fischer, W. Haefs, and Y.-G. Mix, 285–302. Munich: Beck.

Böhm, Wolfgang. 1990. Die Anfänge des Feldversuchswesens in Deutschland. *Zeitschrift für Agrargeschichte und Agrarsoziologie.* 38, no. 2, 155–75.

Böhme, Hartmut, and Gernot Böhme. 1983. *Das Andere der Vernuft: Zur Entwicklung von Rationalitätsstrukturen am Beispiel Kants.* Frankfurt am Main: Suhrkamp.

Bornhardt, Wilhelm. 1931. Geschichte des Rammelsberger Bergbaus von seinen Anfängen bis zur Gegenwart. *Archiv für Lagerstättenkunde* 52.

Bouvet, Maurice. 1937. *Histoire de la pharmacie en France des origines à nos jours.* Paris: Editions Occitania.

Braudel, Fernand. 1979. *Civilization and Capitalism, 15th–18th Century.* Vol. 2, *The Wheels of Commerce,* trans. S. Reynolds. New York: Harper & Row.

Brédif, Josette. 1989. *Classic Printed Textiles from France 1760–1843: Toiles de Jouy.* London: Thames & Hudson.

Brennan, Thomas. 1988. *Public Drinking and Popular Culture in Eighteenth-Century Paris.* Princeton, NJ: Princeton University Press.

Brightwell, Cecelia. 1859. *Palissy the Potter: Huguenot, Artist and Martyr.* New York: Carlton & Porter.

Brock, William H. 1992. *The Fontana History of Chemistry.* London: Fontana.

Brockway, Lucile H. 1979. *Science and Colonial Expansion: The Role of the British Royal Botanical Gardens.* New York: Academic Press.

Broman, Thomas. 1998. The Habermasian Public Sphere and "Science in the Enlightenment." *History of Science* 36:123–49.

Bruland, Kristine. 2004. New Technologies and the Industrial Revolution: Some Unresolved Issues. In *Artisans, industrie: Nouvelles révolutions du Moyen Âge à nos jours,* ed. N. Coquery, L. Hilaire-Pérez, L. Sallmann, and C. Verna, 55–68. Lyon: ENS Editions.

Bryden, D. J. 1993. A 1701 Dictionary of Mathematical Instruments. In *Making Instruments Count: Essays on Historical Scientific Instruments Presented to Gerard L'Estrange Turner,* ed. R. G. W. Anderson, J. A. Bennett, and W. F. Ryan, 365–82. Aldershot, UK: Variorum.

Buchanan, Brenda J. 1995–1996. The Technology of Gunpowder Making in the Eighteenth Century: Evidence from the Bristol Region. *Transactions of the Newcomen Society for the Study of the History of Engineering and Technology* 67:125–59.

———, ed. 1996a. *Gunpowder: The History of an International Technology.* Bath: Bath University Press.

———. 1996b. Meeting Standards: Bristol Powder Makers in the Eighteenth Century. In *Gunpowder: The History of an International Technology,* ed. B. J. Buchanan, 237–52. Bath: Bath University Press.

———. 2005. "The Art and Mystery of Making Gunpowder": The English Experience in the Seventeenth and Eighteenth Centuries. In *The Heirs of Archimedes: Science and the Art of War through the Age of Enlightenment,* ed. B. D. Steele and T. Dorland, 233–74. Dibner Institute Studies in the History of Science and Technology. Cambridge, MA: MIT Press.

———, ed. 2006a. *Gunpowder, Explosives and the State: A Technological History.* Aldershot, UK: Ashgate.

———. 2006b. Saltpetre: A Commodity of Empire. In *Gunpowder, Explosives and the State: A Technological History,* ed. B. J. Buchanan, 67–90. Aldershot, UK: Ashgate.

Bucklow, Spike. 1999. Paradigms and Pigment Recipes: Vermilion, Synthetic Yellows and the Nature of Egg. *Zeitschrift für Kunsttechnologie und Konservierung* 13:140–49.

———. 2000. Paradigms and Pigment Recipes: Natural Ultramarine. *Zeitschrift für Kunsttechnologie und Konservierung* 14:5–14.

———. 2001. Paradigms and Pigment Recipes: Silver and Mercury Blues. *Zeitschrift für Kunsttechnologie und Konservierung* 15: 25–33.

Burmester, Andreas, and Christoph Krekel. 1998. The Relationship between Albrecht Dürer's Palette and Fifteenth/Sixteenth-Century Pharmacy Price Lists: The Use of Azurite and Ultramarine. In *Painting Techniques: History, Materials and Studio Practice,* ed. A. Roy and P. Smith, 89–93. London: International Institute for Conservation of Historic and Artistic Works.

Burt, Roger. 1984. *The British Lead Mining Industry.* Redruth: Cyllansow Truran.

———. 1991. The International Diffusion of Technology during the Early Modern Period: The Case of the Britisch Non-ferrous Mining Industry. *Economic History Review,* 2nd ser., 41 (2): 249–71.

———. 1995. Proto-industrialisation and the British Non-Ferrous Mining Industries. In *Vom Bergbau- zum Industrierevier,* ed. E. Westermann, 317–33. Stuttgart: Franz Steiner Verlag.

Burty, Phillippe. 1886. *Les Artistes Célèbres: Bernard Palissy.* Paris: Librarie de l'Art.

Bynum, William F. 1993. Cullen and the Nervous System. In *William Cullen and the Eighteenth-Century Medical World,* ed. A. Doig et al., 152–62. Edinburgh: Edinburgh University.

Bynum, William F., and Roy Porter, eds. 1988. *Brunonianism in Britain and Europe.* London: Wellcome Institute for the History of Medicine.

Camille, M. 1998. *Mirror in Parchment: The Luttrell Psalter and the Making of Medieval England.* Chicago: University of Chicago Press.

Campbell, Colin. 1987. *The Romantic Ethic and the Spirit of Modern Consumerism.* Oxford: Basil Blackwell.

Camporesi, Piero. 1988. Decay and Rebirth. In *The Incorruptible Flesh,* trans. Tanya Croft-Murray, 67–90. Cambridge: Cambridge University Press.

———. 1994. *Exotic Brew: The Art of Living in the Age of Enlightenment.* Cambridge: Polity.

Carpenter, Kenneth. 1994. *Protein and Energy: A Study of Changing Ideas in Nutrition.* Cambridge: Cambridge University Press.

Carvalho, David N. 1904. *Forty Centuries of Ink.* New York: Banks Law Publishing.

Céard, Jean. 1987. Relire Bernard Palissy. *Revue de L'Art* 78:77–84.

———. 1991. Formes Discursives. In *Précis de Littératre Francais du XVI Siècle: La Renaissance,* ed. R. Aulotte, 155–93. Paris: Presse Universitaire du France.

Chang, Hasok. 2002. Rumford and the Reflection of Radiant Cold: Historical Reflections and Metaphysical Reflexes. *Physics in Perspective* 4:127–69.

Chapman, Stanley, and Serge Chassagne. 1981. *European Textile Printers in the Eighteenth Century. A Study of Peel and Oberkampf.* London: Heinemann.

Chartier, Roger. 1984. Culture as Appropriation: Popular Culture Uses in Early Modern France. In *Understanding Popular Culture: Europe from the Middle Ages to the Nineteenth Century,* ed. S. L. Kaplan, 230–53. Berlin: Mouton.

———. 1987. *Lectures et lecteurs dans la France d'ancien régime.* Paris: Seuil.

Chaudhuri, K. N. 1985. *Trade and Civilisation in the Indian Ocean: An Economic History from the Rise of Islam to 1750.* Cambridge: Cambridge University Press.

Clark, Peter. 1988. The "Mother Gin" Controversy in the Early Eighteenth Century. *Transactions of the Royal Historical Society,* 5th ser., 38:63–84.

Clark, William, Jan Golinski, and Simon Schaffer. 1999. *The Sciences in Enlightened Europe.* Chicago: University of Chicago Press.

Clement, Martin, and Klaus Brennecke. 1975. Über die Entwicklung der Aufbereitung im Harz. In *Zur Zweihundertjahrfeier 1775–1975,* vol. 1, *Die Bergakademie und ihre Vorgeschichte,* ed. Technische Universität Clausthal, 317–30. Clausthal-Zellerfeld: Technische Universität.

Clericuzio, Antonio. 2003. La Chimica della vita: fermenti e fermentazione nella iatrochimica del seicento. *Medicina nei Secoli* 15: 227–45.

Clifton, Gloria C. 1993. The Spectaclemakers' Company and the Origins of the Optical Instrument–Making Trade in London. In *Making Instruments Count: Essays on Historical Scientific Instruments Presented to Gerard L'Estrange Turner,* ed. R. G. W. Anderson, J. A. Bennett and W. F. Ryan, 341–64. Aldershot, UK: Variorum.

Clow, Archibald, and Nan L. Clow. 1952. *The Chemical Revolution: A Contribution to Social Technology.* London: Batchworth.

Cocroft, Wayne. 1996. "William Congreve (1743–1814) Experimenter and Manufacturer." Presented at the 24th Symposium of ICOHTEC, Budapest (unpublished "speaker's text").

Cole, Michael W. 1999. Cellini's Blood. *Art Bulletin* 81:215–35.

———. 2002. *Cellini and the Principles of Sculpture.* Cambridge: Cambridge University Press.

Coley, Noel G. 1973. *From Animal Chemistry to Biochemistry.* Amersham, UK: Bucks.

———. 1982. Physicians and the Chemical Analysis of Mineral Waters in Eighteenth-Century England. *Medical History* 26:123–44.

Connor, Robin D., and Allen D. C. Simpson. 2004. *Weights and Measures in Scotland: A European Perspective.* Edinburgh: National Museum of Scotland.

Cook, Harold J. 1990. The Rose Case Reconsidered: Physic and the Law in Augustan England. *Journal of the History of Medicine* 45:527–55.

———. 2007. *Matters of Exchange: Commerce, Medicine, and Science in the Dutch Golden Age.* New Haven, CT: Yale University Press.

Cooper, Alix. 2004. "The Possibilities of the Land": The Inventory of "Natural Riches" in the Early Modern German Territories. In *Oeconomies in the Age of Newton,* ed. N. De Marchi and M. Schabas, 129–53. *History of Political Economy* annual supp., vol. 34. Durham: Duke University Press.

———. 2007. *Inventing the Indigenous: Local Knowledge and Natural History in Early Modern Europe.* Cambridge: Cambridge University Press.

Cooter, Roger, and Stephen Pumfrey. 1994. Separate Spheres and Public Places: Reflections on the History of Science Popularization and Science in Popular Culture. *History of Science* 32:237–67.

Coquery, Natacha. 1998. *L'hôtel aristocratique: le marché du luxe à Paris au XVIIIe siècle.* Paris: Publications de la Sorbonne.

———. 2003. Mode, commerce, innovation: la boutique parisienne au XVIIIe siècle. Aperçu sur les stratégies de séduction des marchands parisiens de luxe et de demi-luxe. In *Les chemins de la nouveauté: innover, inventer au regard de l'histoire,* ed. L. Hilaire-Pérez and A.-F. Garçon, 187–206. Paris: Editions du CTHS.

Corbin, Alain. 1994. *The Lure of the Sea: The Discovery of the Seaside in the Western World, 1750–1840.* Berkeley: University of California Press.

Court, Susan, and William A. Smeaton. 1979. Fourcroy and the Journal de la Société des Pharmaciens de Paris. *Ambix* 26:39–55.

Coutts, Howard. 2001. *The Art of Ceramics: European Ceramic Design 1500–1830.* New Haven, CT: Yale University Press.

Cowan, Ruth Schwartz. 1983. *More Work for Mother: The Ironies of Household Technology from the Open Hearth to the Microwave.* New York: Basic Books.

———. 1997. *A Social History of American Technology.* New York: Oxford University Press.

Cowen, David L. 1957. The Edinburgh Pharmacopoeia. *Medical History* 1:123–39, 340–51.

———. 1974. *The Spread and Influence of British Pharmacopoeial and Related Literature: An Historical and Bibliographic Study.* Veröffentlichungen der Internationalen Gesellschaft für Geschichte der Pharmazie 41.

———. 1982. The Influence of the Edinburgh Pharmacopoeia and the Edinburgh Dispensatories. *Pharmaceutical Historian* 12:2–4.

———. 1985. Expunctum est mithridatium. *Pharmaceutical Historian* 15:2–3.

———. 2001. *Pharmacopoeias and Related Literature in Britain and America, 1618–1847.* Aldershot, UK: Ashgate.

Craddock, Paul T. 1995. *Early Metal Mining and Production.* Edinburgh: Edinburgh University Press.

Craig, Mary Elizabeth. 1931. *The Scottish Periodical Press, 1750–1789.* Edinburgh: Oliver & Boyd.

Cullen, L. M. 1998. *The Brandy Trade under the Ancien Régime: Regional Specialisation in the Charente.* Cambridge: Cambridge University Press.

Cunningham, Andrew, and Perry Williams. 1993. De-Centring the "Big-Picture": The Origins of Modern Science and the Modern Origins of Science. *British Journal for the History of Science* 26:407–32.

Curth, Louise Hill. 2006. Introduction: Perspectives on the Evolution of the Retailing of Pharmaceuticals. In *From Physick to Pharmacology: Five Hundred Years of Drug Retailing*, ed. L. Curth. Aldershot, UK: Ashgate.

Damrosch, Leo. 2005. *Jean-Jacques Rousseau: Restless Genius*. Boston: Houghton Mifflin.

Darnton, Robert. 1979. *The Business of Enlightenment: A Publishing History of the Encyclopédie, 1775–1800*. Cambridge, MA: Harvard University Press.

———. 1984. *The Great Cat Massacre and Other Episodes in French Cultural History*. New York: Basic Books.

Daston, Lorraine. 2000a. Introduction: The Coming into Being of Scientific Objects. In *Biographies of Scientific Objects*, ed. L. Daston, 1–14. Chicago: University of Chicago Press.

———, ed. 2000b. *Biographies of Scientific Objects*. Chicago: University of Chicago Press.

———, ed. 2004. *Things That Talk. Object Lessons from Art and Science*. New York: Zone.

Daston, Lorraine, and Katharine Park. 1998. *Wonders and the Order of Nature: 1150–1750*. New York: Zone.

Daumas, Maurice. 1953. *Les instruments scientifiques aux XVIIe et XVIIIe siècles*. Paris: Presses Universitaires de France.

Debus, Allen G. 1991. *The French Paracelsians: The Chemical Challenge to Medical and Scientific Tradition in Early Modern France*. Cambridge: Cambridge University Press.

———. 2001. *Chemistry and Medical Debate: Van Helmont to Boerhaave*. Canton, MA: Science History Publications.

Delamain, Robert. 1935. *Histoire du cognac*. Preface by Gaston Chérau. Paris: Librairie Stock.

Denis, Gilles. 2001. Du physicien agriculteur du dix-huitième à l'agronome des dix-neuvième et vingtième siècles: Mise en planche d'un champ de recherche et d'enseignement. *Comptes rendues de l'Académie d'agriculture de France* 87 (4): 81–103.

Dijksterhuis, Fokko Jan. 2007. Constructive Thinking: A Case for Dioptrics. In *The Mindful Hand: Inquiry and Invention from the Late Renaissance to Early Industrialisation*, ed. L. Roberts, S. Schaffer, and P. Dear, 59–82. Amsterdam: Edita.

Dimier, Louis. 1934. Bernard Palissy Rocailleur, Fontenier et Décorateur de Jardins. *Gazette des Beaux-Arts* 12.

Dion, Roger. 1959. *Histoire de la vigne et du vin en France des origines au XIXe siècle*. Paris: Flammarion.

Dittrich, Erhard. 1974. *Die deutschen und österreichischen Kameralisten*. Darmstadt: Wissenschaftliche Buchgesellschaft.

Dobbs, B. J. T. 1975. *The Foundations of Newton's Alchemy*. Cambridge: Cambridge University Press.

Doering-Manteuffel, Sabine, Josef Mancal, and Wolfgang Wüst, eds. 2001. *Pressewesen der Aufklärung. Periodische Schriften im Alten Reich*. Berlin: Akademie-Verlag.

Donnelly, James. 1994. Consultants, Managers, Testing Slaves: Changing Roles for Chemists in the British Alkali Industry, 1850–1920. *Technology and Culture* 35:100–128.

Donovan, Arthur. 1993. *Antoine Lavoisier: Science, Administration, and Revolution*. Cambridge: Cambridge University Press.

Dornier, Carole. 1997. Le Vin, cette liqueur traîtresse. . . . In *Le Vin*, ed. J. Bart and E. Wahl, special issue, *Dix-huitième siècle* 29:167–84.

Douglas, Mary, and Baron Isherwood. 1996. *The World of Goods: Towards an Anthropology of Consumption*. With new intro. London: Routledge.

Dräger, Ulf. 2004. Abriss der Stolberger Münzgeschichte von Mittelalter bis zur frühen Neuzeit. In *„die Mark zu 13 Reichstaler und 8 Groschen beibehalten werde". Die Alte Münze in Stol-*

*berg (Harz)*, ed. M. Lücke and U. Dräger, 9–18. Stolberg: Numismatischer Verlag Leipziger Münzhandlung und Auktion Heidrun Höhn.

Drayton, Richard. 2000. *Nature's Government: Science, Imperial Britain, and the "Improvement" of the World.* New Haven, CT: Yale University Press.

Duhem, Pierre. 1906. Léonard de Vinci, Cardan et Bernard Palissy. *Bulletin Italien* 6 (4): 289–320.

Duhem, Pierre. 1909. Léonard De Vinci et Les Origines de la Géologie. In *Etudes sur Léonard de Vinci: Second Série*, 281–342. Paris: Librarie Scientifique A. Herman et Fils.

Dujardin, Jules. 1900. *Recherches rétrospectives sur l'art de la distillation. Historique de l'Alcool de l'Alambic et de l'Alcoométrie.* Paris: J. Dujardin.

Dupuy, Ernest. 1902. *Bernard Palissy: l'homme, l'artiste, le savant, l'écrivain.* Paris: Société Française d'Imprimerie et de Librairie.

Durie, Alastair. 2003. Medicine, Health and Economic Development: Promoting Spa and Seaside Resorts in Scotland c. 1750–1830. *Medical History* 47:195–216.

Eamon, William. 1994. *Science and the Secrets of Nature: Books of Secrets in Medieval and Early Modern Culture.* Princeton, NJ: Princeton University Press.

Eddy, Matthew D. 2001. The Doctrine of Salts and Rev John Walker's Analysis of a Scottish Spa. *Ambix* 48:137–60.

———. 2002. Scottish Chemistry, Classification and the Early Mineralogical Career of the "Ingenious" Rev. Dr. John Walker. *British Journal for the History of Science* 35:382–422.

———. 2004. Scottish Chemistry, Classification and the Late Mineralogical Career of the "Ingenious" Professor John Walker (1779–1803). *British Journal for the History of Science* 37:373–99.

———. 2005. Set in Stone: The Medical Language of Mineralogy in Scotland. In *Science and Beliefs: From Natural Philosophy to Natural Science*, ed. D. M. Knight and M. D. Eddy, 77–94. Aldershot, UK: Ashgate.

———. 2007. The Aberdeen Agricola: James Anderson and the Medico-Chemistry of Georgics and Geology, 1770–1800. In *New Narratives in Eighteenth-Century Chemistry*, ed. L. M. Principe, 139–56. Dordrecht: Springer.

———. 2008. *The Language of Mineralogy: John Walker, Chemistry and the Edinburgh Medical School, 1750–1800.* Aldershot: Ashgate.

———. 2009. Natural Philosophy and Natural History Books in Eighteenth-Century Scotland. In *The Edinburgh History of the Book in Scotland*, vol. 2, *1707–1800*, ed. S. Brown and W. McDougall. Edinburgh: University of Edinburgh Press.

Edelstein, Sidney M. 1974. *Historical Notes on the Wet-Processing Industry.* New York: American Dyestuff Reporter.

Edgerton, David. 2006. *The Shock of the Old: Technology and Global History since 1900.* London: Profile.

Eichler, Helga. 1978. Die Leipziger Ökonomische Sozietät im 18. Jahrhundert. *Jahrbuch für Geschichte des Feudalismus* 2:357–86.

Eliade, Mircea. 1980. *Schmiede und Alchemisten.* Stuttgart: Klett-Cotta Verlag.

Elliot, Paul, and Stephen Daniels. 2006. The "School of True, Useful and Universal Science"? Freemasonry, Natural Philosophy and Scientific Culture in Eighteenth-Century England. *British Journal for the History of Science* 39:207–29.

Elsner, John, and Roger Cardinal, eds. 1994. *The Cultures of Collecting.* London: Reaktion.

Emerson, Roger L. 2002. The Scientific Interests of Archibald Campbell, 1st Earl of Ilay and 3rd Duke of Argyll (1682–1761). *Annals of Science* 59:21–56.

Estes, J. Worth. 1990. *Dictionary of Protopharmacology: Therapeutic Practices, 1700–1850.* Canton, MA: Science History Publications.

———. 1991. Quantitative Observations of Fever and Its Treatment before the Advent of Clinical Thermometers. *Medical History* 35:189–216.

Fairchilds, Cissie. 1993. The Production and Marketing of Populuxe Goods in Eighteenth-Century Paris. In *Consumption and the World of Goods,* ed. J. Brewer and R. Porter, 228–48. London: Routledge.

Fairlie, Susan. 1964–1965. Dyestuffs in the Eighteenth Century. *Economic History Review* 17:488–510.

Farrington, Benjamin. 1979. *Francis Bacon: Philosopher of Industrial Science.* New York: Octagon.

Fessner, Michael, Angelika Friedrich, and Christoph Bartels. 2002. *"gründliche Abbildung des uralten Bergwerks"—Eine virtuelle Reise durch den historischen Harzbergbau.* Montanregion Harz, vol. 3. Bochum: Deutsches Bergbau-Museum. (CD and textbook.)

Figala, Karin. 1998. Quecksilber. In *Alchemie, Lexikon einer hermetischen Wissenschaft,* ed. C. Priesner and K. Figala, 295–300. Munich: C. H. Beck.

Fildes, Valerie. 1988. *Wet Nursing: A History from Antiquity to the Present.* New York: Basil Blackwell.

Findlen, Paula. 1990. Jokes of Nature and Jokes of Knowledge: The Playfulness of Scientific Discourse in Early Modern Europe. *Renaissance Quarterly* 43:293–331.

———. 1994. *Possessing Nature: Museums and Collecting in Early Modern Italy.* Berkeley: University of California Press.

———. 2002. Inventing Nature: Commerce, Nature and Art in Early Modern Cabinets of Curiosities. In *Merchants and Marvels: Commerce, Science and Art in Early Modern Europe,* ed. P. Findlen and P. Smith, 297–324. New York: Routledge Press.

Fine, Ben, and Ellen Leopold. 1990. Consumerism and the Industrial Revolution. *Social History* 15 (2): 151–79.

Fischer, Ernst, Wilhelm Haefs, and York-Gothart Mix, eds. 1999. *Von Almanach bis Zeitung. Ein Handbuch der Medien in Deutschland 1700–1800.* Munich: Beck.

Fish, Stanley Eugene. 1980. *Is There a Text in This Class? The Authority of Interpretive Communities.* Cambridge, MA: Harvard University Press.

Fissell, Mary, and Roger Cooter. 2003. Exploring Natural Knowledge: Science and the Popular. In *The Cambridge History of Science,* vol. 4, *Eighteenth-Century Science,* ed. R. Porter, 129–58. Cambridge: Cambridge University Press.

Flahaut, Jean. 1995. Lavoisier et les pharmaciens parisiens de son temps. *Revue d'histoire de la pharmacie* 42:349–55.

Flandrin, Jean-Louis. 1989. Distinction through Taste. In *A History of Private Life,* vol. 3, *Passions of the Renaissance,* ed. R. Chartier, 267–307. Cambridge, MA: Harvard University Press, Belknap Press.

———. 1996a. Assaisonnement, cuisine et diététique. In *Histoire de l'alimentation,* ed. J.-L. Flandrin and M. Montanari, 491–509. Paris: Fayard.

———. 1996b. Choix alimentaires et art culinaire. In *Histoire de l'alimentation,* ed. J.-L. Flandrin and M. Montanari, 657–81. Paris: Fayard.

Fleischer, Manfred P. 1981. The First German Agricultural Manuals. *Agricultural History* 55:1–15.

Fleischmann, Wilhelm. 1910. Geschichte über Milch und Milchzucker. *Archiv für die Geschichte der Medizin* 4 (1): 1–19.

Forbes, Archibald. 1929. *A History of the Army Ordnance Service.* 3 vols. London: Medici Society.

Forbes, Robert J. 1948. *Short History of the Art of Distillation from the Beginnings up to the Death of Cellier Blumenthal.* Leiden: E. J. Brill.

Forrester, John M. 2000. The Origins and Fate of James Currie's Cold Water Treatment for Fever. *Medical History* 44:57–74.

Forssman, Erik. 1956. Renaissance, Manierismus und Nürnberger Goldschmiedekunst. In *Säule und Ornament. Studien zum Problem des Manierismus in den nordischen Säulenbüchern und Vorlageblättern des 16. und 17. Jahrhunderts.* Stockholm: Almqvist & Wiksell.

Foster, Michael. 1970. The Physiology of Digestion in the Eighteenth Century. In *Lectures on the History of Physiology during the Sixteenth, Seventeenth and Eighteenth Centuries.* New York: Dover.

Fox, Robert, and Agustí Nieto-Galan. 1999. *Natural Dyestuffs and Industrial Culture in Europe, 1750–1880.* Canton, MA: Science History Publications.

Fox, Robert, and Anthony Turner, eds. 1998. *Luxury Trades and Consumerism in Ancien Régime Paris: Studies in the History of the Skilled Workforce.* Aldershot, UK: Ashgate.

Fragonard, Marie-Madeleine. 1996. Introduction to *Œuvres Complètes de Bernard Palissy.* Ed. K. Cameron and M.-M. Fragonard. Mont-de-Marsan: Editions Interuniversitaires.

Frängsmyr, Tore, John L. Heilbron, and Robin Rider. 1990. *The Quantifying Spirit in the Eighteenth Century.* Berkeley: University of California Press.

Freytag, Nils, and Wolfgang Piereth. 2002. Städtische Holzversorgung im 18. und 19. Jahrhundert—Dimensionen und Perspektiven eines Forschungsfeldes. In *Städtische Holzversorgung. Machtpolitik, Armenfürsorge und Umweltkonflikte in Bayern und Österreich (1750–1850),* ed. W. Siemann, N. Freytag, and W. Piereth, 1–8. Munich: C. H. Beck.

Fries, Albert. ed. 1981. *Albertus Magnus. Ausgewählte Texte,* mit einer Kurzbiographie von Willehad Paul Eckert. Darmstadt: Wissenschaftliche Buchgesellschaft.

Fruton, Joseph S. 1999. *Proteins, Enzymes, Genes: The Interplay of Chemistry and Biology.* New Haven, CT: Yale University Press.

Fujimura, Joan H. 1992. Crafting Science: Standardized Packages, Boundary Objects, and "Translation." In *Science as Practice and Culture,* ed. A. Pickering, 168–211. Chicago: University of Chicago Press.

Fussell, George E. 1976. Agricultural Science and Experiment in the Eighteenth Century. *Agricultural History Review* 24:44–47.

Gage, John. 1998. Colour Words in the High Middle Ages. In *Looking through Paintings: The Study of Painting Techniques and Materials in Support of Art Historical Research (Leids Kunsthistorisch Jaarboek XI),* ed. E. Hermens, 35–48. Baarn: Uitgeverij de Prom.

———. 1999. *Color and Meaning: Art, Science and Symbolism.* New York: Thames & Hudson.

Garnsey, Peter. 1999. *Food and Society in Classical Antiquity.* Cambridge: Cambridge University Press.

Gastinel-Coural, Chantal. 1997. Chevreul à la Manufacture des Gobelins. In *Michel-Eugène Chevreul. Un savant, des couleurs!* ed. G. Roque, B. Bodo, and F. Vienot, 67–80. Paris: Editions du Muséum national d'histoire naturelle.

Gazley, John G. 1973. *The Life of Arthur Young 1741–1820.* Philadelphia: American Philosophical Society.

Gélis, Jacques, Mireille Laget, and Marie-France Morel. 1980. *Der Weg ins Leben. Geburt und Kindheit in früherer Zeit.* Munich: Beck.

Gerber, Michael Rüdiger. 1995. *Die schlesischen Provinzialblätter 1785–1849.* Sigmaringen: Thorbecke.

Gerber-Visser, Gerrendina. In press. "Die Beschaffenheit des Landes"—Topographische Beschreibungen der Oekonomischen Gesellschaft Bern. In *Landschaften agrarisch-ökonomischen Wissens. Regionale Fallstudien zu landwirtschaftlichen und gewerblichen Themen in Zeitschriften und Sozietäten des 18. Jahrhunderts,* ed. M. Popplow. Cottbuser Studien zur Geschichte von Technik, Arbeit und Umwelt. Münster: Waxmann.

Gerhard, Hans-Jürgen, ed. 1997. *Struktur und Dimension. Festschrift für Karl Heinrich Kaufhold zum 65. Geburtstag.* Vol. 1. Stuttgart: Franz Steiner Verlag.

Gettens, Rutherford J., Robert L. Feller, and W. T. Chase. 1994. Vermilion and Cinnabar. In *Artists' Pigments: A Handbook of Their History and Characteristics,* ed. A. Roy. 2 vols. Oxford: Oxford University Press.

Giaufret Colombani, Hélène, and Maria Teresa Mascarello. 1997. De Dictionnaires en Encyclopédie: le savoir oenologique et sa diffusion. In *Le Vin,* ed. J. Bart and E. Wahl, special issue, *Dix-huitième siècle* 29:51–68.

Gillispie, Charles C. 1957. The Natural History of Industry. *Isis* 48:398–407.

———. 1959. *A Diderot Pictorial Encyclopedia of Trades and Industry.* 2 vols. New York: Dover.

———. 1980. *Science and Polity in France at the End of the Old Regime.* Princeton, NJ: Princeton University Press.

Ginzburg, Carlo. 1986. Clues: Roots of an Evidential Paradigm. In *Myths, Emblems, Clues,* trans. J. Tedeschi and A. C. Tedeschi, 96–125. London: Hutchinson Radius.

Gittins, L. 1979. Innovations in Textile Bleaching in Britain in the Eighteenth Century. *Business History Review* 53:194–204.

Goddard, Nicholas. 1989. Agricultural Literature and Societies. In *The Agrarian History of England and Wales,* vol. 6, *1750–1850,* ed. J. Thirsk, 361–83. Cambridge: Cambridge University Press.

Golden, Janet. 1996. *A Social History of Wet Nursing in Amercia: From Breast to Bottle.* Cambridge: Cambridge University Press.

Golinski, Jan. 1992. *Science as Public Culture: Chemistry and Enlightenment in Britain, 1760–1820.* Cambridge: Cambridge University Press.

Gooday, Graeme. 2000. The Flourishing of History of Technology in the United Kingdom: A Critique of Antiquarian Complaints of "Neglect." *History of Technology* 22:189–201.

Goodman, Dena. 1992. Public Sphere and Private Life: Towards a Synthesis of Recent Historiographical Approaches to the Old Regime. *History and Theory* 31:1–20.

Graefe, Christa, ed. 1989. *Staatsklugheit und Frömmigkeit. Herzog Julius zu Braunschweig-Lüneburg, ein norddeutscher Landesherr des 16. Jahrhunderts.* Weinheim: VCH Verlagsgesellschaft.

Grafton, Anthony, April Shelford, and Nancy Siraisi, eds. 1992. *New Worlds, Ancient Texts: The Power of Tradition and the Shock of Discovery.* Cambridge, MA: Harvard University Press, Belknap Press.

Grafton, Antony, and Nancy Siraisi, eds. 1999. *Natural Particulars: Nature and the Disciplines in Renaissance Europe.* Cambridge, MA: MIT Press.

Grenby, Matthew O. 2002. Adults Only? Children and Children's Books in British Circulating Libraries, 1748–1848. *Book History* 5:19–38.

Grossmann, Henryk. 1935. Die gesellschaftlichen Grundlagen der mechanistischen Philosophie und die Manufaktur. *Zeitschrift für Sozialforschung* 4 (2): 161–231.

———. 1987. The Social Foundations of Mechanistic Philosophy and Manufacture. *Science in Context* 1:129–80.

Grote, Andreas, ed. 1994. *Macroscosmos in Microcosmo—Die Welt in der Stube: Zur Geschichte des Sammelns, 1450 bis 1800.* Opladen: Leske und Budrich.

Habermas, Jürgen. 1989. *The Structural Transformation of the Public Sphere: An Inquiry into a Category of Bourgeois Society.* Trans. T. Burger. Cambridge: Polity.

Hacking, Ian. 1983. *Representing and Intervening: Introductory Topics in the Philosophy of Natural Science.* Cambridge: Cambridge University Press.

———. 2002. *Historical Ontology.* Cambridge, MA: Harvard University Press.

Hahn, Roger. 1963. The Application of Science to Society: The Societies of Arts. *Studies on Voltaire and the Eighteenth Century* 25:829–36.

———. 1971. *The Anatomy of a Scientific Institution: The Paris Academy of Sciences, 1666–1803.* Berkeley: University of California Press.

Halimi, Suzy. 1988. La Bataille du gin en Angleterre dans la première moitié du XVIIIe siècle. In *Toxicomanies—Alcool, Tabac, Drogues,* special issue, *Histoire, économie et société* 7:461–73.

Hall, Bert S. 1997. *Weapons and Warfare in Renaissance Europe.* Baltimore: Johns Hopkins University Press.

Hall, Rupert. 1969. The Scholar and the Craftsman in the Scientific Revolution. In *Criticial Problems in the History of Science,* ed. M. Clagett, 3–23. Madison: University of Wisconsin Press.

Halleux, Robert. 1996. The Reception of Arabic Alchemy in the West. In *Encyclopedia of the History of Arabic Science,* ed. R. Rashed, 3:886–902. London: Routledge.

Halporn, Barbara C. 2000. *The Correspondence of Johann Amerbach: Early Printing in Its Social Context.* Trans. B. C. Halporn. Ann Arbor: University of Michigan Press.

Hammermayer, Ludwig. 1995. Zur Publizistik von Aufklärung, Reform und Sozietätsbewegung in Bayern. Die Burghausener Sittlich-Ökonomische Gesellschaft und ihr "Baierisch-Ökonomischer Hausvater" (1779–1786). *Zeitschrift für bayerische Landesgeschichte* 58:341–401.

Hannaway, Owen. 1975. *The Chemists and the Word: The Didactic Origins of Chemistry.* Baltimore: John Hopkins University Press.

Hanschmann, Alexander Bruno. 1903. Palissy's Geist und Charakter und seine Bedeutung für die Geschichte der Pädagogischen Erkenntnis. In *Bernard Palissy, der Künstler, Naturforscher und Schriftsteller, als Vater der induktiven Wissenschaftsmethode des Bacon von Verulam: Ein Beitrag zur Geschichte der Naturwissenschaften und der Philosophie.* Leipzig: Dieterich.

Haraway, Donna J. 1991. Situated Knowledges: The Science Question in Feminism and the Privilege of Partial Perspective. In *Simians, Cyborgs, and Women: The Reinvention of Nature,* 183–201. New York: Routledge.

Harkness, Deborah E. 2006. Nosce teipsum: Curiosity, the Humoural Body and the Culture of Therapeutics in Late Sixteenth- and Early Seventeenth-Century England. In *Curiosity and Wonder from the Renaissance to the Enlightenment,* ed. R. J. W. Evans and A. Marr, 171–92. Aldershot, UK: Ashgate.

Harley, Rosamond D. 2001. *Artists' Pigments 1600–1835.* 2nd ed. London: Archetype.

Harwood, Jonathan. 2005. *Technology's Dilemma: Agricultural Colleges between Science and Practice in Germany, 1860–1934.* Bern: Lang.

Hassan, John. 2003. *The Seaside, Health and the Environment in England and Wales since 1800.* Aldershot, UK: Ashgate.

Hecht, Gabrielle, and Michael Thad Allen. 2001. Introduction: Authority, Political Machines, and Technology's History. In *Technologies of Power: Essays in Honor of Thomas Parke Hughes and Agatha Chipley Hughes,* ed. M. T. Allen and G. Hecht, 1–23. Cambridge, MA: MIT Press.

Heesen, Anke te. 2002. *The World in a Box: The Story of an Eighteenth-Century Picture Encyclopedia.* Chicago: University of Chicago Press.

Heimann, Peter M. 1981. Ether and Imponderables. In *Conceptions of Ether: Studies in the History of Ether Theory 1740–1900*, ed. G. N. Cantor and M. S. Hodge, 61–84. Cambridge: Cambridge University Press.

Hein, Wolfgang-Hagen, and Holm-Dietmar Schwarz, eds. 1975–1997. *Deutsche Apotheker-Biographie.* 4 vols. Stuttgart: Wissenschaftliche Verlagsgesellschaft.

Hembry, Phyllis M. 1990. *The English Spa, 1560–1815.* London: Athlone Press.

Henderson, John A. 1907. *Aberdeenshire Epitaphs and Inscriptions.* Aberdeen: Aberdeen Daily Journal.

Henschke, Ekkehard. 1974. *Landesherrschaft und Bergbauwirtschaft. Zur Wirtschafts- und Verwaltungsgeschichte des Oberharzer Bergbaugebietes im 16. und 17. Jahrhundert.* Berlin: Duncker & Humblot.

Herrmann, Bernd. 2006. Zur Historierung der Schädlingsbekämpfung. In *Technik, Arbeit und Umwelt in der Geschichte. Günter Bayerl zum 60. Geburtstag*, ed. T. Meyer and M. Popplow, 317–38. Münster: Waxmann.

Hess, Christel. 1987. *Presse und Publizistik in der Kurpfalz in der zweiten Hälfte des 18. Jahrhunderts.* Frankfurt am Main: Lang.

Hessen, Boris. 1971. The Social and Economic Roots of Newton's "Principia." In *Science at the Cross Roads: Papers Presented to the International Congress of the History of Science and Technology, Held in London from June 29th to July 3rd, 1931*, 151–212. London: F. Cass.

Hickel, Erika. 1973. *Arzneimittel-Standardisierung im 19. Jahrhundert in den Pharmakopöen Deutschlands, Frankreichs, Großbritanniens und der Vereinigten Staaten von Amerika.* Stuttgart: Wissenschaftliche Verlagsgesellschaft.

Hilaire-Pérez, Liliane. 2000. *L'invention technique au siècle des Lumières.* Pref. by Daniel Roche. Paris: Albin Michel.

Hilaire-Pérez, Liliane, and Anne-Françoise Garçon. 2003. *Les chemins de la nouveauté: innover, inventer au regard de l'histoire.* Paris: Editions du CTHS.

Hills, Paul. 1999. Interpreting Renaissance Color. In *The Italian Renaissance in the Twentieth Century*, ed. A. J. Greico, M. Rocke, and F. G. Superbi, 337–50. Florence: Leo S. Olschki.

Hobhouse, Henry. 1985. *Seeds of Change: Five Plants That Transformed Mankind.* London: Sidgwick & Jackson.

Hobsbawm, Eric J. 2000. *Bandits.* New York: New Press.

Hochstrasser, Julie Berger. 2007. *Still Life and Trade in the Dutch Golden Age.* New Haven, CT: Yale University Press.

Hoffmann, Dietrich. 1978. Der Berghauptmann Heinrich Albert von dem Bussche (1664–1731) und die „Goldene Zeit" des Oberharzer Bergbaus. In *Niedersächsisches Jahrbuch für Landesgeschichte.* Hannover: Verlag Hahnesche Buchhandlung, 50:275–310.

Hogg, Oliver F. G. 1963. *The Royal Arsenal: Its Background, Origin, and Subsequent History.* 2 vols. London: Oxford University Press.

Holenstein, André. 2003. *„Gute Policey" und lokale Gesellschaft im Staat des Ancien Régime. Das Fallbeispiel der Markgrafschaft Baden(-Durlach).* Tübingen: Bibliotheca Academica.

Hollister-Short, Graham. 1995. The Steam Engine in Mine Pumping 1712–c.1870: Continuities and Discontinuities. In *Vom Bergbau- zum Industrierevier*, ed. E. Westermann, 335–45. Stuttgart: Franz Steiner Verlag.

———. 2000. Before and after the Newcomen Engine of 1712: Ideas, Gestalts, Practice. In *Konjunkturen im europäischen Bergbau der vorindustriellen Zeit. Festschrift für Ekkehard Westermann zum 60. Geburtstag*, ed. C. Bartels and M. Denzel, 221–36. Stuttgart: Franz Steiner Verlag.

Holmes, Frederic L. 1963. Elementary Analysis and the Origins of Physiological Chemistry. *Isis* 54 (1): 50–81.

———. 1971. Analysis by Fire and Solvent Extractions: The Metamorphosis of a Tradition. *Isis* 62 (2): 128–48.

———. 1975. The Transformation of the Science of Nutrition. *Journal of the History of Biology* 8:135–44.

———. 1989a. Chemists in the Plant Kingdom. In *Eighteenth-Century Chemistry as an Investigative Enterprise*, ed. F. Holmes. Berkeley: University of California at Berkeley, Office for History of Science and Technology.

———. 1989b. *Eighteenth-Century Chemistry as an Investigative Enterprise: 5 Lectures Delivered at the International Summer School in Bologna, August 1988.* Berkeley: Office for History of Science and Technology.

———. 1995. Beyond the Boundaries: Concluding Remarks on the Workshop. In *Negotiating a New Language for Chemistry: Lavoisier in European Context*, ed. B. Bensaude-Vincent and F. Abbri, 267–78. Canton, MA: Science History Publications.

Holmes, Frederic L., and Trevor H. Levere. eds. 2000. *Instruments and Experimentation in the History of Chemistry.* Cambridge, MA: MIT Press.

Holmyard, E. J. 1990. *Alchemy.* New York: Dover. (Reprint of 1957 ed.)

Holub, Robert C. 2003. *Reception Theory: A Critical Introduction.* London: Routledge.

Hoolihan, Christopher. 1985. Thomas Young, M.D. (1726?–1783) and Obstetrical Education in Edinburgh. *Journal of the History of Medicine and Allied Sciences* 40:327–45.

Hooper-Greenhill, Eilean. 1992. *Museums and the Shaping of Knowledge.* London: Routledge.

Hörmann, Johannes. 1898. Die königliche Hofapotheke in Berlin (1598–1898). *Hohenzollern-Jahrbuch*, 208–26.

Howard, Robert H. 1996. The Evolution of the Process of Powder Making from an American Perspective. In *Gunpowder: The History of an International Technology*, ed. B. J. Buchanan, 3–23. Bath: Bath University Press.

Hudson, Derek, and Kenneth W. Luckhurst. 1954. *The Royal Society of Arts, 1754–1954.* London: Murray.

Hufbauer, Karl. 1982. *The Formation of the German Chemical Community (1720–1795).* Berkeley: University of California Press.

Hughes, Thomas P. 1983. *Networks of Power: Electrification in Western Society, 1880–1930.* Baltimore: Johns Hopkins University Press.

———. 1987. The Evolution of Large Technological Systems. In *The Social Construction of Technological Sysems: New Directions in the Sociology and History of Technology*, ed. W. E. Bijker, T. P. Hughes, and T. Pinch, 51–82. Cambridge, MA: MIT Press.

Hünemörder, Kai. In press. Die Celler Landwirtschaftsgesellschaft und das Hannoverische Magazin: Schnittstellen der Ökonomischen Aufklärung in Kurhannover (1750–1789). In *Landschaften agrarisch-ökonomischen Wissens. Regionale Fallstudien zu landwirtschaftlichen und gewerblichen Themen in Zeitschriften und Sozietäten des 18. Jahrhunderts*, ed. M. Popplow. Cottbuser Studien zur Geschichte von Technik, Arbeit und Umwelt 30. Münster: Waxmann.

Hunter, Michael, and Edward B. Davis, eds. 2000. *Unpublished Writings, 1645–1670.* Vol. 13 of *The Works of Robert Boyle.* London: Pickerings & Chatto.

Hunter, Michael, and Simon Schaffer, eds. 1989. *Robert Hooke: New Studies.* Woodbridge: Boydell Press.

Ingold, Tim. 2000. *The Perception of the Environment: Essays in Livelihood, Dwelling and Skill.* London: Routledge.

Jacob, Margaret C. 1997. *Scientific Culture and the Making of the Industrial West.* New York: Oxford University Press.

Jacob, Margaret C., and Larry Stewart. 2004. *Practical Matter. Newton's Science in the Service of Industry and Empire, 1687–1851.* Cambridge, MA: Harvard University Press.

Jacobus, Mary. 1995. *First Things. The Maternal Imaginery in Literature, Art, and Psychoanalysis.* New York: Routledge.

Jacomy, Bruno. 1990. *Une histoire des techniques.* Paris: Seuil.

Jarck, Horst-Rüdiger, and Gerhard Schildt. 2000. *Die Braunschweigische Landesgeschichte. Jahrtausendrückblick einer Region.* Braunschweig: Appelhans.

Jeanneret, Michel. 2001. *Perpetual Motion: Transforming Shapes in the Renaissance from Da Vinci to Montaigne.* Trans. N. Poller. Baltimore: Johns Hopkins University Press.

Jenner, Mark S. R., and Patrick Wallis. 2007. *Medicine and the Market in England and Its Colonies, c. 1450–c. 1850.* Houndsmill, UK: Palgrave Macmillan.

Johns, Adrian. 2000. The Physiology of Reading. In *Books and the Sciences in History,* ed.y M. Frasca-Spada and N. Jardine. Cambridge: Cambridge University Press.

———. 2007. The Identity Engine: Printing, Publishing and the Birth of the Knowledge Economy. In *The Mindful Hand: Inquiry and Invention from the Late Renaissance to Early Industrialization,* ed. L. Roberts, S. Schaffer and P. Dear, 403–28. Amsterdam: Edita.

Johnston, Dorothy B. 1987. All Honourable Men: The Award of Irregular Degrees in King's and Marischal College in the Eighteenth Century. In *Aberdeen in the Enlightenment,* ed. J. J. Carter and J. H. Pittock. Aberdeen: Aberdeen University Press.

Jones, Colin. 1991. Bourgeois Revolution Revivified: 1789 and Social Change. In *Rewriting the French Revolution,* ed. C. Lucas, 69–118. Oxford: Clarendon.

———. 1996. The Great Chain of Buying: Medical Advertisement, the Bourgeois Public Sphere, and the Origins of the French Revolution. *American Historical Review* 101:13–40.

Jones, Jean. 1985. James Hutton's Agricultural Research and His Life as a Farmer. *Annals of Science* 42:573–601.

Julien, Pierre. 1992. Sur les relations entre Macquer et Baumé. *Revue d'histoire de la pharmacie* 39:65–77.

Justin, Emile. 1935. *Les sociétés royales d'agriculture au XVIII siècle (1757–1793).* Paris and Saint-Lô: Imprimerie Barbaroux.

Kahane, Ernest. 1978. *Parmentier: ou la dignité de la pomme de terre, essai sur la famine.* Paris: A. Blanchard.

Kaplan, David M., ed. 2004. *Readings in the Philosophy of Technology.* Lanham, MD: Rowman & Littlefield.

Kemp, Martin. 1999. Palissy's Philosophical Pots: Ceramics, Grottoes and the Matrice of the Earth. In *Le Origini Della Modernita (II),* ed. W. Tega. Milan: Leo S. Olschki.

———. 2001. *Visualizations: The Nature Book of Art and Science.* Berkeley: University of California Press.

Kennedy, Emmet. 1989. *A Cultural History of the French Revolution.* New Haven, CT: Yale University Press.

Keyser, Barbara W. 1990. Between Science and Craft: The Case of Berthollet and Dyeing. *Annals of Science* 47:213–60.

Kim, Mi Gyung. 2003. *Affinity, That Elusive Dream: A Genealogy of the Chemical Revolution.* Cambridge, MA: MIT Press.

King, Steven. 2006. Accessing Drugs in the Eighteenth-Century Regions. In *From Physic to Pharmacy: Five Hundred Years of Drug Retailing,* ed. L. H. Croft. Aldershot, UK: Ashgate.

Kinini, Angélique. 1999. Les principales fabriques des cotons filés rouges de la Grèce à la fin du XVIIIème siècle en Théssalie: le cas des manufactures de la ville d'Ampélakia. In *Natural Dyestuffs and Industrial Culture in Europe, 1750–1880,* ed. R. Fox and A. Nieto-Galan, 71–100. Canton, MA: Science History Publications.

Klein, Ursula. 1994. *Verbindung und Affinität. Die Grundlegung der neuzeitlichen Chemie an der Wende vom 17. bis zum 18. Jahrhundert.* Basel: Birkhäuser.

———. 1996. The Chemical Workshop Tradition and the Experimental Practice: Discontinuities within Continuities. In *Fundamental Concepts of Early Modern Chemistry,* ed. Wolfgang Lefèvre, special issue, *Science in Context* 9:251–87.

———. 2003a. *Experiments, Models and Tools. Cultures of Organic Chemistry in the Nineteenth Century.* Stanford, CA: Stanford University Press.

———, ed. 2003b. Spaces of Classification. Preprint 240, Max Planck Institute for the History of Science, Berlin.

———. 2005a. Experiments at the Intersection of Experimental History, Technological Inquiry, and Conceptually Driven Analysis: A Case Study from Early Nineteenth-Century France. *Perspectives on Science* 13 (1): 1–48.

———. 2005b. Shifting Ontologies, Changing Classifications: Plant Materials from 1700 to 1830. *Studies in History and Philosophy of Science* 36 A (2): 261–329.

———. 2005c. Technoscience avant la lettre. In *Technoscientific Productivity,* ed. Ursula Klein, special issue, *Perspectives on Science* 13 (2/3): 226–66.

———. 2007a. Apothecary's Shops, Laboratories and Manufacture in Eighteenth-Century Germany. In *The Mindful Hand: Inquiry and Invention from the Late Renaissance to Early Industrialisation,* ed. L. Roberts, S. Schaffer and P. Dear, 246–76. Amsterdam: Edita.

———. 2007b. Apothecary-Chemists in Eighteenth-Century Germany. In *New Narratives in Eighteenth-Century Chemistry,* ed. L. M. Principe, 97–137. Dordrecht: Springer.

———. 2008a. Die technowissenschaftlichen Laboratorien der Frühen Neuzeit. *NTM* 16:5–39.

———. 2008b. The Laboratory Challenge: Some Revisions of the Standard View of Early Modern Experimentation. *Isis* 99 (4): 769–782.

Klein, Ursula, and Wolfgang Lefèvre. 2007. *Materials in Eighteenth-Century Science: A Historical Ontology.* Cambridge, MA: MIT Press.

Klemm, Volker. 1992. *Agrarwissenschaften in Deutschland. Geschichte—Tradition. Von den Anfängen bis 1945.* St. Katharinen: Scriptae Mercaturae.

Knoeff, Rina. 2002. *Hermann Boerhaave (1668–1738): Calvinist Chemist and Physician.* Amsterdam: Koninklijke Nederlandse Akademie van Wetenschappen.

Knoll, Ilona. 2003. *Der Mannheimer Botaniker Friedrich Casimir Medicus (1736–1808): Leben und Werk.* Heidelberg: Palatina Verlag.

Knorr-Cetina, Karin. 1999. *Epistemic Cultures: How the Sciences Make Knowledge.* Cambridge, MA: Harvard University Press.

Koch, Johannes Hugo. 1984. *Mit Model, Krapp und Indigo.* Hamburg: Christians Verlag.

Koch, Manfred. 1963. *Geschichte und Entwicklung des Bergmännischen Schrifttums.* Goslar: Huebener Verlag.

Koerner, Lisbet. 1999. *Linnaeus: Nature and Nation.* Cambridge, MA: Harvard University Press.

Konersmann, Frank. 2005. "Ueber die Nutzbarkeit des Predigtamtes." Pfarrer als Agrarschriftsteller und Landwirte in der Pfalz (1770–1852). *Aufklärung* 17:5–33.

Kopp, Hermann. 1966. *Geschichte der Chemie.* 4 vols. Hildesheim: Olms.

Kopsidis, Michael. 2006. *Agrarentwicklung. Historische Agrarrevolutionen und Entwicklungsökonomie.* Stuttgart: Steiner.

Koyré, Alexandre. 1943. Galileo and Plato. *Journal of the History of Ideas* 4 (4): 400–428.

———. 1968. *Metaphysics and Measurement.* Cambridge, MA: Harvard University Press.

Krafft, Fritz. 2002. Johann Christian Wiegleb und seine Rolle bei der Verwissenschaftlichung der Chemie. In *Apotheker und Universität. Die Vorträge der Pharmaziehistorischen Biennale in Leipzig vom 12. Bis 14. Mai 2000,* ed. C. Friedrich and W.-D. Müller-Jahncke, 151–95. Stuttgart: Wissenschaftliche Verlagsgesellschaft.

Kraschewski, Hans-Joachim. 1978. *Wirtschaftspolitik im deutschen Territorialstaat des 16. Jahrhunderts: Herzog Julius von Braunschweig-Wolfenbüttel (1528–1589).* Cologne: Böhlau Verlag.

———. 1989. Der ökonomische Fürst. Herzog Julius als Unternehmer-Verleger der Wirtschaft seines Landes, besonders des Harz-Bergbaus. In *Staatsklugheit und Frömmigkeit. Herzog Julius zu Braunschweig-Lüneburg, ein norddeutscher Landesherr des 16. Jahrhunderts,* ed. C. Graefe, 41–57. Weinheim: VCH Verlagsgesellschaft.

Kremers, Edward, and George Urdang. 1951. *History of Pharmacy: A Guide and a Survey.* Philadelphia: Lippincott.

Kris, Ernst. 1926. Der Stil Rustique. Die Verwendung des Naturabgusses bei Wenzel Jamnitzer und Bernard Palissy. *Jahrbuch der Kunsthistorisches Sammlungen in Wien* 22:137–208.

Kroker, Werner. 1972. Aspekte der Entwicklung des Markscheidewesens am Oberharz. *Technikgeschichte* 39:280–301.

Krüger, Mechthild. 1968. *Zur Geschichte der Elixire, Essenzen und Tinkturen.* Braunschweig: Technische Hochschule.

Kruse, Christiane. 2000. Fleisch werden–Fleisch malen. Malerei als "incarnazione." Mediale Verfahren des Bildwerdens im Libro dell'Arte von Cennino Cennini. *Zeitschrift für Kunstgeschichte* 63:305–25.

Kubátová, Ludmilla, Hans Prescher, and Werner Weisbach. 1994. *Lazarus Ercker (1528/30–1554). Probierer, Berg- und Münzmeister in Sachsen, Braunschweig und Böhmen.* Leipzig: Deutscher Verlag für Grundstoffindustrie.

Küpper-Eichas, Claudia, and Gesche Löning. 1994. Der Betrieb der Clausthaler Münzstätte 1617–1849. In *Die Münze zu Clausthal. Beiträge zur Geschichte der Münzstätte,* ed. B. Gisevius, C. Küpper-Eichas, G. Löning, W. Schütze, and C. Wiechmann, 1–68. Clausthal-Zellerfeld: Oberharzer Geschichts- und Museumsverein.

Kushner, Eva. 1981. Le Dialogue de 1580–1630: Articulations et Fonctions. In *L'Automne de la Renaissance 1580–1630,* ed. J. Lafond and A. Stegmann. Paris: Librairie Philosophe J. Vrin.

La Berge, Ann Elizabeth Fowler. 1992. Mission and Method: The Early Nineteenth-Century French Public Health Movement. Cambridge: Cambridge University Press.

La Force, Clayburn. 1965. *The Development of the Spanish Textile Industry, 1750–1800.* Berkeley: University of California Press.

La Rocque, Aurèle. 1957. Introduction to *The Admirable Discourses of Bernard Palissy.* Ed. A. La Rocque, 1–19. Urbana: University of Illinois Press.

Labouvie, Eva. 1998. *Andere Umstände. Eine Kulturgeschichte der Geburt.* Cologne: Böhlau.

Lachiver, Marcel. 1988. *Vins, vignes et vignerons: histoire du vignoble français.* Paris: Librairie Arthème Fayard.

Lastinger, Valerie. 1996. Re-Defining Motherhood: Breast-Feeding and the French Enlightenment. *Women's Studies: An Interdisciplinary Journal* 25:603–18.

Latour, Bruno. 1986. Visualization and Cognition: Thinking with Eyes and Hands. *Knowledge and Society* 6:1–40.

———. 1993. *We Have Never Been Modern.* Trans. C. Porter. New York: Harvester Wheatsheaf.

———. 1997. On Actor Network Theory: A Few Clarifications. Contribution to nettime, www.desk.nl/~nettime/, 2 May.

———. 1999. *Pandora's Hope: Essays in the Reality of Science Studies.* Cambridge, MA: Harvard University Press.

———. 2000. On the Partial Existence of Existing and Nonexisting Objects. In *Biographies of Scientific Objects,* ed. L. Daston, 247–69. Chicago: University of Chicago Press.

Latour, Bruno, and Steve Woolgar. 1979. *Laboratory Life: The Social Construction of Scientific Facts.* Beverly Hills, CA: Sage.

Lave, Jean. 1988. *Cognition in Practice: Mind, Mathematics and Culture in Everyday Life.* Cambridge: Cambridge University Press, 1988.

Layton, Edwin T., Jr. 1974. Technology as Knowledge. *Technology and Culture* 15 (1): 31–41.

Leake, Chauncey D. 1925. The Historical Development of Surgical Anesthesia. *Scientific Monthly* 20 (3): 304–28.

Lefèvre, Wolfgang. 2000. Galileo Engineer: Art and Modern Science. *Science in Context* 13 (3/4): 281–97.

———. 2005. Science as Labor. *Perspectives on Science* 13 (2): 194–225.

———, ed. 2007. *Inside the Camera Obscura: Optics and Art under the Spell of the Projected Image.* Preprint, Max Planck Institute for the History of Science, Berlin.

Legrand, Homer E. 1973. The "Conversion" of C.-L. Berthollet to Lavoisier's Chemistry. *Ambix* 22:58–70.

Levere, Trevor H. 1994. *Chemists and Chemistry in Nature and Society 1770–1878.* Aldershot, UK: Ashgate-Variorum.

Levere, Trevor H., and G. L'E. Turner, eds. 2002. *Discussing Chemistry and Steam: The Minutes of a Coffee House Philosophical Society 1780–1787.* Oxford: Oxford University Press.

Littré, Emile. 1882. *Dictionnaire de la Langue Française.* Paris: Hachette.

Long, Pamela O. 2001. *Openness, Secrecy, Authorship: Technical Arts and the Culture of Knowledge from Antiquity to the Renaissance.* Baltimore: Johns Hopkins University Press.

Loux, Francoise. 1980. *Das Kind und sein Körper in der Volksmedizin. Eine historisch-ethnographische Studie.* Stuttgart: Klett-Cotta.

Löw, Reinhard. 1979. *Pflanzenchemie zwischen Lavoisier und Liebig.* Münchner Hochschulschriften, series: Naturwissenschaften 1. Straubing: Donau-Verlag.

Lowood, Henry E. 1991. *Patriotism, Profit, and the Promotion of Science in the German Enlightenment: The Economic and Scientific Societies, 1760–1815.* New York: Garland.

———. 1995. The New World and the European Catalog of Nature. In *America in European Consciousness 1493–1750,* ed. K. O. Kupperman, 295–323. Chapel Hill: University of North Carolina Press.

Lowry, S. Todd. 2004. The Agricultural Foundation of the Seventeenth-Century English Economy. In *Oeconomies in the Age of Newton,* ed. M. Schabas and N. De Marchi, 74–100. *History of Political Economy* annual supp., vol. 34. Durham, NC: Duke University Press.

Lücke, Monika, and Ulf Dräger, eds. 2004. *„Die Mark zu 13 Reichstaler und 8 Groschen beibehalten werde". Die Alte Münze in Stolberg (Harz).* Stolberg: Numismatischer Verlag Leipziger Münzhandlung und Auktion Heidrun Höhn.

Lüdecke, Cornelia. In press. Von der Kanoldsammlung (1717–1726) bis zu den Ephemeriden der Societas Meteorologica Palatina (1781–1792)—Meteorologische Quellen zur Umweltgeschichte des 18. Jahrhunderts. In *Landschaften agrarisch-ökonomischen Wissens. Regionale*

*Fallstudien zu landwirtschaftlichen und gewerblichen Themen in Zeitschriften und Sozietäten des 18. Jahrhunderts*, ed. M. Popplow. Cottbuser Studien zur Geschichte von Technik, Arbeit und Umwelt 30. Münster: Waxmann.

Lysaght, Patricia, ed. 1994. *Milk and Milk Products from Medieval to Modern Times: Proceedings of the Ninth International Conference on Ethnological Food Research.* Edinburgh: Canongate Academic.

Maehle, Andreas-Holger. 1999. *Drugs on Trial: Experimental Pharmacology and Therapeutic Innovation in the Eighteenth Century.* Amsterdam: Rodopi.

———. 2007. Experience, Experiment and Theory: Justifications and Criticisms of Pharmacotherapeutic Practices in the Eighteenth Century. In *Medical Theory and Therapeutic Practice in the Eighteenth Century: A Transatlantic Perspective*, ed. J. Helm and R. Wilson. Stuttgart: Franz Steiner Verlag.

Mani, Nikolaus. 1961. Darmresorption und Blutbildung im Lichte der experimentellen Physiologie des 17. Jahrhunderts. *Gesnerus. Vierteljahresschrift für Geschichte der Medizin und Naturwissenschaften* 18:85–141.

Mark, Overton. 1996. *Agricultural Revolution in England: The Transformation of the Agrarian Economy, 1500–1850.* Cambridge: Cambridge University Press.

Marr, Alexander. 2006. Gentile curiosité: Wonder-Working and the Culture of Automata in the Late Renaissance. In *Curiosity and Wonder from the Renaissance to the Enlightenment*, ed. R. J. W. Evans and A. Marr, 149–70. Aldershot, UK: Ashgate.

Martin, Morag. 1999. Consuming Beauty: The Commerce of Cosmetics in France, 1750–1800. Ph.D. diss., University of California, Irvine.

Mathias, Peter. 1991. Resources and Technology. In *Innovation and Technology in Europe: From the Eighteenth Century to the Present Day*, ed. M. Peter and J. A. Davis, 18–42. Oxford: Basil Blackwell.

Matthee, Rudi. 1985. Exotic Substances: The Introduction and Global Spread of Tobacco, Coffee, Tea, and Distilled Liquor, 16th to 18th Centuries. In *Drugs and Narcotics in History*, ed. R. Porter and M. Teich, 24–51. Cambridge: Cambridge University Press.

Mauskopf, Seymour H. 1988. Gunpowder and the Chemical Revolution. In *The Chemical Revolution: Essays in Reinterpretation*, ed. A. Donovan, special issue, *Osiris* 4:93–118.

———. 1990. Chemistry and Cannon: J.-L. Proust and Gunpowder Analysis. *Technology and Culture* 31:398–426.

———. 1996. From Rumford to Rodman: The Scientific Study of the Physical Characteristics of Gunpowder in the First Part of the Nineteenth Century. In *Gunpowder: The History of an International Technology*, ed. B. J. Buchanan, 277–93. Bath: Bath University Press.

———. 1998. Explosives, Instruments to Test the Ballistic Force of. In *Instruments of Science*, ed. R. Bud and D. J. Warner, 234–36. New York: Garland.

———. 2005. Chemistry in the Arsenal: State Regulation and Scientific Methodology of Gunpowder in Eighteenth-Century England and France. In *The Heirs of Archimedes: Science and the Art of War through the Age of Enlightenment*, ed. Brett D. Steele and Tamera Dorland, 293–330. Cambridge, MA: MIT Press.

———. 2006. Pellets, Pebbles and Prisms: British Munitions for Larger Guns, 1860–1885. In *Gunpowder, Explosives and the State: A Technological History*, ed. B. J. Buchanan, 202–29. Aldershot, UK: Ashgate.

McCallum, R. Ian. 1999. *Antimony in Medical History: An Account of the Medical Uses of Antimony and Its Compounds since Early Times to the Present.* Edinburgh: Pentland Press.

McKendrick, Neil, John Brewer, and John H. Plumb. 1982. *The Birth of a Consumer Society: The Commercialization of Eighteenth-Century England.* Bloomington: Indiana University Press.

McKenzie, Donald Francis. 1966. *The Cambridge University Press, 1696–1712: A Bibliographical Study.* 2 vols. Cambridge: Cambridge University Press.

Meinel, Christoph. 1983. Theory or Practice? The Eighteenth-Century Debate on the Scientific Status of Chemistry. *Ambix* 30:121–32.

Merton, Robert K. 1970. *Science, Technology and Society in Seventeenth-Century England.* New York: Harper & Row.

Metzler, Guido. 2001. Markgraf Karl Friedrich von Baden und die französischen Physiokraten. *Francia* 28 (2): 35–63.

Meyer, Torsten. 1999. *Natur, Technik und Wirtschaftswachstum im 18. Jahrhundert.* Cottbuser Studien zur Geschichte von Technik, Arbeit und Umwelt, 12. Münster: Waxmann.

———. 2003. Von der begrenzten zur unbegrenzten Ausrottung: "Schädlinge" als "natürliches Risiko" im 18. Jahrhundert. In *Die Veränderung der Kulturlandschaft. Nutzungen—Sichtweisen—Planungen,* ed. G. Bayerl and T. Meyer, 61–73. Cottbuser Studien zur Geschichte von Technik, Arbeit und Umwelt, 22. Münster: Waxmann.

Meyer, Torsten, and Marcus Popplow. 2004. "To Employ Each of Nature's Products in the Most Favorable Way Possible"—Nature as a Commodity in Eighteenth-Century German Economic Discourse. *Historical Social Research* 29:4–40.

———, eds. 2006. *Technik, Arbeit und Umwelt in der Geschichte. Günter Bayerl zum 60. Geburtstag.* Münster: Waxmann.

Mintz, Sidney W. 1985. *Sweetness and Power: The Place of Sugar in Modern History.* New York: Elisabeth Sifton, Viking.

Mockridge, Colonal A. H. 1950. The Proving of Ordnance and Propellants. *Journal of the Royal Artillery* 77.

Mokyr, Joel. 2002. *The Gifts of Athena: Historical Origins of the Knowledge Economy.* Princeton, NJ: Princeton University Press.

———. 2005. The Intellectual Origins of Modern Economic Growth. *Journal of Economic History* 65:285–351.

Möller, R. 1962. Chemiker und Pharmazeut der Goethezeit. Eine Skizze des Lebens und Schaffens Johann Gotttfried August Göttlings. *Die Pharmazie* 17:624–34.

Moran, Bruce T. 2005. *Distilling Knowledge: Alchemy, Chemistry and the Scientific Revolution.* Cambridge, MA: Harvard University Press.

———. 2007. *Andreas Libavius and the Transformation of Alchemy: Separating Chemical Cultures with Polemic Fire.* Sagamore Beach, MA: Watson.

Morley, Henry. 1853. *Palissy the Potter: In Two Volumes.* Boston: Ticknor, Reed & Fields.

Morrison-Low, A. D. 2007. *Making Scientific Instruments in the Industrial Revolution.* Aldershot, UK: Ashgate.

Moxon, Joseph. 1962. [1683–1684]. *Mechanick Exercises on the Whole Art of Printing.* 2nd ed. Ed. H. David and H. Carter. London: Oxford University Press.

Müller, Hans-Heinrich. 1975. *Akademie und Wirtschaft im 18. Jahrhundert. Agrarökonomische Preisaufgaben und Preisschriften der Preußischen Akademie der Wissenschaften (Versuch, Tendenzen und Überblick).* Berlin: Akademie Verlag.

Müller-Wille, Staffan. 2004. Nature as a Marketplace: The Political Economy of Linnaean Botany. In *Oeconomies in the Age of Newton,* ed. N. De Marchi and M. Schabas, 155–73. *History of Political Economy* annual supp., vol. 34. Durham, NC: Duke University Press.

Multhauf, Robert P. 1966. *The Origins of Chemistry.* London: Oldbourne Book Company.

Mumford, Lewis. 1986. *The Future of Technics & Civilization.* London: Freedom.

———. 2000. *Art and Technics.* Introduction by Casey Nelson Blake. New York: Columbia University Press.

Musgrave, Alan. 1976. Why Did Oxygen Supplant Phlogiston? Research Programmes in the Chemical Revolution. In *Method and Appraisal in the Physical Sciences: The Critical Background to Modern Science, 1800–1905,* ed. C. Howson, 181–209. Cambridge: Cambridge University Press.

Musson, Albert Edward, ed. 1972. *Science, Technology, and Economic Growth in the Eighteenth Century.* London: Methuen.

Musson, Albert Edward, and Eric Robinson. 1969. *Science and Technology in the Industrial Revolution.* Manchester: Manchester University Press.

Myers, Fred R. 2001. Introduction: The Empire of Things. In *The Empire of Things: Regimes of Value and Material Culture,* ed. F. Myers, 3–61. Santa Fe, NM: School of American Research Press/James Currey.

Nahoum-Grappe, Véronique. 1987. Les usages sociaux de l'alcool: les mots et les conduites en France entre 1750 et 1850. *Information sur les sciences sociales* 26 (2): 435–49.

Nathan, Frederic L. 1909. The Royal Gunpowder Factory: Waltham Abbey. In *The Rise and Progress of the British Explosives Industry: Published under the Auspices of the VIIth International Congress of Applied Chemistry by Its Explosives Section,* 316–23. London: Whittaker and Co.

Nathans, Benjamin. 1990. Habermas's "Public Sphere" in the Era of the French Revolution. *French Historical Studies* 16:620–44.

Newman, William R. 1994. *Gehennical Fire: The Lives of George Starkey, an American Alchemist in the Scientific Revolution.* Cambridge, MA: Harvard University Press.

———. 2004. *Promethean Ambitions: Alchemy and the Quest to Perfect Nature.* Chicago: University of Chicago Press.

———. 2006. *Atoms and Alchemy: Chymistry and the Experimental Origins of the Scientific Revolution.* Chicago: University of Chicago Press.

Newman, William R., and Anthony Grafton, eds. 2001. *Secrets of Nature: Astrology and Alchemy in Early Modern Europe.* Cambridge, MA: MIT Press.

Newman, William R., and Lawrence M. Principe. 2002. *Alchemy Tried in the Fire: Starkey, Boyle, and the Fate of Helmontian Chymistry.* Chicago: University of Chicago Press.

Nieto-Galan, Agustí. 1995. The French Chemical Nomenclature in Spain: Critical Points, Rhetorical Arguments and Practical Uses. In *Negotiating a New Language for Chemistry: Lavoisier in European Context,* ed. B. Bensaude-Vincent and F. Abbri, 173–91. Canton, MA: Science History Publications.

———. 2000. From the Workshop to the Print: Bancroft, Berthollet and the Textbooks on the Art of Dyeing in the Late Eighteenth-Century. In *Communicating Chemistry: Textbooks and Their Audiences, 1789–1939,* ed. B. Bensaude-Vincent and A. Lundgren, 275–304. Canton, MA: Science History Publications.

———. 2001. *Colouring Textiles: A History of Natural Dyestuffs in Industrial Europe.* Dordrecht: Kluwer.

———. 2002. La tecnología química: el caso de la tintura. In *Historia de la Ciencia y la Técnica en la Corona de Castilla,* vol. 4, *Ilustración,* ed. J. L. Peset, 631–52. Valladolid: Junta de Castilla y León.

O'Brien, Patrick. 1991. The Mainsprings of Technological Progress in Western Europe 1750–1850. In *Innovation and Technology in Europe: From the Eighteenth Century to the Present Day,* ed. P. Mathias and J. A. Davis, 6–17. Oxford: Basil Blackwell.

Oakley, Kenneth P. 1965. Folklore of Fossils. *Antiquity* 39:9–16, 117–25.

Oldenziel, Ruth. 1999. *Making Technology Masculine.* Amsterdam: Amsterdam University Press.

Orland, Barbara. 2003. Turbo-Cows: Producing a Competitive Animal in the Nineteenth and Early Twentieth Centuries. In *Industrializing Organisms: Introducing Evolutionary History,* ed. P. Scranton and S. Schrepfer, 167–90. London: Rutgers University Press.

———. 2004. Alpine Milk: Dairy Farming as a Pre-modern Strategy of Land Use. *Environment and History* 10:327–64.

Osborne, Michael A. 1994. *Nature, the Exotic, and the Science of French Colonialism.* Bloomington: Indiana University Press.

Panne, Kathrin., ed. 2002. *Albrecht Daniel Thaer—Der Mann gehört der Welt.* Celle: Stadtarchiv.

Park, Katherine. 1985. *Doctors and Medicine in Early Renaissance Florence.* Princeton, NJ: Princeton University Press.

Park, Katherine, and Lorraine Daston. 2006. Introduction: The Age of the New. In *Early Modern Science,* vol. 3 of *The Cambridge History of Science,* ed. K. Park and L. Daston, 1–17. Cambridge: Cambridge University Press.

Partington, James R. 1961–1970. *A History of Chemistry.* 4 vols. London: Macmillan

Pearce, Susan M., ed. 1989. *Museum Studies in Material Culture.* Leicester, UK: Leicester University Press.

———. 1990. *Objects of Knowledge.* London: Athlone.

———. 1994. *Interpreting Objects and Collections.* London: Routledge.

Pearce, Susan M., and Kenneth Arnold, eds. 2000. *The Collector's Voice: Critical Readings in the History of Collecting.* Vol. 2, *Early Voices.* Hampshire, UK: Ashgate.

Pelzer, Marten. 2004. "Was die Schule für das heranwachsende Geschlecht ist, das ist der landwirtschaftliche Verein für die älteren Landwirte. . . . " Bildungsanspruch und -wirklichkeit landwirtschaftlicher Vereine im 19. Jahrhundert. *Zeitschrift für Agrargeschichte und Agrarsoziologie* 52:41–58.

Pennell, Sara. 1999. Consumption and Consumerism in Early Modern England. *Historical Journal* 42 (2): 549–64.

Pérez-Ramos, Antonio. 1988. *Francis Bacon's Idea of Science and the Maker's Tradition.* Oxford: Clarendon Press.

Perrig, Alexander. 1980. Leonardo: Die Anatomie der Erde. *Jahrbuch der Hamburger Kunstsammlungen* 25:51–80.

Perrot, Philippe. 1995. *Le Luxe: une richesse entre faste et confort, XVIIIe–XIXe siècle.* Paris: Editions du Seuil.

Phillips, J. W. 1998. *Printing and Bookselling in Dublin, 1670–1800.* Dublin: Irish Academic Press.

Pickering, Andrew. 1989. Living in the Material World. In *The Uses of Experiment: Studies in the Natural Sciences,* ed. D. Gooding, T. Pinch, and S. Schaffer, 275–97. Cambridge: Cambridge University Press.

———. 1995. *The Mangle of Practice: Time, Agency, and Science.* Chicago: University of Chicago Press.

———. 2005. Decentering Sociology: Synthetic Dyes and Social Theory. *Perspectives on Science* 13 (3): 352–405.

Pickstone, John V. 2000. *Ways of Knowing: A New History of Science, Technology and Medicine.* Manchester: Manchester University Press.

Plagnol-Diéval, Marie-Emmanuelle. 1997. Vin canaille et vin moral sur les scènes privées. In *Le Vin*, ed. J. Bart and E. Wahl, special issue, *Dix-huitième siècle* 29:237–53.

Planiscig, Leo. 1927. *Andrea Riccio*. Vienna: Schroll.

Plant, Marjorie. 1974. *The English Book Trade*. London: Allen & Unwin.

Poirier, Jean-Pierre. 1996. *Lavoisier: Chemist, Biologist, Economist*. Trans. R. Balinski. Philadelphia: University of Pennsylvania Press.

———. 2007. Lavoisier's Friends. http://historyofscience.free.fr/Lavoisier-Friends/a_chap2_lavoisier.html, 9 February.

Polanyi, Michael. 1994. *Personal Knowledge: Towards a Post-Critical Philosophy*. Chicago: University of Chicago Press.

Pollard, Mary. 2000. *A Dictionary of Members of the Dublin Book Trade, 1550–1800*. London: Bibliographical Society.

Pomeranz, Kenneth. 2000. *The Great Divergence. China, Europe, and the Making of the Modern World Economy*. Princeton, NJ: Princeton University Press.

Pomian, Krzysztof. 1987. *Collectioneurs, amateurs et curieux: Paris, Venise: XVIe–XVIIIe siècle*. Paris: Gallimard.

Poni, Carlo. 1998. Mode et innovation: les stratégies des marchands en soie de Lyon au XVIIIe siècle. *Revue d'histoire moderne et contemporaine* 45:589–625.

Popplow, Marcus. 2006. Hoffnungsträger "Unächter Acacien=Baum": Zur Wertschätzung der Robinie von der Ökonomischen Aufklärung des 18. Jahrhunderts bis zu aktuellen Konzepten nachhaltiger Landnutzung. In *Technik, Arbeit und Umwelt in der Geschichte. Günter Bayerl zum 60. Geburtstag*, ed. T. Meyer and M. Popplow, 297–316. Münster: Waxmann.

———, ed. In press. *Landschaften agrarisch-ökonomischen Wissens. Regionale Fallstudien zu landwirtschaftlichen und gewerblichen Themen in Zeitschriften und Sozietäten des 18. Jahrhunderts*. Cottbuser Studien zur Geschichte von Technik, Arbeit und Umwelt 30. Münster: Waxmann.

Porter, Dorothy, and Roy Porter. 1989. *Patient's Progress: Doctors and Doctoring in Eighteenth-Century England*. Stanford, CA: Stanford University Press.

Porter, Roy. 1985. Lay Medical Knowledge in the Eighteenth Century. *Medical History* 29:138–68.

———, ed. 1990. *The Medical History of Waters and Spas*. London: Wellcome.

———. 1991. Introduction to *The English Malady (1733)*. Ed. G. Cheyne. London: Tavistock.

———. 1992. *Doctor of Society: Thomas Beddoes and the Sick Trade in Late-Enlightenment England*. London: Routledge.

———. 2003. *Flesh in the Age of Reason*. New York: Norton.

Porter, Roy, Simon Schaffer, Jim Bennett, and Olivia Brown. 1985. *Science and Profit in 18th-Century London*. Cambridge: Whipple Museum of the History of Science.

Potocki, Jan. 1996. *The Manuscript Found in Saragossa*. Trans. I. MacLean. London: Penguin.

Poulot, Dominique. 1997. Une nouvelle histoire de la culture matérielle? *Revue d'histoire moderne et contemporaine* 44 (2): 344–57.

Prass, Reiner. 1997. *Reformprogramm und bäuerliche Interessen. Die Auflösung der traditionellen Gemeindeökonomie im südlichen Niedersachsen, 1750–1883*. Göttingen: Vandenhoek und Ruprecht.

Priesner, Claus. 1986. Spiritus Aethereus—Formation of Ether and Theories of Etherification from Valerius Cordus to Alexander Williamson. *Ambix* 33:129–52.

Prown, Jules. 1982. Mind in Matter: An Introduction to Material Culture Theory and Method. *Winterthur Portfolio* 17:1–19.

Prown, Jonathan, and Richard Miller. 1996. The Rococo, the Grotto, and the Philadelphia High Chest. *American Furniture,* 105–36.

Public Sphere. 1992. In *The Public Sphere in the Eighteenth Century,* special issue, *French Historical Studies* 17.

Pyle, Cynthia M. 2000. Art as Science: Scientific Illustration, 1490–1670, in Drawing, Woodcut and Copper Plate. *Endeavour* 24:69–75.

Quintili, Paolo. 2004. Introduction to *Denis Diderot: Eléments de physiologie.* Ed. P. Quintili, 11–102. Paris: Champion.

Rammelsberger Bergbaumuseum Goslar, ed. 2004. *Der Riechenberger Vertrag.* Verlag Goslarsche Zeitung Karl Krause.

Reddy, William M. 1984. *The Rise of Market Culture: The Textile Trade and French Society, 1750–1900.* Cambridge: Cambridge University Press; Paris: Editions de la Maison des Sciences de l'Homme.

Renn, Jürgen, and Matteo Valleriani. 2001. Galileo and the Challenge of the Arsenal. *Nuncius* 16 (2): 481–503.

Reudenbach, Bruno. 2003. Praxisorientierung und Theologie. Die Neubewertung der Werkkünste in „De diversibus artibus" des Theophilus Presbyter. In *Helmarshausen. Buchkultur und Goldschmiedekunst im Hochmittelalter,* ed. I. Baumgärtner, 199–218. Kassel: Euregioverlag.

Reynolds, Terry S. 1983. *Stronger Than a Hundred Men: A History of the Vertical Water Wheel.* Baltimore: Johns Hopkins University Press.

Rheinberger, Hans-Jörg. 1997. *Toward a History of Epistemic Things: Synthesizing Proteins in the Test Tube.* Stanford, CA: Stanford University Press.

Ribeiro, Aileen. 1995. *The Art of Dress: Fashion in England and France, 1750 to 1820.* New Haven, CT: Yale University Press.

Richardson, R. C. 2004. Social Engineering in Early Modern England: Masters, Servants, and the Godly Discipline. *Clio* 33:163–87.

Rickard, T. A. 1932. *Man and Metals: A History of Mining in Relation to the Development of Civilisation.* Vol. 1. London: Whittlesley Home.

Riddle, John M., and James A. Mulholland. 1980. Albert on Stones and Minerals. In *Albertus Magnus and the Sciences: Commemorative Essays,* ed. J. A. Weisheipl 203–34. Toronto: Pontifical Institute of Mediaeval Studies,.

*The Rise and Progress of the British Explosives Industry: Published under the Auspices of the VIIth International Congress of Applied Chemistry by Its Explosives Section.* 1909. London: Whittaker and Co.

Risse, Guenter B. 1986. *Hospital Life in Enlightenment Scotland: Care and Teaching at the Royal Infirmary of Edinburgh.* Cambridge: Cambridge University Press.

———. 1992. Medicine in the Age of Enlightenment. In *Medicine in Society: Historical Essays,* ed. A. Wear, 149–95. Cambridge: Cambridge University Press.

———. 2005. *New Medical Challenges during the Scottish Enligthenment.* New York: Rodopi.

Roberts, Lissa. 1999. Science Becomes Electric: Dutch Interactions with the Electrical Machine during the Eighteenth Century. *Isis* 90:680–714.

Roberts, Lissa, Simon Schaffer, and Peter Dear, eds. 2007. *The Mindful Hand: Inquiry and Invention from the Late Renaissance to Early Industrialisation.* Amsterdam: Edita; www.knaw.nl/publicaties/pdf/20041102.pdf.

Roberts, Michael. 1998. "To Bridle the Falsehood of Unconscionable Workmen, and for Her Own Satisfaction": What the Jacobean Housewife Needed to Know about Men's Work, and Why. *Labour History Review* 63:4–30.

Robinson, Harry. 1976. *A Geography of Tourism*. London: Macdonald and Evens.

Robinson, Roger J. 1996. The Madness of Mrs Beattie's Family: The Strange Case of the "Assassin" of John Wilkes. *British Journal for Eighteenth-Century Studies* 19:183–97.

———. 2004. *The Correspondence of James Beattie*. 4 vols. Bristol: Thoemmes.

Roche, Daniel. 1978a. Négoce et culture dans la France du XVIIIe siècle. *Revue d'histoire moderne et contemporaine* 25:375–95.

———. 1978b. *Le siècle des lumières en province*. 2 vols. Paris: École des hautes études en sciences sociales.

———. 1994. *The Culture of Clothing: Dress and Fashion in the Ancien Régime*. Cambridge: Cambridge University Press.

———. 2000a. *A History of Everyday Things: The Birth of Consumption in France, 1600–1800*. Cambridge: Cambridge University Press.

———. 2000b. Preface to *L'invention technique au siècle des Lumières*. Ed. L. Hilaire-Pérez, 11–23. Paris: Albin Michel.

Ron, Moshe. 1991. *Biblioteca Tinctoria: Annotated Catalogue of the Sidney M. Edelstein Collection in the History of Bleaching, Dyeing, Finishing and Spot Removing*. Jerusalem: Jewish National and University Library.

Rosenberg, Charles E. 1983. Medical Text and Social Context: Explaining William Buchan's "Domestic Medicine." *Bulletin of the History of Medicine* 57:22–42.

Roseneck, Reinhard. 2001. *Der Rammelsberg. Tausend Jahre Mensch-Natur-Technik*. 2 vols. Goslar: Verlag Goslarsche Zeitung Karl Krause.

Rosenfeld, Louis. 1999. *Four Centuries of Clinical Chemistry*. New York: CRC Press.

Rossi, Paolo. 1970. *Philosophy, Technology, and the Arts in the Early Modern Era*. New York: Harper & Row.

———. 1984. *The Dark Abyss of Time: The History of the Earth and the History of Nations from Hooke to Vico*. Trans. L. G. Cochrane. Chicago: University of Chicago Press.

Rousseau, George S. 2004. *Nervous Acts: Essays on Literature, Culture, and Sensibility*. New York: Palgrave Macmillan.

*The Royal Gunpowder Factory, Waltham Abbey, Essex: An RCHME Survey*. 1994. London: Royal Commission on the Historical Monuments of England.

Rudgley, Richard. 1993. *The Alchemy of Culture: Intoxicants in Society*. London: British Museum Press.

Rudwick, Martin. 1972. *The Meaning of Fossils*. New York: American Elsevier.

Rustock, Andrea. 2002. *Vital Accounts: Quantifying Health and Population in Eighteenth-Century England and France*. Cambridge: Cambridge University Press.

Sandl, Marcus. 1999. *Ökonomie des Raumes. Der kameralwissenschaftliche Entwurf der Staatswirtschaft im 18. Jahrhundert*. Cologne: Böhlau.

Sargentson, Carolyn. 1996. *Merchants and Luxury Markets: The Marchands Merciers of Eighteenth-Century Paris*. London: Victoria and Albert Museum.

Sarton, George. 1931. *The History of Science and the New Humanism*. New York: Holt.

Scagliola, Robert. 1943. *Les Apothicaires de Paris et les Distillateurs*. Clermont-Ferrand: Imprimerie Générale Jean de Bussac.

Schabas, Margaret, and Neil De Marchi, eds. 2003. *Oeconomies in the Age of Newton*. Durham, NC: Duke University Press.

Schaffer, Simon. 1989. Glass Works: Newton's Prisms and the Uses of Experiment. In *The Uses of Experiment: Studies in the Natural Sciences*, ed. D. Gooding, T. Pinch, and S. Schaffer, 67–104. Cambridge: Cambridge University Press.

———. 1997a. The Earth's Fertility as a Social Fact in Early Modern Britain. In *Nature and Society in Historical Context,* ed. M. Teich, R. Porter, and B. Gustafsson, 124–47. Cambridge: Cambridge University Press.

———. 1997b. Experimenters' Techniques, Dyers' Hands, and the Electric Planetarium. *Isis* 88 (3): 456–83.

———. 1999. Enlightened Automata. In *The Sciences in Enlightened Europe,* ed. W. Clark, J. Golinski, and S. Schaffer. Chicago: University of Chicago Press, 126–65.

———. 2004. A Science Whose Business Is Bursting: Soap Bubbles as Commodities in Classical Physics. In *Things That Talk: Object Lessons from Art and Science,* ed. L. Daston, 147–92. New York: Zone.

Scheicher, Elisabeth. 1985. The Collection of Archduke Ferdinand II at Schloss Ambras. In *The Origins of Museums: The Cabinet of Curiosities in Sixteenth- and Seventeenth-Century Europe,* ed. O. Impey and A. MacGregor. Oxford: Clarendon Press.

Schendel, A. F. E. van. 1972. Manufacture of Vermilion in 17th-Century Amsterdam, the Pekstok Papers. *Studies in Conservation* 17:70–82.

Schiebinger, Londa. 1993. *Nature's Body. Gender in the Making of Modern Science.* Boston: Beacon.

Schiebinger, Londa, and Claudia Swan, eds. 2005. *Colonial Botany: Science, Commerce, and Politics in the Early Modern World.* Philadelphia: University of Pennsylvania Press.

Schießl, Ulrich. 1989. *Die deutschsprachige Literatur zu Werkstoffen und Techniken der Malerei von 1530 bis ca. 1950.* Worms: Wernersche Verlagsgesellschaft.

Schindler, Norbert, and Wolfgang Bonß. 1980. Praktische Aufklärung—Ökonomische Sozietäten in Süddeutschland und Österreich im 18. Jahrhundert. In *Deutsche patriotische und gemeinnützige Gesellschaften,* ed. R. Vierhaus, 255–353. Munich: Kraus.

Schivelbusch, Wolfgang. 1993. *Tastes of Paradise: A Social History of Spices, Stimulants, and Intoxicants.* Trans. D. Jacobson. New York: Vintage.

Schlögl, Rudolf. 1993. Die patriotisch-gemeinnützigen Gesellschaften. Organsiation, Sozialstruktur, Tätigkeitsfelder. In *Aufklärungsgesellschaften,* ed. H. Reinalter, 61–81. Frankfurt am Main: Lang.

Schnapper, Antoine. 1988. *Le Géant, La Licorce et La Tulipe: Collections et Collectionneurs dans la France du XVIIe Siecle.* Tours: Mame Imprimeurs.

Schneider, Wolfgang. 1968–1975. *Lexikon zur Arzneimittelgeschicht: Sachwörterbuch zur Geschichte der pharmazeutischen Botanik, Chemie, Mineralogie, Pharmakologie, Zoologie.* 7 vols. Frankfurt am Main: Govi.

———. 1972. *Geschichte der pharmazeutischen Chemie.* Weinheim: Verlag Chemie.

Schöne, Andreas. 2001. Die Leipziger Ökonomische Sozietät. In *Sächsische Aufklärung,* ed A. Klingenberg et al. , 73–91. Leipzig: Leipziger Universitätsverlag.

Secord, James A. 2003. *Victorian Sensation: The Extraordinary Publication, Reception, and Secret Authorship of Vestiges of the Natural History of Creation.* Chicago: University of Chicago Press.

———. 2004. Knowledge in Transit. *Isis* 95:654–72.

Segers-Glocke, Christiane, ed. 2000. *Auf den Spuren einer frühen Industrielandschaft. Naturraum-Mensch-Umwelt im Harz.* Arbeitshefte zur Denkmalpflege in Niedersachen 21. Hameln: CW Niemeyer Buchverlag.

Serres, Michel. 1982. *Parasite.* Baltimore: Johns Hopkins University Press.

———. 1989. *Statues: le second livre des fondations.* Paris: Flammarion.

Sgard, Jean. 1991. *Dictionnaire de la presse. Part 1: Dictionnaire des journaux.* 2 vols. Paris: Universitas.

Shafer, Robert Jones. 1958. *The Economic Societies in the Spanish World, 1763–1821.* Syracuse, NY: Syracuse University Press.

Shammas, Carole. 1990. *The Pre-Industrial Consumer in England and America.* Oxford: Clarendon.

Shapin, Steven, and Simon Schaffer. 1985. *Leviathan and the Air-Pump: Hobbes, Boyle, and the Experimental Life.* Princeton, NJ: Princeton University Press.

Shell, Hanna Rose. 2004. Casting Life, Recasting Experience: Bernard Palissy's Occupation between Maker and Nature. *Configurations* 12:1–40.

Sherwood, Joan. 1993. The Milk Factor: The Ideology of Breast-feeding and Post-partum Illnesses, 1750–1850. *Canadian Bulletin of Medical History* 10:25–47.

Sieglerschmidt, Jörn. 1999. Die virtuelle Landschaft der Hausväterliteratur. In *Natur-Bilder. Wahrnehmungen von Natur und Umwelt in der Geschichte,* ed. H. Breuninger and R. P. Sieferle, 223–54. Frankfurt am Main: Campus.

Simmons, Anna. 2006. Medicines, Monopolies and Mortars: The Chemical Laboratory and Pharmaceutical Trade at the Society of Apothecaries in the Eighteenth Century. *Ambix* 53:221–36.

Simmons, W. H. 1963. *A Short History of the Royal Gunpowder Factory at Waltham Abbey.* Privately published, Controllate of the Royal Ordnance Factories.

Simon, Christian. 1994. Labour Relations at Manufactures in the Eighteenth Century: The Calico-Printers in Europe. *International Review of Social History* 39:115–44.

Simon, Jonathan. 2005. *Chemistry, Pharmacy and Revolution in France, 1777–1809.* Aldershot, UK: Ashgate.

Simon, Thomas. 2004. *„Gute Policey". Ordnungsleitbilder und Zielvorstellungen politischen Handelns in der Frühen Neuzeit.* Frankfurt am Main: Klostermann.

Simonetto, Michele. 2001. *I lumi nelle campagne. Accademie e agricoltura nella Repubblica di Venezia, 1768–1797.* Treviso: Edizioni Fondazione Benetton Ricerche.

Simpson, A. D. C. 1989. Robert Hooke and Practical Optics: Technical Support at a Scientific Frontier. In *Robert Hooke: New Studies,* ed. M. Hunter and S. Schaffer, 33–62. Woodbridge, UK: Boydell.

Siraisi, Nancy G. 1990. *Medieval and Early Renaissance Medicine: An Introduction to Knowledge and Practice.* Chicago: University of Chicago Press.

Smeaton, William A. 1957. L'Avant-Coureur: The Journal in Which Some of Lavoisier's Earliest Research Was Reported. *Annals of Science* 13:219–34.

———. 1962. *Fourcroy, Chemist and Revolutionary, 1755–1809.* Cambridge: Heffer & Sons.

Smith, John G. 1979. *The Origins and Early Development of the Heavy Chemical Industry in France.* Oxford: Clarendon.

Smith, Pamela H. 2004. *The Body of the Artisan: Art and Experience in the Scientific Revolution.* Chicago: University of Chicago Press.

———. 2005. Making as Knowing: Craft as Natural Philosophy. Lecture delivered at the conference "Ways of Making and Knowing: The Material Culture of Empirical Knowledge," Victoria & Albert Museum, London, July 2005.

———. 2006a. Art, Science, and Visual Culture in Early Modern Europe. *Isis* 96:83–100.

———. 2006b. Laboratories. In *Early Modern Science,* vol. 3 of *The Cambridge History of Science,* ed. K. Park and L. Daston, 290–305. Cambridge: Cambridge University Press.

Smith, Pamela H., and Paula Findlen, eds. 2002. *Merchants and Marvels: Commerce, Science, and Art in Early Modern Europe*. New York: Routledge.

Smith, W. D. A. 1982. *Under the Influence: A History of Nitrous Oxide and Oxygen Anaesthesia*. London: Macmillan.

Smith, Woodruff D. 1992–1993. Complications of the Commonplace: Tea, Sugar, and Imperialism. *Journal of Interdisciplinary History* 23:259–78.

———. 2002. *Consumption and the Making of Respectability, 1600–1800*. New York: Routledge.

Soukup, Rudolf Werner. 2007. *Chemie in Österreich: Bergbau, Alchemie und frühe Chemie. Von den Anfängen bis zum Ende des 18. Jahrhunderts*. Vienna: Böhlau.

Sournia, Jean-Charles. 1990. *A History of Alcoholism*. Introduction by Roy Porter; trans. N. Hindley and G. Stanton. Oxford: Basil Blackwell.

Spary, E. C. 1996. Political, Natural and Bodily Economies. In *Cultures of Natural History*, ed. N. Jardine, J. A. Secord, and E. C. Spary. Cambridge: Cambridge University Press, 178–96.

———. 2000. *Utopia's Garden: French Natural History from Old Regime to Revolution*. Chicago: University of Chicago Press.

———. 2003. "Peaches Which the Patriarchs Lacked": Natural History, Natural Resources, and the Natural Economy in Eighteenth-Century France. In *Oeconomies in the Age of Newton*, ed. M. Schabas and N. D. Marchi, 14–41. Durham, NC: Duke University Press.

———. n.d. Eating the Enlightenment (manuscript).

Star, Susan Leigh, and James R. Griesemer. 1989. Institutional Ecology, "Translation" and Boundary Objects: Amateurs and Professionals in Berkeley's Museum of Vertebrate Zoology, 1907–39. *Social Studies of Science* 19 (3): 387–420.

Stearn, Roger T. n.d. Congreve, Sir. William, second baronet (1772–1828), rocket designer. In *Oxford Dictionary of National Biography*, electronic edition, www.oxforddnb.com. Oxford University Press.

———. n.d. Napier, George (1751–1804), army officer. In *Oxford Dictionary of National Biography*, electronic edition, www.oxforddnb.com. Oxford University Press.

Steele, Brett D. 1994. Muskets and Pendulums: Benjamin Robins, Leonhard Euler, and the Ballistics Revolution. *Technology and Culture* 35:348–82.

———. 2006. Rational Mechanics as Enlightenment Engineering: Leonhard Euler and Interior Ballistics. In *Gunpowder, Explosives and the State: A Technological History*, ed. B. J. Buchanan, 281–302. Aldershot, UK: Ashgate.

Steiner, Philippe. 1996. Les revues économiques de langue française au XVIIIème siècle (1751–1776). In *Les revues d'économie en France. Genèse et actualité 1751–1994*, ed. L. Marco, 33–77. Paris: L'Harmattan.

Stewart, Larry. 1989. Global Pillage: Science, Commerce, and Empire. In *The Agrarian History of England and Wales*, vol. 6, *1750–1850*, ed. J. Thirsk, 825–44. Cambridge: Cambridge University Press.

———. 1992. *The Rise of Public Science: Rhetoric, Technology and Natural Philosophy in Newtonian Britain*. Cambridge: Cambridge University Press.

Stillman, John M. 1960. *The Story of Alchemy and Early Chemistry*. New York: Dover. (Reprint of 1924 ed.)

Stocking, George W., Jr. 1985. Essays on Museums and Material Culture. In *Objects and Others: Essays on Museums and Material Culture*, 3–14. Madison: University of Wisconsin Press.

Stöllner, Thomas, Rainer Slotta, and Abdolrasool Vatandoust. 2004. *Persiens Antike Pracht. Bergbau—Handwerk—Archäologie*. Vol. 1. Bochum: Deutsches Bergbau-Museum.

Stroup, Alice. 1990. *A Company of Scientists: Botany, Patronage, and Community at the Seventeenth-Century Parisian Royal Academy of Sciences.* Berkeley: University of California Press.

Stuber, Martin. In press. „ . . . dass gemeinnüzige wahrheiten gemein gemacht werden"—Zur Publikationstätigkeit der Oekonomischen Gesellschaft Bern 1759–1798. In *Landschaften agrarisch-ökonomischen Wissens. Regionale Fallstudien zu landwirtschaftlichen und gewerblichen Themen in Zeitschriften und Sozietäten des 18. Jahrhunderts,* ed. M. Popplow. Cottbuser Studien zur Geschichte von Technik, Arbeit und Umwelt 30. Münster: Waxmann.

Sturdy, David J. 1995. *Science and Social Status: The Members of the Académie des Sciences, 1666–1750.* Woodbridge, Suffolk: Boydell Press.

Stürzbecher, Manfred. 1966. *Beiträge zur Berliner Medizingeschichte: Quellen und Studien zur Geschichte des Gesundheitswesens vom 17. bis zum 19. Jahrhundert.* Berlin: Walter de Gruyter.

Suhling, Lothar. 1976. *Der Seigerhüttenprozess. Die Technologie des Kupferseigerns nach dem frühen metallurgischen Schrifttum.* Stuttgart: Dr. Riederer Verlag.

———. 1983. *Aufschließen, Gewinnen und Fördern. Geschichte des Bergbaus.* Reinbeck bei Hamburg: Rowohlt.

———. 2000. "Artzschmeltzen zu grossenstein" im Jahre 1540: Technologisches aus einem Fuggerschen Hüttenwerk in Tirol. In *Konjunkturen im europäischen Bergbau der vorindustriellen Zeit. Festschrift für Ekkehard Westermann zum 60. Geburtstag,* ed. C. Bartels and M. Denzel, 189–201. Stuttgart: Franz Steiner Verlag.

———. 2004. Hüttentechnische Verfahren zur Gewinnung von Silber, Blei und Kupfer als Kuppelprodukte im ostalpinen Bergbau um 1500. In *Der Tiroler Bergbau und die Depression der europäischen Montanwirtschaft im 14. und 15. Jahrhundert. Akten der internationalen bergbaugeschichtlichen Tagung Steinhaus,* ed. R. Tasser and E. Westermann, 227–39. Innsbruck: Studien Verlag.

Suleiman, Susan, and Inge Crosman, eds. 1980. *The Reader in the Text: Essays on Audience and Interpretation.* Princeton, NJ: Princeton University Press.

Sumner, James. 2001. John Richardson, Saccharometry and the Pounds-Per-Barrel Extract: The Construction of a Quantity. *British Journal for the History of Science* 34:255–73.

Sussman, George D. 1982. *Selling Mother's Milk: The Wet-Nursing Business in France, 1715–1914.* Urbana: University of Illinois Press.

Taylor, Georgette. 2006. Unification Achieved: William Cullen's Theory of Heat and Phlogiston as an Example of His Philosophical Chemistry. *British Journal for the History of Science* 39:477–501.

Technische Universität Bergakademie, Freiberg, ed. 1994. *Lazarus Ercker. Sein Leben und seine Zeit. Zur Geschichte des Montan- und Münzwesens im mittleren Europa.* Annaberg-Buchholz: Landesstelle für Erzgebirgische und Vogtländische Volkskultur.

Thomas, Nicholas. 1991. *Entangled Objects.* Cambridge, MA: Harvard University Press.

Thompson, Jr., Daniel V. 1933–1934. Artificial Vermilion in the Middle Ages. *Technical Studies in the Field of the Fine Arts* 2:62–70.

Timm, Albrecht. 1964. *Kleine Geschichte der Technologie.* Stuttgart: W. Kohlhammer.

Toraude, Léon Gabriel. 1907. *J.-F. Demachy, Histoires et Contes précédés d'une Etude Historique, Anecdotique et Critique sur sa vie et ses oeuvres.* Paris: Charles Carrington.

Torlais, Jean. 1961. *Un esprit encyclopédique en dehors de "L'Encyclopédie": Réaumur. D'après des documents inédits.* 2nd ed. Paris: Albert Blanchard.

Torrens, Hugh. 1985. Early Collecting in the Field of Geology. In *The Origins of Museums: The Cabinet of Curiosities in Sixteenth- and Seventeenth-Century Europe,* ed. O. Impey and A. MacGregor. Oxford: Clarendon Press.

Travis, Anthony S. 1994. From Manchester to Massachusetts via Mulhouse: The Transatlantic Voyage of Aniline Black. *Technology and Culture* 35:70–99.

Tribe, Keith. 1988a. *Governing Economy: The Reformation of German Economic Discourse, 1750–1840.* Cambridge: Cambridge University Press.

———. 1988b. Die Kameral-Hohe Schule zu Lautern und die Anfänge der ökonomischen Lehre in Heidelberg (1774–1822). In *Die Institutionalisierung der Nationalökonomie an deutschen Universitäten,* ed. N. Waszek, 162–91. St. Katharinen: Scripta Mercaturae.

Tröhler, Ulrich. 2000. *"To Improve the Evidence of Medicine": The 18th Century British Origins of a Critical Approach.* Edinburgh: Royal College of Physicians.

Turner, Antony. 1987. *Early Scientific Instruments: Europe 1400–1800.* London: Sotheby's Publications.

Turner, Bryan S. 1996. *The Body and Society.* 2nd ed. London: Sage.

Turner, Gerard L'Estrange. 1990. *Scientific Instruments and Experimental Philosophy: 1550–1850.* Brookfield: Variorum.

Valleriani, Matteo. 2007. Galileo Engineer. PhD thesis, Humboldt University Berlin, Berlin.

Valverde, Mariana. 1998. *Diseases of the Will: Alcohol and the Dilemma of Freedom.* Cambridge: Cambridge University Press.

Van Creveld, Martin L. 1989. *Technology and War: From 2000 B.C. to the Present.* London: Collier Macmillan.

Van Helden, Albert. 1977. *The Invention of the Telescope.* Philadelphia: American Philosophical Society.

Van Helden, Albert, and Thomas L. Hankins, eds. 1994. *Instruments,* special issue, *Osiris* 9.

Vatin, Francois. 1990. *L'Industrie du Lait. Essai d'Histoire économique.* Paris: Editions L'Harmattan.

Vickery, Amanda. 1993. Women and the World of Goods: A Lancashire Consumer and Her Possessions, 1751–1781. In *Consumption and the World of Goods,* ed. J. Brewer and R. Porter, 274–301. London: Routledge, 274–301.

Vierhaus, Rudolf, ed. 1980. *Deutsche patriotische und gemeinnützige Gesellschaften.* Munich: Kraus.

Voet, Andries. 1952. *Ink and Paper in the Printing Process.* New York: Interscience.

Wagenbreth, Otfried, and Eberhard Wächtler, eds. 1990. *Bergbau im Erzgebirge. Technische Denkmale und Geschichte.* Leipzig: Deutscher Verlag für Grundstoffindustrie.

Wallert, Arie. 1990. Alchemy and Medieval Art Technology. In *Alchemy Revisited,* ed. Z. R. W. M. von Martels. Leiden: Brill.

Walvin, James. 1997. *Fruits of Empire: Exotic Produce and British Taste, 1660–1800.* Basingstoke, UK: Macmillan.

Warde, Paul. 2006. *Ecology, Economy, and State Formation in Early Modern Germany.* Cambridge: Cambridge University Press.

Warner, Jessica. 1994. Resolv'd to Drink No More: Addiction as a Preindustrial Construct. *Journal of Studies on Alcohol* 55 (6): 685–91.

Watson-Verran, Helen, and David Turnbull. 1995. Science and Other Indigenous Knowledge Systems. In *Handbook of Science and Technology Studies,* ed. S. Jasanoff, G. E. Marble, J. C. Peterson, and T. Pinch, 115–39. London: Sage.

Weatherill, Lorna. 1988. *Consumer Behaviour and Material Culture in Britain, 1660–1760.* London: Routledge.

Webster, Charles. 1982. Paracelsus and Demons: Science as a Synthesis of Popular Belief. In *Scienze Credenze Occulte Livelli di Cultura,* ed. P. Zambelli, 3–20. Florence: Leo S. Olschke.

Weisgerber, Gerd. 1987. Montanarchäologie—ein Weg zum Verständnis früher Rohstoffversorgung. In *Die Grossen Abenteuer der Archäologie*, vol. 9, ed. R. Pörtner and H. G. Niemeyer, 3503–40, Salzburg: Anderas Verlag.

Weisheipl, James A., ed. 1980. *Albertus Magnus and the Sciences: Commemorative Essays.* Toronto: Pontifical Institute of Mediaeval Studies.

West, Jenny. 1991. *Gunpowder, Government and War in the Mid-Eighteenth Century.* Woodbridge, Suffolk: Boydell Press [Royal Historical Society].

Westermann, Angelika. 2005a. Bergrecht. In *Enzyklopädie der Neuzeit,* 2:33–39. Stuttgart: Metzler.

———. 2005b. Bergregal. In *Enzyklopädie der Neuzeit,* 2:39–41. Stuttgart: Metzler.

Westermann, Ekkehard. 1986. Zur Silber- und Kupferproduktion Mitteleuropas vom 15. bis zum frühen 17. Jahrhundert. Über Bedeutung und Rangfolge der Reviere von Schwaz, Mansfeld und Neusohl. *Der Anschnitt* 38:187–211.

Wiborg, Frank Bestow. 1926. *Printing Ink: A History.* New York: Harper & Brothers.

Willer, Wilfried. 1967. Die Bibliothek der churpfälzisch physikalisch-ökonomischen Gesellschaft (1770–1804). *Bibliothek und Wissenschaft* 4:240–302.

Wilmot, Sarah. 1990. *The Business of Improvement: Agriculture and Scientific Culture in Britain, c. 1770–c. 1870.* Bristol: Historical Geography Research Society.

Wilson, Catherine. 1995. *The Invisible World: Early Modern Philosophy and the Invention of the Microscope.* Princeton, NJ: Princeton University Press.

Winter, Alison. 1998. *Mesmerized: Powers of Mind in Victorian Britain.* Chicago: University of Chicago Press.

Wood, Paul. 1993. *The Aberdeen Enlightenment.* Aberdeen: Aberdeen University Press.

Yalçin, Ünsal, Cemal Pulak, and Rainer Slotta, eds. 2005. *Das Schiff von Uluburun. Katalog der Ausstellung des Deutschen Bergbau-Museums Bochum vom 15. Juli 2005 bis 16. Juli 2006.* Bochum: Deutsches Bergbau-Museum.

Zilsel, Edgar. 2000. *The Social Origins of Modern Science.* Ed. D. Raven, W. Krohn, and R. S. Cohen. Dordrecht: Kluwer.

Zorach, Rebecca. 2006. *Blood, Milk, Ink, Gold: Abundance and Excess in the French Renaissance.* Chicago: University of Chicago Press.

# *Contributors*

CHRISTOPH BARTELS is a historian of technology and research scholar at the *Deutsches Bergbau-Museum* (Bochum). He is the editor and author of numerous books and essays on the history of medieval and early modern mining and metallurgy in Germany.

MATTHEW D. EDDY is a senior lecturer in the history of science and culture at Durham University. He is the author of *The Language of Mineralogy: John Walker, Chemistry, and the Edinburgh Medical School, 1750–1800* (2008) and (with David M. Knight) he has edited William Paley's *Natural Theology* (2006; 2008) and a volume of essays entitled *Science and Beliefs* (2005). He is currently writing a book about print culture and natural history during the Scottish Enlightenment.

ADRIAN JOHNS is a professor in the Department of History and the Committee on Conceptual and Historical Studies of Science at the University of Chicago. He is the author of *The Nature of the Book: Print and Knowledge in the Making* and *Piracy: The Intellectual Property Wars from Gutenberg to Gates.* He is currently completing a book on pirate radio.

URSULA KLEIN is a senior research scholar at the Max Planck Institute for the History of Science (Berlin) and a professor of history and philosophy of science at the University of Konstanz. Her current research projects are concerned with the intersection of chemistry and the arts and crafts in eighteenth-century Germany and with the history of the laboratory. She is the author of *Experiments, Models, Paper,* and (with W. Lefèvre) *Materials in Eighteenth-Century Science: A Historical Ontology.*

SEYMOUR MAUSKOPF is a professor of history at Duke University. His fields of research interest are the history of chemistry (*Crystals and Compounds,* 1976; *Chemical Sciences in the Modern World,* 1993) and the history of marginal science (*The Elusive*

*Science,* with Michael R. McVaugh, 1980). In 1998, he received the Dexter Award for Outstanding Contributions to the History of Chemistry from the American Chemical Society.

AGUSTÍ NIETO-GALAN is a senior lecturer of history of science at the Universitat Autònoma de Barcelona. He has written widely on the history of chemistry and natural dyestuffs in the eighteenth and nineteenth centuries. Among his most important publications in this field are (ed. with Robert Fox) *Natural Dyestuffs and Industrial Culture in Europe, 1750–1880* (1999); and *Colouring Textiles* (2001). He has also published *La seducción de la máquina* (2001); *Cultura industrial, historia y medio ambiente* (2004); (ed. with José Ramón Bertomeu-Sánchez) *Chemistry, Medicine, and Crime: Mateu J.B. Orfila (1787–1853) and His Times* (2006); and (ed. with Faidra Papanelopoulou and Enrique Perdiguero) *Popularizing Science and Technology in the European Periphery, 1800–2000* (2009).

BARBARA ORLAND is a historian of technology and science and a senior scholar at the Science Studies Program, University of Basel. Her current research interests cover different fields of the history of biomedical technologies and the history of agriculture, nutrition, and food technologies. She is the author of numerous essays on the history of the body, gender, biotechnology, and popular science. She is finishing a book on the cultural history of nutritional sciences.

MARCUS POPPLOW is a historian of technology, science, and the environment in medieval and early modern Europe. Since 2005 he has worked as an independent scholar for academic and private institutions. He is the editor of several books and the author of numerous essays, including "Why Draw Pictures of Machines," in Wolfgang Lefèvre, ed., *Picturing Machines 1400–1700* (2004). He is completing a project on the economic enlightenment.

HANNA ROSE SHELL is an assistant professor of Science, Technology and Society (STS) at the Massachusetts Institute of Technology and a junior fellow at the Harvard Society of Fellows. As a historian of science and technology, she is the author of *Hide and Seek: Camouflage, Photography and the Media of Reconnaissance* (2010); and editor of William Temple Hornaday's *Extermination of the American Bison* (2002 [1889]). As a filmmaker, she has directed *Locomotion in Water* (2005), *Secondhand (Pepe)* (2007), and *Blind* (2009).

PAMELA H. SMITH is a professor of history at Columbia University and the author of *The Business of Alchemy: Science and Culture in the Holy Roman Empire* (1994) and *The Body of the Artisan: Art and Experience in the Scientific Revolution* (2004). In her present research, she attempts to reconstruct the vernacular knowledge of early modern European metalworkers.

E. C. SPARY is a lecturer in the history of eighteenth-century medicine at the Wellcome Trust Centre for the History of Medicine at University College, London. Her interests span the domains of health, nutrition, medical chemistry, the Parisian cooperations, the French food and drug trade, and the relations between medicine and the Enlightenment. She is the author of *Utopia's Garden: French Natural History from Old Regime to Revolution* (2000).

# Index